Current Topics in Primate Vocal Communication

Current Topics in Primate Vocal Communication

Edited by

Elke Zimmermann

German Primate Center
Göttingen, Germany

John D. Newman

National Institute of Child Health and Human Development
Poolesville, Maryland

and

Uwe Jürgens

German Primate Center
Göttingen, Germany

Springer Science+Business Media, LLC

Library of Congress Cataloging-in-Publication Data

Current topics in primate vocal communication / edited by Elke Zimmermann, John D. Newman, and Uwe Jürgens.
p. cm.
"Proceedings of the XIV Congress of the International Primatological Society, held August 16-21, 1992, in Strasbourg, France, and of the XV Congress of the International Primatological Society, held August 3-8, 1994, in Kuta-Bali, Indonesia"--T.p. verso.
Includes bibliographical references and index.
ISBN 978-1-4757-9932-3 ISBN 978-1-4757-9930-9 (eBook)
DOI 10.1007/978-1-4757-9930-9

1. Primates--Behavior--Congresses. 2. Animal communication--Congresses. I. Zimmermann, Elke. II. Newman, John D. III. Jürgens, Uwe. IV. International Primatological Society. Congress (14th : 1992 : Strasbourg, France) V. International Primatological Society. Congress (15th : 1994 : Kuta-Bali, Indonesia)
QL737.P9C885 1995
599.8'0459--dc20 95-18860
CIP

Proceedings of the XIV Congress of the International Primatological Society, held August 16–21, 1992, in Strasbourg, France, and of the XV Congress of the International Primatological Society, held August 3–8, 1994, in Kuta-Bali, Indonesia

ISBN 978-1-4757-9932-3

Originally published by Plenum Press, New York in 1995
Softcover reprint of the hardcover 1st edition 1995

10 9 8 7 6 5 4 3 2 1

PREFACE

More than 25 years ago, the first major review of primate communication appeared (Altmann, 1967). Since then, information on the communicative abilities of primates increased rapidly, resulting, 15 years later, in the appearance of the first volume in which signaling systems were analyzed in a broader variety of primate groups within an evolutionary perspective (Snowdon, Brown and Petersen, 1982). Seven years later, the first volume dedicated solely to primate vocal communication appeared (Todt, Goedeking and Symmes, 1988) and another four years later a volume followed in which nonverbal vocal communication in non-human primates and human infants was compared (Papousek, Jürgens and Papousek, 1992).

None of these volumes, however, provided information about current technical advances in the field of bioacoustics, especially in digital sound analyzing systems, which offer primatologists, anthropologists and linguists nowadays a variety of rapid methods for analyzing human speech and non-human primate vocalizations in a quantitative and comparative way. Choosing the right method is difficult if a synopsis of these tools is lacking. Furthermore, information was particularly lacking on the natural signaling systems of two important primate groups, the prosimians and the apes. Likewise, new and unexpected insights into the ontogeny and evolution of vocal communication were gained during the past few years by the use of highly sophisticated sound analysis and statistical techniques.

The scope of this volume, therefore, is to reflect recent technical advances for analyzing vocalizations and to compile and synthesize important aspects of the current state-of-the-art of evolutionary orientated research on primate vocal communication from prosimians to humans. The contributions will be grouped around four different themes: 1. Technical advances in the study of primate vocal signals, 2. Social and ecological determinants of vocal communication, 3. Morphological and neuronal substrates for vocal communication, and 4. Cognitive and evolutionary aspects.

The *first section* is concerned with *technical advances in the study of primate vocal signals*. It introduces major analyzing techniques now commercially available in the field of bioacoustics and points out their main advantages and disadvantages. In this section *Owren and Linker* present a critical summary of acoustical techniques used to analyze animal sounds, especially primate vocal signals. Likewise, they provide some examples of their successful usage. *A. Zimmermann* introduces the principles of the artificial neuronal network theory and points out examples of its application for continuous and speaker-independent recognition. Applying it to non-human primate vocal signals, this procedure will open up the possibility for automatic classification of primate vocal repertoires, for characterizing ontogenetic changes and for unravelling potential learning processes.

The *second section* examines *how ecological and social factors affect the structure and usage of primate vocal signals* in various primate groups, from nocturnal prosimians to apes. Using highly sophisticated acoustical and statistical analysing techniques combined with a detailed analysis of sound-associated behaviors, new and unexpected insights into the ontogeny and evolution of primate communication systems are yielded. In this section, *E. Zimmermann* analyses the rich diversity of advertisement calls in nocturnal prosimians extending from the low frequency up to the ultrasonic range and proposes a hypothesis for their evolution. Dialect-like vocal differentiations are revealed. Deme- and individually specific vocal patterns in adults emerge from plastic infant calls and seem to be influenced by vocal patterns of older conspecifics. *Newman* presents a comprehensive review of our current knowledge of vocal ontogeny in macaques and marmosets. Focusing on a short-term isolation paradigm, he demonstrates similar developmental lines in both groups. The rich variability in vocal morphology during early infancy is gradually lost with increasing age and may provide a basis for vocal learning. That the vocal output of non-human primates is more influenced by conspecifics than previously expected is supported by studies on two primate taxa with a longterm history in bioacoustic research, squirrel monkeys and Japanese macaques. *Biben and Bernhards* conducted a meticulous quantitative acoustical analysis of the ontogeny of close-range affiliative calls in squirrel monkey youngsters and revealed changes in vocal structure which cannot be explained solely by inheritance. *Sugiura and Masataka* investigated vocal exchanges of coo-calls in free-ranging Japanese monkeys. They discerned that the acoustic pattern of vocal responses is influenced by the preceding calls of a conspecific, an indication of vocal accomodation. The last two contributions of this section provide novel information on the vocal communicative abilities in macaques and apes. *Todt and collaborators* worked on the vocal behavior of a macaque species with a quite unusual infant care system, Barbary macaques. Their detailed description of the vocal repertoire of this species highlighting peculiarities in their vocal system will allow a comparison with related macaque species showing different modes of infant care. *Hohmann and Fruth* focus on an important aspect of ape vocal communication, the regulation of social relationships by loud calls. They include first data on bonobos from their own longterm field study in Zaire. Based on a data set on the distribution, usage and function of loud calls in apes and its comparison with a comprehensive data set from monkey species displaying similar cohesion systems they hypothesize that sex differences are brought about during evolution by differences in migration patterns.

Morphological and neuronal substrates for vocal communication are examined in the *third section*. *Schön Ybarra* summarizes what is currently known about the morphology of the non-human primate vocal tract from a comparative point of view and relates morphological differentiations to differences in vocal production. *Jürgens* contrasts the neuronal basis for innate, voluntarily controlled and learned vocalizations by comparing the neuronal pathways for vocal production in squirrel monkeys, macaques and humans. He introduces a neurobiological hypothesis of why these non-human primate groups should lack the voluntary fine motor control of the vocal folds. *Heffner and Heffner* compare lesioning effects in auditory cortex on the perception of communicative vocalizations in macaques and humans. They conclude that cortical mechanisms for processing conspecific sounds in macaques are analogous to cortical speech perception mechanisms in humans. *Bieser* reports on an important feature of human speech and non-human communicative sound recognition, amplitude envelope encoding. An electrophysiological single-unit study of the auditory cortex of unanaesthesized squirrel monkeys revealed field-specific neuronal coding mechanisms for temporal information processing.

The *final section* discusses *cognitive prerequisites for* and raises *theories about the evolution of speech and language in humans. Gouzoules and collaborators* review studies that have reported external reference in the natural communication of non-human primates and illuminate conditions under which such signaling is likely to evolve. Based on the available evidence for representational signaling, they propose a fuzzy logic system as an appropriate way to explain how information about the external world is categorized. *Locke* characterizes three phases of early language learning in human infants and compares them with the development of vocal communication in non-human primates. Based on the revealed similarities and divergencies between them, he presents a biolinguistic theory on how and why infants acquire the linguistic capacity for spoken language. *Lieberman* stresses similarities between human speech and primate vocalizations. He proposes that voluntary human speech evolved in stages from emotionally controlled non-human primate vocalizations.

This volume is based in part on updated contributions presented orally at the XIV and XV Congress of the International Primatological Society 1992 in Strasbourg, France, and 1994 in Kuta, Indonesia, respectively. Contributions from investigators working in similar areas of research were solicited to provide a broader perspective. Each contribution was peer-reviewed by one specialist in the field and one biologist.

We think that this volume provides a useful compilation of current research topics on primate vocal communication and offers stimulating ideas for further research. Its content will be of interest to students and professionals in zoology, psychology, animal behavior, anthropology, primatology, linguistics, neurobiology and comparative anatomy trying to understand human communicative behavior from an evolutionary perspective.

The help of numerous reviewers to improve the quality of this volume by expert comments is gratefully acknowledged. In particular, we thank: Kim Beeman, Dorothey Cheney, Bertrand Deputte, Robin Dunbar, Hubert Fastl, Anne Fernald, Heinrich Güttinger, Marc Hauser, Don Kroodsma, Hans-Jürg Kuhn, Peter McNeilage, Günther Palm, Gerhard Schuller and Heather Williams.

Last but not least we thank Marlies Heinemann-Jahn for her editorial assistance and Eileen Bermingham of Plenum Press for continuous support.

Elke Zimmermann, John Newman, and Uwe Jürgens

Göttingen and Poolesville

References

Altmann, S.A. (ed.), 1967, "Social Communication among Primates," Chicago Univ. Press, Chicago.

Papousek, H., Jürgens, U., and Papousek, M. (eds.), 1992, "Nonverbal Vocal Communication: Comparative and Developmental Approaches," Cambridge Univ. Press, Cambridge.

Snowdon, C.T., Brown, C.H., and Petersen, M.R. (eds.), 1982, "Primate Communication," Cambridge Univ. Press, Cambridge.

Todt, D., Goedeking, P., and Symmes, D. (eds.), 1988, "Primate Vocal Communication," Springer Verlag, Heidelberg.

CONTENTS

SECTION I: Technical advances in the study of primate vocal signals

SECTION II: Social and ecological determinants of vocal communication

SECTION III: Morphological and neuronal substrates for vocal communication

SECTION IV: Cognitive and evolutionary aspects

SOME ANALYSIS METHODS THAT MAY BE USEFUL TO ACOUSTIC PRIMATOLOGISTS

Michael J. Owren and Christopher D. Linker

Department of Psychology
University of Colorado at Denver
Denver, CO 80217-3364, USA

INTRODUCTION

Throughout the course of modern primatology, researchers have shown an abiding interest in characterizing the acoustic properties of vocalizations produced by their study species. While the number of primatologists specializing in such work (which will be called acoustic primatology) has remained small, it has been recognized that vocalization-related data can have important implications for many fundamental questions about non-human primate behavior (e.g., Cheney and Seyfarth, 1990). The acoustic primatologist's task is not, however, a straightforward one. Acoustics is a complex, technologically intensive discipline and effective acoustic analysis requires the use of a variety of specialized procedures. Unfortunately, relatively little in the way of a common set of methods or standards has been established for analysis of non-human primate calls and few techniques have been specifically developed for use with these signals.

The overall immaturity of the field of acoustic primatology appears to be directly traceable to the relatively small scope of this discipline, even in comparison to other areas of bioacoustics. To illustrate the latter, we examined a series of articles by P. Hansen (published in the first 4 volumes of the specialty journal *Bioacoustics*) that provide a comprehensive listing of recent bioacoustics-related references. Only 192 (8%) of the 2529 articles listed were found to be related to non-human primates. In contrast, fully 1099 (43%) of the citations concerned avian species. Of the 44 empirical articles published in these same volumes of *Bioacoustics* (1988 to 1993), only one concerned acoustic primatology.

Given the existence of more than two hundred non-human primate species, the total number and diversity of vocalizations of potential interest presents a daunting analysis challenge. While a proportionate array of specialized techniques and approaches is therefore warranted, the tools used are typically those most readily available to bioacousticians from commercial vendors. These vendors, in turn, serve a market that is most strongly influenced by the needs of researchers working in areas other than primatology. While some very effective analyses of non-human primate vocalizations have been and continue to be

Current Topics in Primate Vocal Communication
Edited by E. Zimmermann *et al.*, Plenum Press, New York, 1995

performed on this basis, there appears to be an overall willingness on the part of researchers to allow the capabilities of the most commonly available bioacoustics-related tools to influence the analyses undertaken rather than the other way around.

This chapter is meant to encourage acoustic primatologists to routinely adopt the expectation that more optimal analysis techniques can and should be available for use with non-human primate vocalizations. As substantive technological developments in acoustic analysis cannot be expected to take place within primatology, however, techniques and approaches will continue to be borrowed from other sources. We propose that the area of human speech research has been generally underutilized as such a source, due in part to the relatively inflexible and narrow implementation of many speech-related analysis tools. The study of speech is marked by tremendous productivity and technological innovation, and it can be expected that acoustic analysis breakthroughs of the past (including the invention of the sound spectrograph, development of numerous digital signal processing algorithms, and implementation of these algorithms on low-cost microcomputer platforms) will be matched in the future. In many cases, only a modest degree of effort is required in adapting speech-specific approaches for use with non-human primate calls.

We also argue that acoustic primatology can benefit from adopting a production-based perspective on vocal behavior, paralleling a successful approach taken in speech studies. As the vocal production mechanisms in human and non-human primates share many important characteristics (see Negus, 1949; Ploog, 1988), theoretical models and analysis techniques developed for the former may also be directly applicable to the latter. In contrast, sound production processes in non-human primates are quite different from those of many animals studied in bioacoustics research, like birds, anurans, or insects.

We have tried to prepare the chapter specifically with the non-specialist in mind, but cannot provide a comprehensive presentation of all relevant background material. Instead, compromises have consistently been made between breadth, accessibility, and the limited number of techniques that can be discussed. The topics selected are concerned with measurements of the acoustic properties of vocalizations and, regrettably, important developments in other areas are ignored. We begin with brief presentations of the acoustic theory of speech production, sound spectrography, and digital analysis of acoustic signals. This background sets the stage for describing a few of the analysis techniques we believe to have useful applications in primatology, but that are used either relatively little or not at all. In rough correspondence to the production-based perspective that is developed, these methods are presented under the general headings of measuring frequency-spectral content and characterizing laryngeal energy, respectively. Sources of further information are included whenever possible, with particular emphasis on those that do not require a strong mathematical background.

BACKGROUND

The Source-Filter Approach to Speech Production

In 1960, Gunnar Fant proposed a model in which the speech signal is viewed as a linear combination of an underlying energy source and subsequent filter (see also Chiba and Kajiyama, 1941; Stevens and House, 1955). This "acoustic theory of speech production," or "source-filter theory" suggests an approach to speech analysis in which the acoustic signal is decomposed into independent components related to separable production processes. The energy source for speech is modeled either as a quasi-periodic pulse series corresponding to the puffs of air that emanate from the glottis (the opening between the vibrating vocal folds) or as non-periodic turbulent noise that is produced by constricting or occluding the supra-

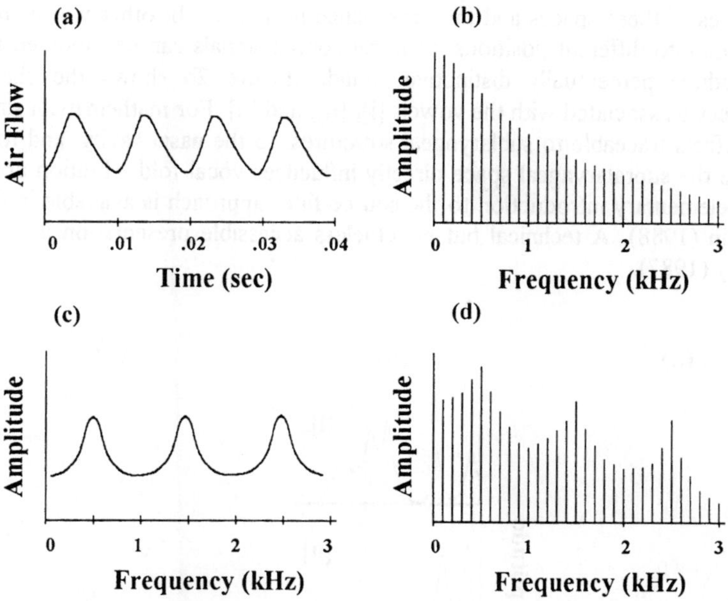

Figure 1. (a) A glottal airflow waveform, (b) the frequency spectrum of this waveform, (c) a filtering or transfer function associated with a neutral, relaxed vocal tract, and (d) the frequency spectrum of the resulting speech sound.

laryngeal vocal tract. The pressure wave thus created has both temporal and frequency characteristics. In vowel production, the rate of vibration of the vocal folds is termed the fundamental frequency (F_0, or first harmonic). Vibratory movement in the vocal folds produces signals whose frequency spectra include energy not only at the F_0, but also at higher harmonics--frequencies that are integer multiples of the base rate of vibration. A possible glottal pressure waveform and its frequency spectrum are illustrated in Figures 1a and 1b, respectively, showing a signal with an F_0 of approximately 100 Hz (slightly lower than the typical value for a human adult male). As illustrated, glottal waveform frequency spectra typically show monotonically declining harmonic amplitudes.

The cavities and tissues found above the larynx strongly influence the glottal waveform through their resonance and anti-resonance (damping) properties. Vocal tract resonances, called formants, reinforce energy in specific frequency ranges. The frequency spectra of vowels typically reveal the presence of 4 or 5 vocal tract formants below 5 kHz and it is the relative locations of the lowest 2 or 3 of these formants that is considered to determine the acoustic qualities associated with various vowels (Joos, 1948). Figure 1c shows the characteristic filtering function of a neutral, relaxed vocal tract. Formants also affect speech sounds that are based on turbulent, or aperiodic, energy sources. However, the resulting noisy sounds are typically characterized by only 1 or 2 formants, usually located above 3 kHz. Energy below this range is damped by anti-resonances in the supralaryngeal filter. As shown in Figure 1d, then, a given speech sound reflects the characteristics of both the source energy function and subsequent vocal tract filtering.

In humans, the supralaryngeal cavities are the pharynx, the oral cavity, and the nasal cavity, as shown in Figure 2a. The pharynx and oral cavity are the most important to the filtering process, as movements of the tongue, mandible, and lips can substantially influence both the shapes of these spaces and their resonance properties. In other words, by moving the articulators into different positions, identical source signals can be modified in various ways to produce perceptually distinctive sounds. Figure 2b shows the characteristic frequency spectra associated with the vowels [i], [u], and [a]. For mathematical simplicity, a number of effects traceable to sublaryngeal structures, to the nasal cavity, and to coupling effects where the supralaryngeal space directly influences vocal fold vibration are typically ignored. An elementary introduction to the source-filter approach is available in Lieberman and Blumstein (1988). A technical but nonetheless accessible presentation is provided by O'Shaugnessy (1987).

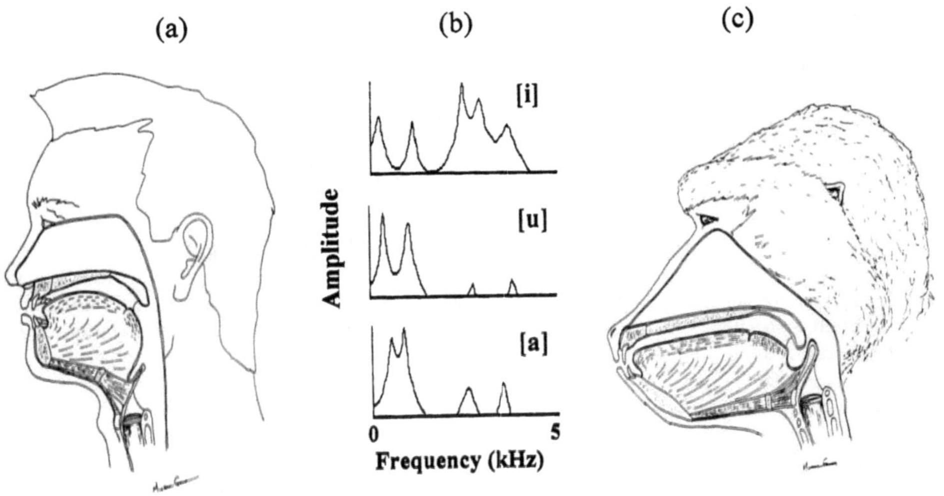

Figure 2. (a) A schematic, midsaggital view of an adult male human (*Homo sapiens*) vocal tract, (b) spectral slices illustrating the frequency energy patterning of the vowels [i], [u], and [a], and (c) a schematic, midsaggital view of an olive baboon (*Papio anubis*) vocal tract. (Drawings by Michael Graham)

Applying the Source-Filter Approach to Non-human Primate Vocalizations

Studies of non-human primates have demonstrated many essential similarities in the sound production structures of non-human primates and humans (e.g., Geist, 1933; Hill and Booth, 1957; Hilloowala, 1975, 1976; Kelemen 1938, 1948, 1969; Kelemen and Sade, 1960; Negus, 1949; Schön, 1971; Schön Ybarra, this volume). These similarities are apparent upon comparing Figures 2a and 2c, where the latter may be considered to illustrate the "standard" non-human primate vocal tract. By extension, then, basic source-filter concepts can be applied to non-human primates as well as to humans. However, important differences in sound production have also been proposed. Lieberman (1968), for instance, suggested that monkeys and apes are unable to maintain precise control over vocal fold vibration and therefore cannot produce highly regular, tonal sounds analogous to human vowels. While supralaryngeal cavities clearly exert a filtering effect on source energy during non-human primate vocal production, it has also been suggested that significant modification of vocal tract shape does not occur (Kelemen, 1948, 1969; Lieberman, 1972; Lieberman, Klatt, and Wilson, 1969; Negus, 1949; Zhinkin, 1963). Compared to humans, non-human primates have longer, thinner tongues, larynges that are positioned higher in the neck, and a relative lack of flexible soft tissues in the supralaryngeal cavities. Due to these differences, non-human primates appear to have less opportunity to alter the resonance properties of their vocal tracts through flexible modification of articulator positions.

Sufficient data are now available to arguably counter both of these assertions. For example, field primatologists can attest to the occurrence of highly tonal vocalizations in many different species--calls that indicate a high level of vocal fold control. Sound spectrograms of such calls show evidence of both regular vocal fold movement and, in many cases, pronounced harmonic structure. F_0's in these vocalizations range significantly and can be lower (e.g., Andrew, 1976), somewhat higher (e.g., Owren, Dieter, Seyfarth, and Cheney, 1991), or many times higher (e.g., Clark and Wrangham, 1993; Hauser, Evans, and Marler, 1993; Zimmermann, 1981) than corresponding rates in humans. Both Gautier (1983) and Brown, Brown, Santos, and Dagenais (C. Brown, personal communication) have found through laryngographic recordings that highly regular vocal fold vibration can occur in non-human primates. Brown et al. also measured minute variations in the periodicity of vocal fold vibration in a Sykes' monkey and a human, finding comparable values for the two subjects. Evidence described by Hauser and Fowler (1991) indicates that monkeys can exhibit predictably declining F_0 values when producing repeated calls, a phenomenon that may be analogous to pitch declination occurring at the end of human speech utterances.

Sound spectrography has also revealed strong evidence of vocal tract filtering in a number of non-human primates. For instance, spectrograms of baboon grunts (Andrew, 1976; Owren, Linker, and Rowe, 1993) reveal vowel-like formant structures. In macaques, highly tonal coo calls also reveal patterns of harmonic energy that can vary both within and between individuals, and that do not show monotonically decreasing amplitude values across harmonics, as might be expected for an unfiltered laryngeal waveform. Some investigators have argued that monkeys and apes explicitly modify vocal tract resonances through various articulatory manoeuvers (e.g., Hauser and Fowler, 1991; Richman, 1976). Owren and Bernacki (1988) showed that two types of noisy snake and eagle alarm calls produced by vervet monkeys differ in overall frequency patterning and subsequently found these patterns to be salient cues for vervets sorting synthetic alarm calls in a laboratory operant-conditioning task (Owren, 1990). Direct evidence of separable contributions of source and filter has been presented by Hauser et al. (1993), who found that variation in mandible position in vocalizing rhesus macaques was correlated with spectral changes in the resulting call in the absence of changes in F_0. Hauser and Schön Ybarra (1994) reported evidence of lip configuration effects on the acoustic realization of rhesus monkey coo calls, but not in scream vocalizations. Brown et al. (C. Brown, personal communication) found direct evidence of modifiable vocal tract resonances in their Sykes' monkey subject by comparing the frequency spectra of laryngeal signals and corresponding calls.

As mentioned earlier, anatomical studies have also shown there to be differences between human and non-human primate vocal tracts, including structures that are not found in humans. While a few species exhibit extralaryngeal resonance structures (Gautier and Gautier, 1977), a wider variety may show a vocal fold specialization that humans also lack. Schön Ybarra's (this volume) examination of a variety of non-human primate cadavers has revealed the presence of a relatively rigid vocal lip on the medial extremity of each fold. This lip appears to provide the ability to produce two independent vocal fold vibration patterns--for instance in combining periodic with aperiodic source energies. His proposal is consistent with observed call characteristics in many non-human primates, and the work also indicates that significant modification of both F_0 and vocal tract position can occur in various species.

Analog Sound Spectrography in Speech Research

Historically, studies of both speech and animal signals relied on the analog sound spectrograph as the single most important acoustic analysis instrument. This device became available to the research community soon after World War II (Flanagan, 1972) and came into use with non-human primate vocal signals approximately a dozen years later (Rowell, 1962; Rowell and Hinde, 1962). The spectrograph (or sonagraph) produces a time-by-frequency plot referred to as a spectrogram (or sonagram), as shown in Figure 3. In this representation, which is produced by repeatedly sweeping an analysis filter across the frequency range, the amount of acoustic energy present at any given point is indicated by the darkness of the corresponding tracing. Tracings are produced by a localized burning effect that occurs when energy is detected within the bandwidth of the analysis filter and indicate the average value of that energy.

Until quite recently, the analog spectrograph enjoyed a perennial popularity in bioacoustics research. In design, however, this instrument was primarily intended for use with speech signals rather than other kinds of sounds. In fact, the analog spectrograph's settings were specifically chosen to allow clear analyses of the F_0 and vocal tract resonances

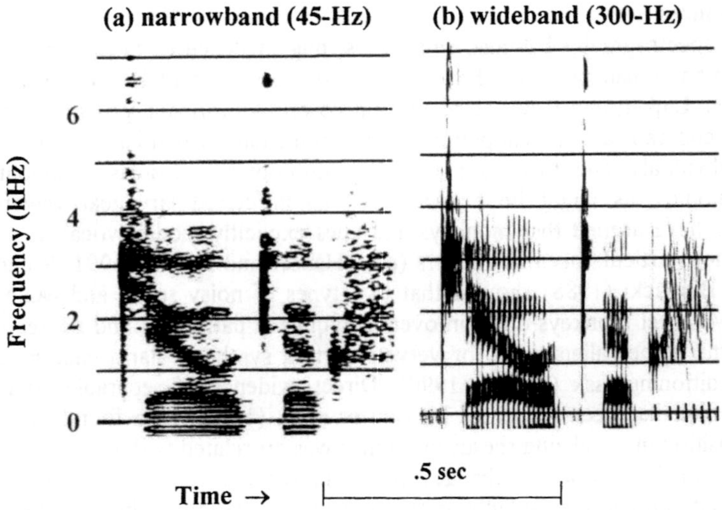

Figure 3. Analog sound spectrograms of author MJO saying the words "Joe took."

of human adult males. Given a typical F_0 of approximately 120 Hz in males, a 45-Hz-wide bandpass filter (the "narrowband" setting on a standard analog spectrograph) reveals individual energy peaks in the frequency spectrum that correspond to the F_0 and higher harmonics (see Figure 3a). The F_0 value is indicated by the frequency difference between adjacent harmonics. The F_0 is also apparent when a wider analysis filter is used, which decreases frequency resolution but increases temporal resolution. The results of applying the normative "wideband" analysis filter of 300 Hz are shown in Figure 3b. Whereas the 25-ms temporal resolution associated with a 45-Hz analysis bandwidth is too coarse to show individual cycles of vocal fold vibration, a 300-Hz bandwidth provides a temporal resolution of 3 ms and individual glottal energy pulses become visible as vertical striations. As is apparent in comparing Figures 3a and 3b, improving resolution in either the temporal or frequency domains produces a requisite resolution decrease in the other domain (see Beecher, 1988 for an extensive discussion of this issue).

The primary purpose of using a wideband filter in sound spectrographic analysis of speech was not, however, to allow measurement of F_0 values but to reveal the locations of vocal tract resonances, or formants. Each of the dark bands evident in Figure 3b shows the effect of a formant present during speech production. Because a wideband filter subsumes adjacent harmonics in adult male speech, a strong averaging or smoothing effect occurs in the resulting spectrogram. Areas where adjacent harmonics are uniformly high in amplitude due to a vocal tract resonance become darkly shaded. Conversely, loss of energy in excess of expected frequency-dependent decreases indicates either that no formant occurs in that frequency range or that an anti-resonance may be present. The overall formant pattern apparent in the wideband spectrogram closely follows the distribution of high and low harmonic amplitudes shown in the narrowband spectrogram. By design, then, the wideband filter setting spans at least 2 harmonics in the typical speech of adult male while still allowing differentiation of closely spaced formants.

The trading relationship between temporal- and frequency-domain resolutions was specifically taken into account in tailoring the analog spectrograph to the speech of adult males. The standard filters work well for such sounds, but less well with higher-pitched speech produced by women and children. In general, wideband filters are not appropriate for signals that exhibit relatively high-frequency F_0's where harmonics are widely spaced. As a rule of thumb, the F_0 value should not exceed one-half the bandwidth of the filter when spectral rather than temporal or F_0 characteristics are of interest.

Analog Sound Spectrography in Acoustic Primatology

Bioacoustics researchers traditionally showed little sensitivity to the importance of filter bandwidth when conducting analog spectrograph analyses. As noted by Beecher (1988), however, avian researchers benefitted from a fortuitous, approximate match between the standard filter options and the frequency-temporal features of many bird songs. Due to the extreme range of variation in frequency modulation, tonality, and F_0 values of non-human primate vocalizations, however, acoustic primatologists were not so fortunate. Many analyses, for instance, involved relatively tonal calls whose characteristics were poorly suited to either of the available bandwidths, with the result that the two filter settings came to be used almost interchangeably. It is not unusual to find measurements made from wideband spectrograms of periodic vocalizations whose F_0's were much higher than one-half the filter bandwidth, the result of which was to produce uninformative smearing of energy from individual harmonics across the width of the filter. Spectrographic analyses of noisy non-human primate sounds with either no F_0 or an unstable F_0 have also been quite uninformative, whether wideband or narrowband. For such sounds, the analog spectrograph is unable to reveal either source characteristics, filter characteristics, or the net distribution of spectral energy.

The limited dynamic range of the analog spectrograph also presented problems and may have spawned at least one poorly conceived measure that is still in use. Due to limitations in the gray scale resolution of burnt paper, the analog spectrograph can only show an amplitude range of 12 dB. A much wider range can be shown using the instrument's "amplitude contour" capability, but this setting has rarely been used by bioacousticians. The analog spectrographic representation is therefore almost useless for measuring relative amplitudes across the frequency range, which encouraged the use of simplistic measures like "dominant frequency" and "bandwidth." The former is the single highest-amplitude energy band appearing in the frequency spectrum. While this variable may be informative in some circumstances, a more useful approach is to characterize the entire spectral energy pattern. Bandwidth is the range of frequencies in the call and has often been based simply on visual inspection of the spectrogram. Given the instrument's limitations, this approach is of questionable value. The measure can also be criticized from a production standpoint, where uncontrolled variation in source amplitude, attenuation of high-frequency components during transmission, and amplitude at the microphone can produce rather complex influences on the range of frequencies that is recorded. One can argue that given a large sample of calls, bandwidth differences between calls or individuals should nonetheless be a useful indicator of differences in acoustic characteristics--all the more robust in fact, given the possible influence of error factors. Unfortunately, the sample sizes customarily available are usually quite small. In such cases, it is important to maximize both the validity and reliability of the measures used in order to avoid producing spurious differences or missing differences that do exist.

Overall, primatologists have historically shown only modest sensitivity to the inherent limitations of the analog sound spectrograph as an acoustic analysis tool, although notes of caution concerning these limitations have long been available in the bioacoustics literature (e.g., Davis, 1964). Because its salient design features have often been misunderstood or ignored, the instrument's output has been subject to much misinterpretation. This particular format nonetheless remains the prototypical visual representation for bioacoustics and speech research (although see Cooke, Beet, and Crawford, 1993) and is routinely simulated in digital processing environments. Fortunately, the latter approach allows researchers both to move beyond the spectrographic representation to its underlying quantitative characteristics and to apply other more specialized processing techniques.

Digital Signal Processing (DSP)

Over the last decade, the development of powerful, affordable microcomputers, coupled with the implementation of digital signal processing (DSP) algorithms in commercial software packages, has put sophisticated acoustic analysis capabilities within the reach of virtually all researchers. These changes are of particular importance in acoustic primatology, where the broad range of signals of interest requires a flexible approach to signal analysis. On the one hand, simply moving to the digital domain improves the objectivity and accuracy of measurement. On the other hand, DSP allows dramatic increases in the range and power of the tools that can be applied, given a sufficient development effort. The challenge presented to acoustic primatologists, however, is to work cooperatively in mastering the conceptual basis of DSP and agreeing upon common analysis procedures. Although the number of accessible presentations of DSP fundamentals is quite small, some are available. Stoddard (1990), for example, provides an excellent review of the components and basic operations of a microcomputer-based digital processing system. A relatively elementary introduction to the mathematics underlying DSP is available in Steiglitz (1974). Rabiner and Gold (1975) cover these mathematical topics more thoroughly and extensively, while O'Shaugnessy's (1987) comprehensive textbook on speech communication includes many DSP applications. Baken's (1987) overview of measurement of speech and voice characteristics from a clinical perspective is thorough and very accessible. Wakita (1976) discusses a number of digital processing techniques used in speech research in a brief but informative book chapter.

Digital Sampling

The necessary first step in DSP is to create a digital representation of the waveform of interest. This process entails using an analog-to-digital conversion (known as A/D or ADC) system that produces a series of discrete voltage values representing the continously varying waveform. These values are amplitude samples of the signal recorded as they appear as input to the A/D system. Examples appear in Figure 4, which shows the sampled waveforms of a baboon grunt and vervet monkey snake alarm call, respectively. The apparently continuous form of these waveforms illustrates the common practice of connecting the sample points with lines in graphical representations and does not reflect continuity in the underlying waveform files.

The sampling procedure is well-described by Stoddard (1990) and will not be examined here except to emphasize two points. First, it must be ensured that the sampled waveform represents the amplitude variation of the original signal as completely as possible. As a rule of thumb, the signal of interest should on average fill approximately three-quarters

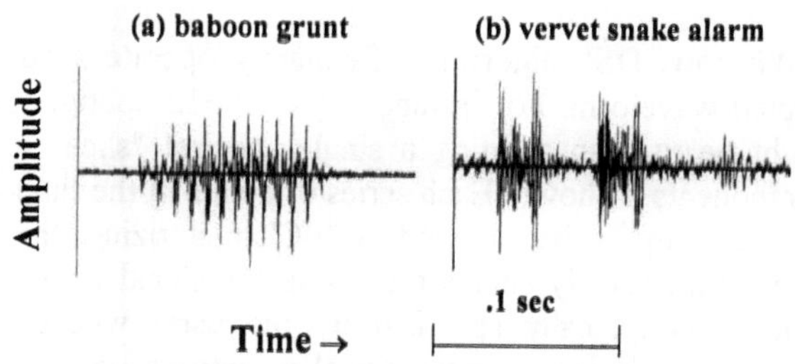

Figure 4. The digitally based time-domain waveform representations of (a) a highly tonal chacma (*Papio ursinus*) baboon grunt and (b) a very noisy vervet monkey (*Cercopithecus aethiops*) snake alarm call.

of the available amplitude range. However, the sampled signal cannot be allowed to exceed this range. "Peak clipping," in which one or more amplitude peaks are truncated in the digitizing process, produces distortions in the signal's frequency spectrum and must be avoided. Second, energy at frequencies in the original signal that exceed one-half the digital sampling rate must be removed or strongly attenuated using a low-pass filter. If samples are read from the analog waveform 20,000 times per second, for instance, energy above 10 kHz (known as the Nyquist, or folding frequency) must be removed either before or while the signal is being sampled. Sampling at 20 kHz, then, allows representation of frequency components only up to 10 kHz. A more conservative procedure is to filter the signal at a somewhat lower frequency, for instance at four-tenths the sampling rate. Failure to remove energy above the Nyquist frequency with an "anti-aliasing" low-pass filter produces an "aliasing" effect in which higher-frequency components are inaccurately represented as energy occurring at lower frequencies.

Some Characteristics of DSP Analysis Routines

The Structure of Speech-Related Routines. As noted earlier, speech-related research has provided the impetus for a significant proportion of the technological advances that have occurred in digital processing of acoustic signals. Accordingly, many DSP software or hardware packages provide both unspecialized, "generic" DSP tools and more specialized routines designed for speech analysis. The latter are typically constructed around a generic algorithm that has been optimized for automated processing of speech. While the algorithm that forms the core of the routine may be applicable to virtually any signal, other aspects of the routine often render it unsuitable for use beyond the intended application.

Typically, a speech-related DSP routine consists of three parts. A "preprocessor" component conditions the sampled waveform, the "basic extractor" routine executes the measurement of interest, and a "postprocessor" performs tasks ranging from error correction to providing well-formatted results (Hess, 1983). Unfortunately, a preprocessor usually conditions an input signal in accordance with parameter settings that are tied closely to typical speech values. Applied to non-human primate vocalizations, the algorithm may "ignore" or even remove critical aspects of the signal. The preprocessor can thus do harm to the signal instead of aiding in analysis and must be modified or circumvented. Analogously, the postprocessor may ignore or remove measurement values of interest because they fall outside the relevant speech range. The basic extractor algorithm, in contrast, can usually be readily used with non-human primate vocalizations. In some cases, a speech-related routine may be made more usable by adjusting the analysis settings. More often, however, these settings cannot be modified sufficiently and a more flexible package in which the extractor algorithm itself is available may be required. In both cases, the algorithm and its customary implementation must be well-understood by the user.

The Analysis Window. DSP algorithms frequently operate sequentially over short segments of the sampled waveform. For instance, 256 or 512 contiguous datapoints from the time domain might be used in creating a single spectral "slice" in which energy at various frequency components is shown. Each series of points in the time-domain waveform metaphorically constitutes an analysis "window." Characterizing the entire waveform segment of interest therefore usually entails traversing the signal in stepwise fashion. The size of analysis window is typically specified by the user, whose most fundamental consideration is the rate at which important signal properties are changing. Ideally, the waveform segment falling under the window is stationary, as analysis of non-stationary segments produces inaccurate results. The "shape" of the analysis window must also be specified, which is equivalent to setting an independent weighting factor for each sample point during the analysis operation. The simplest case is to weight each point equally, which is described as using a rectangular window. The most common true weighting window is a cosine-shaped tapering function that ranges from 0 at each end to 1 in the middle. Windows of this kind (which include Hanning, Hamming, Blackman, and others) help prevent frequency-domain distortions that can result from time-domain amplitude discontinuities occurring at the endpoints of a rectangular window.

CHARACTERIZING THE FREQUENCY SPECTRUM

The Fourier Transform

Fourier analysis is arguably the most widely used technique for characterizing the frequency spectrum of an acoustic signal. The transform involved is based on Fourier's theorem, which proposes that any periodic waveform in the time domain can be exactly modeled as the sum of an infinite series of sinusoidal functions. Acoustic waveforms of interest are "decomposed" into individual frequency components whose frequency, amplitude, and phase values are known. The discrete Fourier transform (DFT) produces a frequency (or power) spectrum representation based on a short-time analysis window. As the window defines a signal of finite duration, an inverse relationship exists between time- and frequency-domain resolutions, as discussed earlier for analog spectrography. Fourier analysis was made practicable by Cooley and Tukey (1965), who described an efficient method of computing the DFT based on restricting window lengths to powers of 2. This fast Fourier transform (FFT) is found in virtually every acoustic analysis package.

FFT-based digital representations that are designed to resemble analog spectrograms have also become quite common. The approach involved is to sequentially transform short segments of the signal of interest, producing a series of power spectra. Amplitude values are represented at n points across the frequency spectrum, where n is one-half the number of points in the time-domain analysis window. The spectrogram's values are stored in a matrix of rows and columns in which each cell contains the amplitude value calculated for a given frequency at a particular time. Averaging or interpolating values in the time and frequency dimensions allows digital spectrograms to approximate the smooth appearance of analog spectrograms. Figure 5a shows a digital spectrogram of the vervet snake alarm shown in

Figure 4b, while a power spectrum based on the first 256 points of this same call is shown in Figure 5b.

Digital analysis systems typically allow the user significantly more flexibility in parameter selection (e.g., analysis window length and sampling rate) than do analog spectrographs. Due to the inverse relationship between time and frequency resolutions, these parameters should be adjusted so as to optimize the characterization of the particular feature of interest. The primary consideration is the amount and rate of change in the time and frequency domains (see Beecher, 1988). A stationary signal can be analyzed with a long analysis window, producing the greatest possible frequency resolution. With a non-stationary signal, however, a shorter analysis window encompassing only relatively unchanging waveform segments must be used. Optimal parameter selection therefore represents a compromise between achieving maximal frequency resolution and the rapidity of change in the particular signal under analysis. It should be noted, though, that this resolution trade-off is not inevitable in spectral analysis. Rather, it is a consequence of the "quasi-stationary," or "window-analyze-shift" approach that is employed both in Fourier analysis and a number of other methods (Loughlin, Atlas, and Pitton, 1993). Several techniques that can represent both time and frequency characteristics accurately have become available, but each has its own limitations and disadvantages. As these difficulties are overcome, the Fourier transform may be replaced by other analysis approaches.

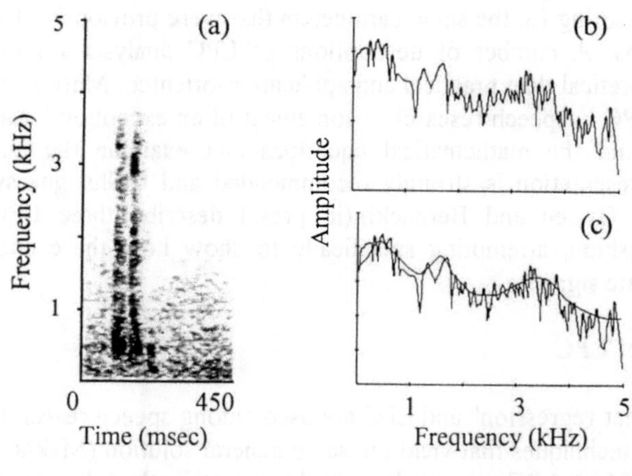

Figure 5. (a) A spectrogram of the vervet snake alarm call previously shown in Figure 4b, based on 512-point FFT's. A 256-point FFT derived from the beginning of the call appears by itself in (b) and is overlaid by a 7-coefficient, 256-point LPC model calculated over the same segment in (c).

A Need for Data Reduction

A salient characteristic of Fourier analysis is that all data from the time domain are preserved. In other words, time-domain data are transformed to the frequency domain without alteration, except for differential weighting effects produced by the analysis window. While Fourier analysis characterizes each spectral component in terms of its frequency, amplitude, and phase values, only frequency and amplitude are explicitly represented in the spectrogram. If phase information is included, the spectral representation can be returned to the time domain through an inverse transform, thereby recreating the original waveform segment. Fourier analysis, then, does not reduce the data being processed. A single transform, incorporating for instance 256 sampled points, produces a spectrum with 128 frequency component values. Ignoring phase information, the number of data points has technically been reduced by one-half. However, the transformation has not helped the investigator in deciding which aspects of the resulting spectrum are of greatest importance. While such fidelity can be important and useful, the point of conducting the analysis is usually to extract the spectral characteristics that are most likely to serve as distinctive features to the species in question. If the search for distinctive spectral features can be grounded in specific knowledge of the species' production mechanisms or perceptual processing, those features can be targetted in processing and the sheer quantity of data to be processed can be significantly reduced. As noted earlier, the source-filter model has provided an important theoretical foundation for characterizing the distinctive features of human speech and the analysis approaches that have resulted can arguably be used with non-human primate calls as well.

Linear Predictive Coding (LPC)

Following the introduction and subsequent unification of several DSP formulations under the collective heading of "linear predictive coding" in the 1960's and 70's, LPC gained considerable prominence both as an analysis and synthesis tool in human speech research (Wakita, 1976). One strength of LPC is that it provides accurate estimates of formants and can be useful in measuring F_0, the same parameters that were previously characterized using analog spectrograms. A number of descriptions of LPC analysis are available, but are generally more theoretical than practical and application-oriented. Markel and Gray's (1976) classic primer on LPC in speech research is somewhat of an exception inasmuch as one can fruitfully read around the mathematical equations and examine the numerous figures. Wakita's (1976) presentation is strongly recommended and O'Shaugnessy (1987) should also be consulted. Owren and Bernacki (in press) describe three LPC techniques in nonmathematical fashion, attempting specifically to show how these approaches can be applied to bioacoustic signals.

Two Approaches of LPC

The terms "linear regression" and LPC are used among speech researchers to designate any of a number of techniques that yield the same general solution (Makhoul, 1975; Markel and Gray, 1976). Most LPC approaches produce a smoothed frequency spectrum (or spectral envelope) that includes only peaks. Spectral "dips" that may be present in the original spectrum are disregarded in these approaches, which are collectively referred to as "autoregressive" (AR) models. Two of these models are discussed here. An LPC-based smoothed spectrum is illustrated in Figure 5c, where the model is derived from the vervet monkey snake alarm call and has been superimposed on the corresponding, but independently extracted power spectrum. Figure 6 shows FFT- and LPC-based spectrograms of the baboon grunt that is also shown in Figure 4a, along with a single LPC smoothed spectrum model and the frequency peak values associated with it.

Linear Regression. The linear regression approach is based on linear least-squares estimation techniques developed in the late 18th century (Markel and Gray, 1976). Thus, the historical roots of LPC were roughly contemporaneous with the development of the Fourier transform. Linear regression is performed in the temporal domain, where the strategy is to approximate successive points of a digitally sampled signal as linear combinations of past values. The result is a smoothed waveform whose simplified frequency characteristics can then be obtained. To illustrate the linear regression concept, consider a sequence of 100 sampled data points where successive sets of points are used to predict the next value. Each "prediction" takes the form of a linear regression equation with M coefficients. If M is equal to 10, the eleventh point of the sequence is predicted first, by weighting each of the preceding sample points with a unique coefficient (a_1 to a_{10}) and adding the resulting products. The twelfth point is predicted as the sum of weighted points 2 through 11, etc. Predicting each sample point in this manner produces a set of simultaneous linear equations, analogous to a problem in multiple regression. The solution is a single regression equation that predicts successive sample points of the time segment on the basis of the previous M points while minimizing overall prediction error in the least-squares sense. Any of several methods can be used to compute the regression coefficients, including autocorrelation- and covariance-based approaches. These methods allow efficient computation by taking advantage of redundancy properties in the regression equations.

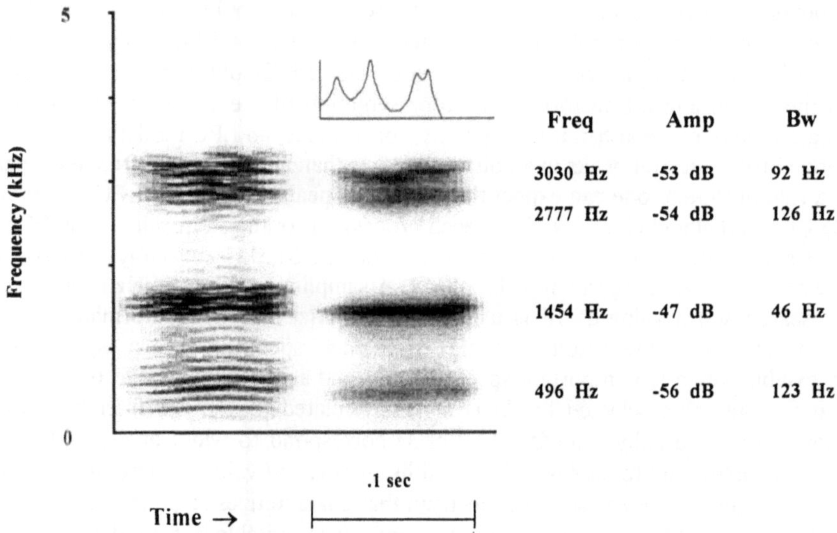

Figure 6. The baboon grunt previously illustrated in Figure 4 is shown here in digital spectrograms based on 256-point FFT's (left) and 14-coefficient, 256-point LPC models (middle), respectively. A single LPC model derived from the midpoint of the call is shown above the latter, with frequency on the horizontal axis and amplitude on the vertical axis. Values listed on the right represent the frequency, amplitude, and bandwidth values for each peak shown both in the spectrogram and in the spectral slice.

Inverse Filtering. A second variety of LPC, the inverse filter approach, is based directly on the source-filter model of sound production. Here it is assumed that the acoustic signal represents a linear combination of sound source and subsequent filter. The source is further assumed to be either a quasi-periodic impulse series or white noise--its characteristics are therefore known. The inverse filter that best recovers the input (the proposed source function) from the output (the sound itself) can then be sought. This filter is represented as a polynomial based on a series of coefficients (or a parameters) where the reciprocal of this filter polynomial approximates the transfer function of the (vocal tract) filter associated with production of the sound. Filter coefficient values are often computed using least-squares-based autocorrelation or covariance methods.

Filters Derived from Regression Coefficients. The connection between these two methods (and other LPC approaches) is that the coefficients of the inverse filter polynomial are identical to the regression coefficients derived through linear prediction. In other words, these two formulations can be used to arrive at the same general solution in spite of having very different theoretical rationales (see Rabiner and Shafer, 1978). The residual or error signal of the linear regression approach corresponds to the source waveform as recovered through inverse filtering. The mathematical equivalence of these approaches allows the coefficients to be derived in an efficient, hybrid manner. Both of the methods described produce a smoothed spectral model--either directly as the inverse of the vocal tract filter model or indirectly, as a transformation of the smoothed waveform. Replicable, quantified estimates of the frequency, amplitude, and bandwidth characteristics of each spectral peak are available either through root-solving the inverse filter polynomial or through direct examination of the smoothed spectrum.

As with other DSP methods, LPC requires that the investigator selects values for important analysis parameters and the particular values selected play a critical role in shaping the resulting spectral envelope. The most critical parameters are the size and shape of the analysis window, the number of coefficients used in the model (M), and a preemphasis factor that can be applied to the smoothed spectrum. Detailed discussion of each is beyond the scope of this chapter and the reader should consult the references mentioned earlier. However, selection of M lies at the heart of LPC analysis and will be briefly considered. Most importantly, the value selected should produce the best "fit" between the model and a transform-based representation of the same waveform segment, depending on the purpose at hand. In Figure 5c, for instance, an LPC model is being used to characterize broad spectral peaks in a noisy call rather than any individual frequency peak.

As a rule of thumb, one can expect the number of peaks "found" by LPC algorithms to be $(M/2) - 1$ (Markel and Gray, 1976). Speech processing routines typically allow M to vary between 3 and 99 or more. For periodic speech sounds, Markel and Gray (1976) suggest that M be set to the sampling rate in kHz, plus 4. A sampling rate of 10 kHz is customary in speech research and a value of 14 is used to characterize the 4 to 5 formant peaks that typically occur within the resulting 5-kHz frequency spectrum. Noisy speech sounds typically exhibit only a few important spectral peaks and an M value of 4 to 6 is suggested. Conceptually, the value selected for M is inversely related to analysis filter bandwidth in analog sound spectrography. Lower values of M correspond to wider analysis filters while higher values correspond to narrower bandwidths. A small M value is therefore appropriate for a producing an "averaged" smooth spectrum that characterizes regions of major spectral energy. A value of M equal to or slightly over $2h$ is suitable for modeling individual harmonic frequency peaks, where h is the number of harmonics present.

Using LPC with Non-Human Primate Vocalizations

Applicability. While the mathematical solutions provided by various LPC approaches can be mathematically equivalent, the interpretations that are appropriate to each can be somewhat different. All approaches share some assumptions, but there are critical differences in the underlying models. Both the linear regression and inverse filter approaches require that each signal segment being analyzed is stationary, though this assumption is almost always violated to some degree. A second assumption is that the signal can be characterized in terms of frequency peaks alone. AR models do not characterize possible dips in the spectrum where energy present in certain portions of the frequency spectrum of the source waveform may have been attenuated by vocal tract filtering. Such anti-resonances clearly occur in non-human primate vocal production, but have been rarely investigated (although see Haimoff, 1983).

The linear regression approach is inherently the most broadly applicable LPC technique because it operates in the temporal domain and is not tied to any particular production model. It has been used effectively with a range of both tonal and noisy speech sounds as well as in other kinds of time-series analysis (e.g., Robinson, 1967). The inverse filter approach, in contrast, explicitly assumes a linear source-filter production system in which the source energy is a quasi-periodic or aperiodic waveform. As noted earlier, non-human primate call production appears to involve a source-filter system, although these animals may have a much greater variety of laryngeal vibration capabilities than do humans.

On the one hand, then, these two LPC approaches (and others not discussed here), differ in their theoretical applicability to various sounds. On the other hand, each yields the same basic mathematical outcomes. Therefore, while LPC is clearly appropriate for non-human primate vocalizations, its use requires some care in interpretation. The approach can safely be used as a smoothing technique for spectral peaks in which no conclusions are drawn concerning the underlying production system. Depending on implementation, the spectral model may be calculated in any number of ways, but the spectrum produced will be fundamentally the same. The extent to which the smoothed spectrum can be regarded as a vocal tract filtering function has not been investigated for non-human primates. The powerful synthesis techniques associated with LPC modeling have been little used in primatology, although Owren (1990) synthesized replica versions of vervet monkey snake and eagle alarm calls based on a source-filter model of production. These sounds were presented to vervets that had been trained to reliably classify the alarm calls in an operant-conditioning-based procedure and were immediately classified at rates comparable to those shown for natural calls. Other examples of applications of LPC to non-human primate vocalizations are described by Kojima (1992), Owren and Bernacki (1988), and Seyfarth, Cheney, Harcourt, and Stewart (1994).

Advantages. As an analysis tool, LPC has several advantages over sound spectrography, whether analog or digital. Once the analysis parameters have been selected, LPC provides automatic, replicable estimates of frequency peak locations, amplitudes, and bandwidths. LPC also provides substantial data reduction, because much of the variability in the original waveform is discarded. The result is a manageable, quantified characterization of the frequency spectrum. In addition, resynthesis of any analyzed signal can be readily performed, allowing the investigator to test empirically whether communicatively significant features of the acoustic signal have been captured by the model.

Detailed spectral attributes of tonal sounds can be automatically and reliably quantified by specifying a large number of coefficients, as shown in Figure 7. This approach is particularly applicable in cases where there is a high F_0 and few or no higher harmonics. Used in this way, the LPC model includes a separate peak for each harmonic and the spectral envelope should not be interpreted as a filter function. A smaller value is selected for M to obtain a smoothed and reduced envelope if the investigator wishes either to simplify the spectrum of a complex signal, whether noisy or tonal, or can assume a linear source-filter production mechanism and is attempting to recover filter characteristics. In these cases, each LPC peak encompasses a spectral region (a set of adjacent harmonics or a broad peak in a noisy spectrum) and may reflect a vocal tract resonance or formant.

Recently, it has been suggested that the spectral peaks captured by traditional formant analysis may not be the most important cues used in speech perception. Instead, overall spectral shape features that encode the global spectrum rather than individual peak characteristics may provide a more complete set of acoustic correlates than do formants. This approach has been applied to both consonants (Nossair and Zahorian, 1991) and vowels (Zahorian and Jagharghi, 1991, 1993) and is reminiscent of the dominant frequency measure discussed earlier. The overall spectral shape measure used by Zahorian and his colleagues is based on coefficients that are applicable to any frequency spectrum and are not tied to any particular production model. A similar approach may be appropriate in acoustic primatology as an extension of the dominant frequency approach that can more completely capture the full spectrum of information available to a listening animal.

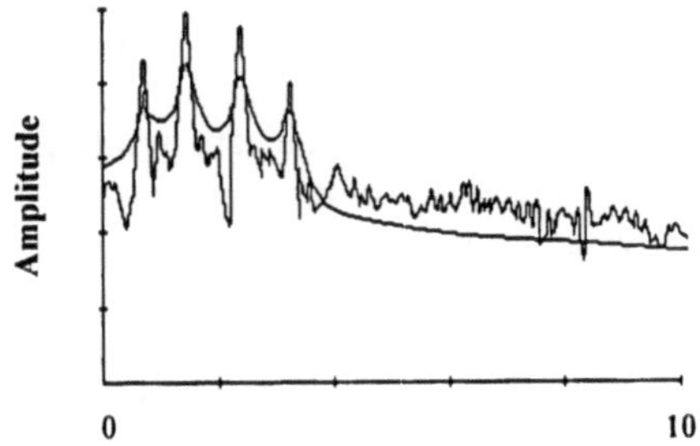

Figure 7. An 256-point FFT of a Japanese macaque coo call showing 4 high-amplitude harmonics characterized using a 9-coefficient, 256-point LPC model.

Spectral Tilt. As noted earlier, vocal fold vibration in human speech production produces a signal whose spectrum exhibits a frequency-dependent energy roll-off (Figure 1). This energy loss is partially but not entirely offset by radiation effects associated with the lips and mouth opening. To compensate further and thereby reveal the locations of possible high-frequency formants in a speech sound, the analog spectrograph included a "high-frequency shaping" circuit that artificially increased spectral energy levels at a rate of 6 dB per octave. In DSP, preemphasis algorithms are used for the same purpose, changing the overall tilt of the spectrum through selection of settings that might range from 0 (no preemphasis) to 100 (full preemphasis). In LPC, the effect of preemphasis is to preselect one of the coefficient values used in the model. If preemphasis is set to 0, this coefficient is free to vary in accordance with the overall tilt of the frequency spectrum and its value can be used as a measure of that spectral tilt. There are, in fact, probably relatively few occasions when preemphasis should be applied to non-human primate calls.

Owren and Bernacki (1988) used the un-preemphasized coefficient as a measure of overall spectral slope (*TILT*) in vervet monkey alarm calls (i.e., *TILT* = C(1) * 100). Using this measure, a generally flat spectral shape results in a *TILT* value close to 0, while frequency spectra that rise or fall across the range of analysis produce positive (0 to 100) and negative (0 to -100) *TILT* values, respectively. A measure like *TILT* provides some advantages when compared to the bandwidth measure discussed earlier. Specifically, while bandwidth results can depend on the absolute levels of high-frequency energy, *TILT* measures the relative shape of the overall spectrum and is therefore less likely to be unduly influenced by variability within and between individuals or by event-specific sound-production and audio-recording factors. Spectral slope can also be measured by exporting the individual points making up the smoothed spectral model to a spreadsheet program and fitting a straight line to the model using linear regression.

CHARACTERIZING THE SOURCE FUNCTION

Characterizing the source energy of a communication signal represents a very basic aspect of analysis, regardless of species. In the context of the primate vocal production system, several different questions are involved. One important question concerns the nature of the source. In human speech production, the vocal folds can produce both periodic and aperiodic energy, and turbulent air flows created elsewhere in the vocal tract can also be used as noisy energy sources. Due to the presence of rigid lips on the vocal folds of many non-human primates, these animals appear to have a much wider variety of laryngeal production capabilities (Schön Ybarra, this volume). Many species have in fact been observed to produce vocalizations that are virtually pure sinewaves or very fast trills, can exhibit frequency upsweeps and downsweeps of several octaves or more, or show a controlled, aperiodic pattern of vocal fold vibration not found in humans.

When a signal's source energy derives from vocal fold vibration, it is of interest to document both momentary F_0 rates and the stability or mode of that vibration. While these tasks may appear to be straightforward in principle, each can be quite complex. Hess (1983) has outlined the many difficulties involved in his comprehensive overview of F_0 determination in speech signals, an endeavor he terms "pitch extraction." Restated somewhat more generally, the problems he describes also apply to non-human primate vocalizations. First, vocalizations are never fully stationary--their characteristics change over time. Measuring a changing signal is significantly more difficult than making the same measurements in a stationary signal and many non-human primate calls probably present more extreme challenges in this regard than do most human speech sounds. Second, there are pronounced differences between individual vocalizers in vocal fold characteristics, supralaryngeal vocal tract anatomy, and vocal tract dynamics during sound production. It is therefore difficult to design a single measure or set of measures that can be used to characterize vibration patterns across individuals producing the same sounds or a single individual producing a variety of sounds. Third, F_0 can vary over a very large range, particularly in non-human primates. Finally, source function vibrations exhibit both uncontrolled and controlled variability. In humans, such variability results from occasional but routine irregularity, use of different vocal registers (e.g., vocal fry and falsetto), temporary changes in vocal fold condition (e.g., hoarseness), as well as pathological vocal fold conditions. As noted, these problems are exacerbated in non-human primates, although clearly pathological callers are probably excluded from most samples.

A small number of both time- and frequency-domain measures used in pitch extraction are described here. Time-domain algorithms are typically designed to locate one or more of the following aspects in the waveform: the fundamental frequency, a quasi-periodic time structure, an alternation of high and low amplitudes, or points of discontinuity. These approaches can vary in complexity, depending on implementation. While each can provide good results, all can also fail. Frequency-domain techniques usually involve analyzing the signal with maximal frequency resolution and identifying the frequency peak that most clearly corresponds to the F_0.

Time-Domain Pitch Extraction

In the time domain, the basic task of the pitch extraction algorithm is to correctly recognize individual pitch periods within the time-domain waveform and to mark each cycle at a corresponding point--typically either where a sampled or interpolated value is zero (zero-crossing approach) or at a high-amplitude point (threshold-crossing approach). One of the most basic time-domain techniques is to simply count the number of zero- or threshold-crossings remaining after preprocessing. An experienced observer can often determine pitch periods in a waveform more accurately by eye than can an automated algorithm, but the process is both tedious and subjective. While a variety of quite sophisticated preprocessing approaches have been developed for isolating F_0 energy in speech sounds, they typically do not perform well with non-human primate vocalizations without modification or disabling of preprocessing components. These components often include low-pass, bandpass, or inverse filtering.

Preprocessing based on low-pass and bandpass filtering removes energy at frequencies outside the F_0 region at an early stage. The inverse filter method is more complex in that frequency-domain modeling precedes waveform filtering. In this approach, a spectral model is generated, then inverted, and finally used to filter the signal. The overall strategy is to attenuate the energy at harmonics above the F_0 that can interfere with zero-crossing or threshold-crossing algorithms. While promising for non-human primate vocalizations as well, the inverse filter method is unlikely to be effective without significant modification. Due to differences in F_0 values and vocal tract filtering characteristics, automated inverse filtering performed by a speech-related routine is likely to remove not only higher harmonics, but also the F_0 itself. A signal that has a rich harmonic structure will be more amenable to the inverse filter approach than will a signal with fewer or more widely spaced harmonics. LPC is often used to produce the inverse filter, which can be an effective strategy but requires use of appropriate parameter settings. Figure 8 shows an expanded and filtered version of the baboon grunt waveform from Figure 4 to illustrate the zero-crossings detection process.

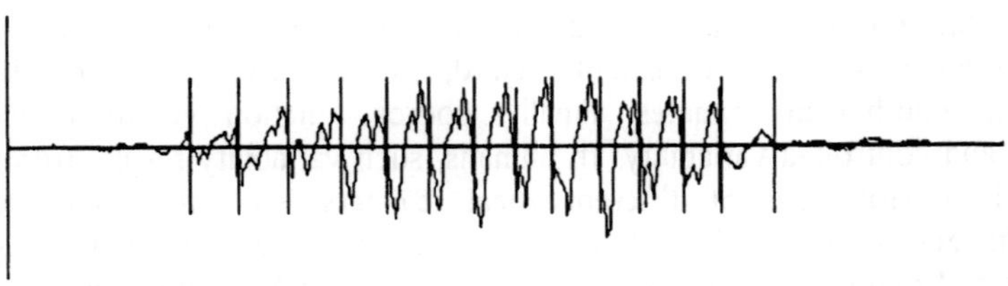

Figure 8. An expanded version of the baboon grunt shown in Figure 4. Both low-pass and inverse filters have been used to isolate F_0-related zero-crossing points, which are marked by vertical lines.

Frequency-Domain Pitch Extraction

Measuring from a Digital Spectrogram. Frequency-domain techniques explicitly rely on spectral analysis in the preprocessing stage. The simplest approach is to produce a digital spectrogram that allows direct measurement of the F_0. If F_0 energy is not present, higher harmonic values can serve the same purpose. This strategy lends itself well to interactive analysis using FFT-based spectrograms (representations produced by other means like LPC are also appropriate). To illustrate this approach, the F_0 energy of a Japanese macaque coo call, whose time-domain waveform is shown in Figure 9a, has been isolated on the digital spectrogram shown in Figure 9b. User-controlled cursors are then used to mark the time and frequency boundaries of F_0 energy, allowing automatic extraction of the highest-amplitude frequency value within each column. A smoothed version of the resulting F_0 contour is shown in Figure 9c. F_0 measurements made in this fashion can be efficient and reliable, but are only as accurate as allowed by the frequency resolution of the underlying power spectra or spectral models. Frequency resolution of F_0 values can be increased significantly in signals that show strong high-amplitude harmonics by measuring the *n*th harmonic and dividing each resulting value by *n*.

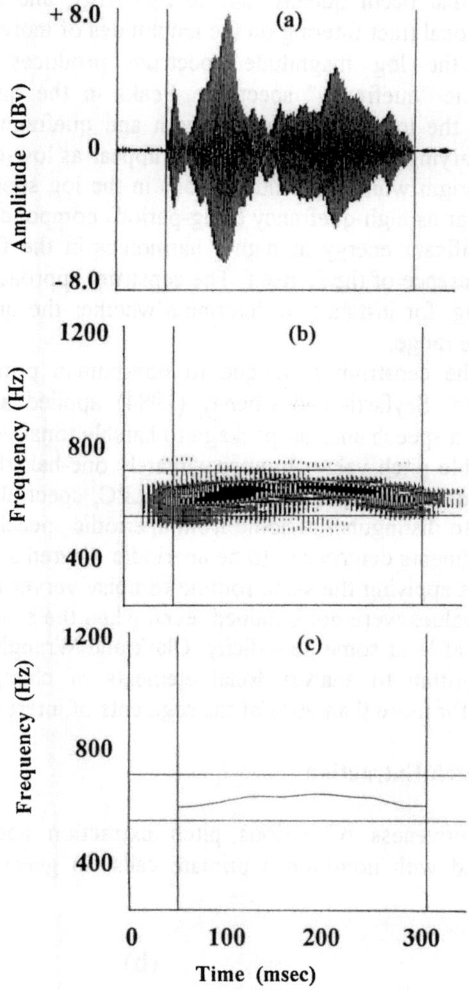

Figure 9. Frequency-domain pitch extraction from a Japanese macaque coo call. The waveform (a) is analyzed using 2048-point FFT's, producing a digital spectrogram (b). The F_0 contour is then manually bracketed using computer screen cursors and F_0 energy is automatically extracted (c).

Cepstral Pitch Extraction. Speech researchers commonly use the "cepstrum," or FFT of the log-magnitude FFT spectrum (Noll, 1964) in pitch extraction. This approach is explicitly based on the source-filter model of vocal production in that the goal is to separate periodic components of a signal (i.e., the F_0 and its harmonics) from non-periodic components traceable to vocal tract effects. First, Fourier analysis is used to calculate a power spectrum over a short segment of the waveform. The analysis window must include at least two pitch periods, but for greatest accuracy should not include changing F_0 values. The second step is to calculate a log transform of the frequency spectrum, which produces a spectrum with greatly compressed amplitude values. This log spectrum itself is then treated as if it were a time-domain waveform. As shown in Figure 10a, this "waveform" exhibits amplitude fluctuations that occur quickly, due to F_0 energy, and slowly, due to the now compressed effects of vocal tract filtering on the amplitudes of individual harmonics. Taking the inverse FFT of the log magnitude spectrum produces a novel time-domain representation called the "quefrency" spectrum. Peaks in the quefrency domain reflect periodicity detected in the log magnitude spectrum and quefrency values correspond to pitch periods. Slowly varying amplitude fluctuations appear as low-quefrency (short-period) components near the origin while rapid fluctuations in the log spectrum (spanning smaller frequency ranges) appear as high-quefrency (long-period) components. Effective use of the cepstrum requires significant energy at higher harmonics in the frequency spectrum, but does not require the presence of the F_0 itself. The cepstrum approach is often supplemented by additional processing, for instance to determine whether the quefrency peaks detected fall within an acceptable range.

Applications of the cepstrum technique to non-human primate vocalizations have produced mixed results. Seyfarth and Cheney (1984) applied a cepstrally based pitch extraction routine from a speech analysis package to harshly tonal vervet monkey grunt calls and recovered reasonable pitch values in approximately one-half the cases. The routine in question included extensive preprocessing based on LPC, spectral tilt, zero-crossing rate, and statistical analysis to distinguish periodic from aperiodic speech segments, returning an F_0 value of zero for segments determined to be aperiodic. Owren and Bernacki (1988) were much less successfull in applying the same routine to noisy vervet monkey snake and eagle alarm calls. Usable F_0 values were not obtained, even when the alarm calls were deemed by other means to exhibit at least some periodicity. Clark and Wrangham (1993) used a more generic cepstrum algorithm to analyze tonal elements of chimpanzee pant hoots and obtained usable values for more than 90% of the segments of interest.

Conclusions About Pitch Extraction

As yet, the effectiveness of various pitch extraction approaches has not been systematically evaluated with non-human primate calls. In general, direct measurements

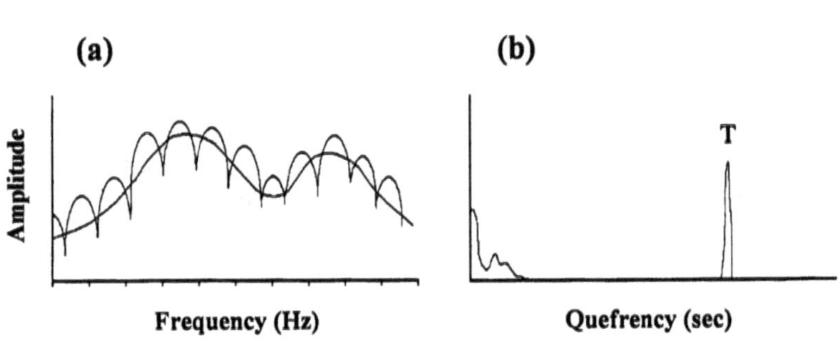

Figure 10. Taking the log magnitude of a Fourier spectrum produces a waveform-like cepstrum (a). Computing the inverse Fourier transform results in a quefrency domain representation (b) where slowly and quickly varying amplitude fluctuations in the cepstrum produce separable periodicity peaks. The F_0 periodicity is represented by quefrency peak at time "T." Redrawn after Noll (1964).

from digital spectrograms probably provide the greatest robustness to noise and spurious signal components. On the other hand, this and other frequency-domain approaches require substantial computation and can be slower than time-domain techniques. Cepstral analysis is perhaps the most computationally intensive, but has been widely used in speech research and can provide very precise results. This approach can be most successfully applied to non-human primate vocalizations when the calls in question are more rather than less periodic and show a larger rather than a smaller number of harmonics. Acoustic primatology would benefit from greater availability of cepstrum-based routines with flexible preprocessing, basic extractor, and postprocessing components. All pitch extraction techniques are found to fail under some circumstances, which places a burden on the investigator to consistently monitor the performance of each routine being used.

Vocal Perturbation Measures

Vocal perturbation refers to the stability of vibration in a periodic (or approximately periodic) energy source. Such stability can be characterized through measuring cycle-to-cycle variation in F_0 in the time-domain waveform or the relative prominence of aperiodic components, or interharmonic energy, in the frequency spectrum (see Baken, 1987). The best known measures of F_0 stability are "jitter" and "shimmer." A number of different approaches have been used to characterize interharmonic energy, including the "spectral noise level," "harmonic-to-noise ratio," and "signal-to-noise ratio" measures.

Jitter and Shimmer. Hypothetically, if a periodic waveform is produced by a perfectly stable vibrating energy source, each cycle will be identical to preceding and following cycles. Naturally produced periodic waveforms, however, inevitably exhibit cycle-to-cycle variability related to the characteristics of both the physical structures that are in motion and the underlying control system regulating that motion. Cycle-to-cycle variation in the rate (or period) of vibration is measured as signal jitter. As lower-frequency (longer-period) F_0's are associated with higher jitter values in speech production (Baken, 1987), normalization strategies are required to allow meaningful comparisons of subjects differing in vocal pitch. Jitter is therefore often computed by dividing the mean perturbation value by the mean frequency (or period). One approach is to add the absolute values of all cycle-to-cycle frequency differences, divide the resulting total by the mean F_0 computed for that segment, and multiplying by 100 to produce a percentage measure (Hollien, Michel, and Doherty, 1973). A second approach is to normalize each cycle-to-cycle difference individually, using the mean F_0 for that pair of cycles (Doherty and Shipp, 1988). Shimmer is analogous to jitter, but concerns cycle-to-cycle variability in peak amplitude. This measure is expressed either as a percentage value or in deciBels and does not vary systematically with F_0 changes in speech production. Shimmer computations are, however, strongly affected by the amplitude changes associated with signal onsets and offsets. As jitter values can also be influenced by F_0 instabilities associated with these transitions, the most reliable results may be obtained if only the central portion (e.g., the middle 60%) of the signal of interest is included in analysis. Figure 11 illustrates the application of jitter and shimmer measurements to coo calls recorded from two adult female Japanese macaques. Both auditorally and visually, the call illustrated on the left in Figure 11 is slightly more tonal than is the call on the right. This subtle difference is also apparent as small differences in the cycle-to-cycle stability of the waveforms from which the spectrograms were produced.

Measuring Interharmonic Energy. While cycle-to-cycle variability in the glottal source energy of human speech is known to contribute to vocal perturbation, it is not the sole source of the aperiodicities that can both be perceptually salient and appear as interharmonic energy on spectrograms. Emanuel and Sansone's (1969) spectral noise level

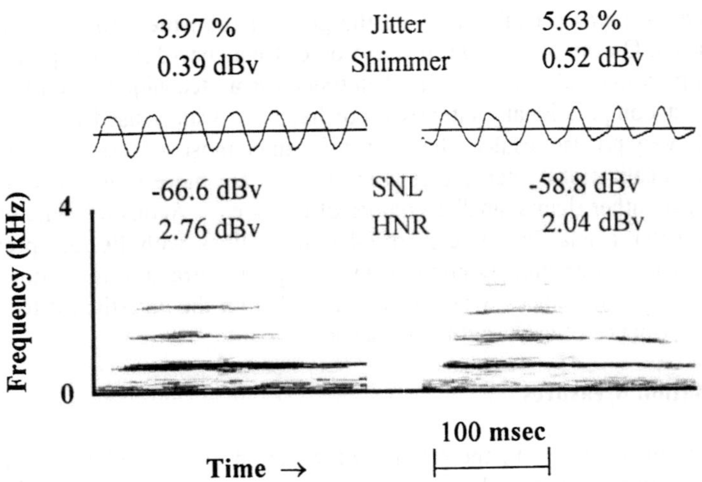

Figure 11. Spectrograms of coo calls from two adult female Japanese macaques that were selected based on a slight difference in vocal clarity. The slightly more tonal call shown on the left produced smaller vocal perturbation and interharmonic noise vlaues than did the hoarser call on the right. Representative 10-msec waveform segments are shown above the spectrograms.

(SNL) measure is based on the premise that perfectly periodic signals should show no interharmonic energy and that the relative prominence of the noise that inevitably does occur in naturally produced speech can be characterized relative to this ideal state by measuring minimum amplitude levels occurring in the frequency spectrum. SNL is computed by first producing a series of high-resolution FFT spectra for the signal of interest and determining the minimum amplitude levels over contiguous 100-Hz segments of each. SNL values can then be calculated over the entire FFT series based on individual frequency segments or as a single, grand mean. This measure has been tested with a variety of human subjects and has been found to be correlated with perceptual measures of vocalic roughness (see Baken, 1987; Newman and Emanuel, 1991). The coo-call frequency spectra shown in Figure 11 show a slight difference both in visible interharmonic energy and overall SNL values.

A difficulty with the SNL measure is that only interharmonic noise levels are taken into account, while energy levels in the harmonics themselves are ignored. Yumoto and his colleagues have therefore described a technique for quantifying both aperiodic components and periodic components (Yumoto, Gould, and Baer, 1982; Yumoto, Sasaki, and Okamura, 1984). In their harmonic-to-noise ratio (HNR) approach, a sound segment is treated as a combination of two independent components--perfectly periodic energy and unpatterned, random noise. The noise component is further assumed to vary randomly around a mean value of zero, producing cancellation of these components if the signal is averaged over a significant time interval. In contrast, the mean of periodic components will reveal reliable positive and negative amplitudes. If the waveform is segmented into equal lengths, then, the overall mean of the resulting large number of waveform segments should be noise-free. If the averaged waveform is subtracted on a point-by-point basis from each of the original waveform segments, the "remainder" waveforms retain only the energy attributable to the noise component associated with each. The HNR value, expressed in deciBels, is the ratio of the amplitude of the average (noise-free) waveform and the mean amplitude of the remainder waveforms. Yumoto et al. (1982, 1984) found normal subjects to exhibit significantly lower HNR values than did subjects with laryngeal pathologies and an analogous result occurs when the HNR is calculated for the coo calls shown in Figure 11.

Klingholz's (1987) signal-to-noise ratio (SNR) measure may be the most elegant currently available approach to measuring interharmonic energy. This conceptually simple but computationally sophisticated technique takes advantage of the random characteristics of noise components and the deterministic nature of harmonic components by approximating the latter using a spectral synthesis-by-analysis, or reconstruction algorithm. Describing the procedure in the simplest possible terms, FFT's are used to first identify the frequencies, amplitudes, and bandwidths of all harmonics present in the spectrum. The harmonic spectrum is then synthesized using these values and checked against the actual spectrum. Given an acceptable match, the SNR is computed as the ratio of the harmonic energy level in the synthetic signal to the spectral noise level of the natural sound. Applying the SNR procedure to synthetic vowels, Klingholz (1987) found it to be unaffected by F_0 variation over a 80- to 360-Hz range. SNR values were then shown to be significantly lower in the natural speech of individuals with laryngeal pathologies and hoarse- or rough-sounding voices.

Conclusions About Vocal Perturbation Measures

The applied, practical nature of most available vocal perturbation measures suggests that these techniques can be readily used with non-human primate calls. As noted earlier, noisiness is a very common feature of non-human primate vocalizations but is difficult to characterize using traditional spectrographic methods, whether analog or digital. Within a given vocal repertoire, calls may show continuous grading along a noisiness dimension, including sounds that are purely tonal, entirely noisy, or intermediate in nature. It is reasonable to suppose that in many cases the degree of noisiness present in a call may bear an important relationship either to call function, the individual voice characteristics of the caller, or both. Due to the difficulties of characterizing noisiness objectively, however, this aspect of non-human primate vocalizations has received very little attention. Measures like jitter, shimmer, and HNR have the potential to detect quite subtle differences in vocal perturbation but require identification of individual waveform cycles in the time domain. They may therefore be best-suited to calls that are unambiguously periodic. Measures like SNL and SNR, on the other hand, are probably very useful with sounds showing significant noise components.

CONCLUSIONS

Given the variety of vocal signals and production mechanisms found among non-human primates, there is no panacea for the many analysis challenges faced by acoustic primatologists. Nonetheless, there is room for significant improvement both in the scope and sophistication of the techniques that are typically used. Given the availability of inexpensive microcomputers, powerful digital signal processing software, and high levels of applicable research activity in related areas of acoustic communication, an opportunity exists to realize such improvement. The challenge to acoustic primatologists, is that we must be willing to educate both ourselves and each other to become well-informed, proactive consumers of digital processing technology.

While some non-human primate vocalizations closely resemble sounds produced by avians, cetaceans, or other non-human animals, most do not. Therefore, any single one of the available bioacoustically oriented analysis packages is unlikely to satisfy all the analysis needs of an acoustic primatologist. On the other hand, many otherwise applicable routines found in speech-oriented analysis packages are too specialized or inflexible for effective use with non-human primate calls. Overall, then, there is ample reason for eclectic selection of both analysis techniques and implementations.

In spite of the difficulties that can arise in using analysis tools meant for speech applications, we have specifically argued that the source-filter approach to speech production is a valuable perspective to adopt in analyzing non-human primate vocalizations. Even if modification or adaptation is required, a number of speech-related techniques that have their roots in this perspective can be fruitfully used. To date, non-human primate vocalization analyses have tended to emphasize features that are visually prominent in sound spectrographic representations--both in the original analog-derived format and in digitally based versions that emulate that form. But there is little evidence to indicate that visually salient features in spectrograms of non-human primate calls are also perceptually important to the species in question. Rather, reliance on the traditional spectrographic representation has tended to encourage the use of overly simplistic measures that may be unlikely candidates for effectively bridging between complex acoustic properties and requisitely sophisticated auditory processing.

In addition to improving the objectivity, precision, and scope of acoustic analysis work already being conducted, the measures described in this chapter (and others like them) may also provide some novel approaches to the study of non-human primate behavior. One issue of great recent interest, for instance, has been whether non-human primate calls (and animal acoustic signals in general) are primarily motivational in nature or can be considered symbolic and referential (see for instance, Marler, Evans, and Hauser, 1992). As discussed by Scherer and Kappas (1988), resolution of this debate is made all the more difficult by the absence of reliable external indicators of internal motivational states. While some notable attempts have been made to examine non-human primate vocalizations produced under different emotional circumstances (e.g., Jürgens, 1979), complications arise both with respect to induction and verification of arousal state, and characterizations of the calls produced. The vocal perturbation measures described here provide one potentially promising approach to indexing arousal states. These measures may be particularly useful in capturing very subtle differences in call acoustics, both within and between different individual callers. Scherer and Kappas (1988) have further suggested that a variety of the frequency spectrum measures developed in speech research can be extended to measuring the acoustic effects of arousal in non-human primate vocalizations as well.

In conclusion, while this chapter has specifically emphasized the weaknesses we perceive in existing analysis methods and practices, it is also important to acknowledge that a wealth of useful information concerning the acoustic characteristics of non-human primate calls has been painstakingly provided by careful work on the part of many investigators. However, a crossroads of sorts has been reached in acoustic primatology in that powerful technologies and analysis approaches developed in other areas are becoming available for use. Taking advantage of this opportunity to improve our field will require a significant investment of time and energy, but is nonetheless the road we must choose.

ACKNOWLEDGEMENTS

Equipment used to illustrate the analyses described in this chapter was supported in part by Faculty Development and Seed Money awards from the University of Colorado at Denver to Michael Owren. Analog spectrograms were produced by Robert Bernacki using the spectrographic facilities of the Speech Research Laboratory at Indiana University (supported in part by NIH grant DC-0111-17) which is directed by David Pisoni. The editors, Marc Hauser, and two anonymous reviewers provided many helpful comments on previous versions of this chapter. Special thanks to Jo-Anne Bachorowski for help in countless aspects of preparing this work, to Michael Graham for the artwork used in Figure 2, and to Jan Mitchell for support and much patience.

REFERENCES

Andrew, R.J., 1976, Use of formants in the grunts of baboons and other nonhuman primates, *in:* "Origins and evolution of language and speech," S.R. Harnad, H.D. Steklis, and J. Lancaster, eds., *Ann. N.Y. Acad. Sci.* 280: 673-693.

Baken, R.J., 1987, "Clinical Measurement of Speech and Voice," Little-Brown, Boston.

Beecher, M.D., 1988, Spectrographic analysis of animal vocalizations: Implications of the "uncertainty principle", *Bioacoust.* 1:187-208.

Cheney, D.L and Seyfarth, R.M., 1990, "How Monkeys See the World," University of Chicago, Chicago.

Chiba, T. and Kajiyama, J., 1941, "The Vowel: Its Nature and Structure," Tokyo-Kaiseikan, Tokyo.

Clark, A. and Wrangham, R.W., 1993, Acoustic analysis of wild chimpanzee pant hoots: Do Kibale forest chimpanzees have an acoustically distinct food arrival pant hoot? *Amer. J. Primatol.* 31:99-109.

Cooke, M., Beet, S., and Crawford, M., eds., 1993, "Visual Representations of Speech Signals," Wiley and Sons, New York.

Cooley, J.W. and Tukey, J.W., 1965, An algorithm for the machine calculation of complex Fourier series, *Math. Comp.* 19:297-301.

Davis, L.I., 1964, Biological acoustics and the use of the sound spectrograph, *Southwest. Nat.* 9:118-145.

Doherty, E.T. and Shipp, T., 1988, Tape recorder effects on jitter and shimmer extraction, *J. Speech Hear. Res.* 31:485-490.

Emanuel, F. W. and Sansone, F.E., Jr., 1969, Some spectral features of "normal" and simulated "rough" vowels, *Folia Phoniat.* 21:401-415.

Fant, G., 1960, "The Acoustic Theory of Speech Production," Mouton, The Hague.

Flanagan, J.L., 1972, "Speech Analysis, Synthesis, and Perception", 2nd ed., Springer-Verlag, Berlin.

Gautier, J.-P., 1983, Qualities and drawbacks of radiotransmitted vocalizations of monkeys through laryngophones: New perspectives in analysis, *Folia Primatol.* 41: 218-230.

Gautier, J.-P. and Gautier, A., 1977, Communication in Old World monkeys, *in:* T.A. Sebeok, ed., "How Animals Communicate," Indiana University, Bloomington.

Geist, F.D., 1933, Nasal cavity, larynx, mouth and pharynx, *in:* "The Anatomy of the Rhesus Monkey," C.G. Hartman and W.L. Straus, Jr., eds., Hafner, New York.

Haimoff, E.H., 1983, Occurrence of anti-resonance in the song of the siamang (*Hylobates syndactylus*), *Amer. J. Primatol.* 5:249-256.

Hauser, M.D. and Fowler, C.A., 1991, Declination in fundamental frequency is not unique to human speech: Evidence from non-human primates, *J. Acoust. Soc. Am.* 91:363-369.

Hauser, M.D. and Schön Ybarra, M., 1994, The role of lip configuration in monkey vocalizations: Experiments using xylocaine as a nerve block, *Brain Lang.*, 46:232-244.

Hauser, M.D. , Evans, C.S., and Marler, P., 1993, The role of articulation in the production of rhesus monkey, *Macaca mulatta*, vocalizations, *Anim. Behav.* 45:423-433.

Hess, W., 1983, "Pitch Determination of Speech Signals," Springer-Verlag, New York.

Hill, W.C.O. and Booth, A.H., 1957, Voice and larynx in African and Asiatic *Colobidae*, *Bombay Nat. Hist. Soc.* 54:309-321.

Hilloowala, R.A., 1975, Comparative anatomical study of the primate hyoid apparatus, *Am. J. Anat.* 142:367-385.

Hilloowala, R.A., 1976, The primate hyolaryngeal apparatus and herbivorous modifications, *Acta Anat.* 95:279-286.

Hollien, H., Michel, J., and Doherty, E.T., 1973, A method for analyzing vocal jitter in sustained phonation, *J. Phonet.* 1:85-91.

Joos, A.M., 1948, Acoustic phonetics, *Language* Supl. 24:1-136.

Jürgens, U., 1979, Vocalization as an emotional indicator. A neuroethological study in the squirrel monkey, *Behaviour* 69:88-117.

Kelemen, G., 1938, Comparative anatomical studies on the junction of larynx and resonant tubes, *Acta Otolaryngol.* 26:276-283.

Kelemen, G., 1948, The anatomical basis of phonation in the chimpanzee, *J. Morph.* 82:229-256.

Kelemen, G., 1969, Anatomy of the larynx and the anatomical basis of vocal performance, *Chimp.* 1:165-186.

Kelemen, G. and Sade, J., 1960, The vocal organs of the howling monkeys (*Alouatta palliata*), *J. Morph.* 107:123-140.
Klingholz, F., 1987, The measurement of the signal-to-noise ratio (SNR) in continuous speech, *Speech Commun.* 6:15-26.
Kojima, S., 1992, Voco-auditory functions of the chimpanzee: Early vocal development. Unpublished manuscript.
Lieberman, P., 1968, Primate vocalizations and human linguistic ability, *J. Acoust. Soc. Am.* 44:1574-1584.
Lieberman, P., 1972, "The Speech of Primates," Mouton, The Hague.
Lieberman, P., 1975, "On the Origins of Language: An Introduction to the Evolution of Human Speech," Macmillan, New York.
Lieberman, P. and Blumstein, S.E., 1988, "Speech Physiology, Speech Perception, and Acoustic Phonetics," Cambridge University, New York.
Lieberman, P., Klatt, D.H., and Wilson, W.H., 1969, Vocal tract limiations on the vowel repertoires of rhesus monkeys and other nonhuman primates, *Science* 164:1185-1187.
Loughlin, P.J., Atlas, L.E., and Pitton, J.W., 1993, Advanced time-frequency representations for speech processing, *in:* "Visual Representations of Speech Signals," M. Cooke, S. Beet, and M. Crawford, eds., Wiley and Sons, New York.
Makhoul, J., 1975, Linear prediction: A tutorial review, *Proc. IEEE* 63:561-580.
Markel, J.D. and Gray, A.D. Jr., 1976, "Linear Prediction of Speech," Springer-Verlag, New York.
Marler, P., Evans, C.S., and Hauser, M.D., 1992, Animal signals: Motivational, referential, or both? *in:* "Nonverbal Vocal Communication: Comparative and Developmental Approaches," H. Papousek, U. Jürgens, and M. Papousek, eds., Cambridge University, New York.
Negus, V.E., 1949, "The Comparative Anatomy and Physiology of the Larynx," Hafner Publishing, New York.
Newman, R.A. and Emanuel, F.W., 1991, Pitch effects on vowel roughness and spectral noise for subjects in four musical voice classifications, *J. Speech Hear. Res.* 34:753-760.
Noll, A.M., 1964, Short-time spectrum and "cepstrum" techniques for vocal-pitch detection, *J. Acoust. Soc. Am.* 36:296-302.
Nossair, Z.B. and Zahorian, S.A., 1991, Dynamic spectral shape features as acoustic correlates for initial stop consonants, *J. Acoust. Soc. Am.* 89: 2978-2991.
O'Shaugnessy, D., 1987, "Speech Communication," Addison-Wesley, Reading, Mass.
Owren, M.J., 1990, Classification of alarm calls by vervet monkeys (*Cercopithecus aethiops*): II. Synthetic calls, *J. Comp. Psychol.* 104:29-41.
Owren, M.J. and Bernacki, R.H., 1988, The acoustic features of vervet monkey alarm calls, *J. Acoust. Soc. Am.* 83:1927-1935.
Owren, M.J. and Bernacki, R.H., in press, Applying linear predictive coding (LPC) to the frequency analysis of animal acoustic signals, *in:* "Recent Advances in the Study of Animal Communication," S.L. Hopp, C.S. Evans, and M.J. Owren, eds., Springer-Verlag, New York.
Owren, M.J., Dieter, J.A., Seyfarth, R.M., and Cheney, D.L., 1991, Food calls of adult female Japanese macaques, their normally raised offspring, and offspring cross-fostered between species, *Behaviour* 120:218-231.
Owren, M.J., Linker, C.D., and Rowe, M.P., 1993, Acoustic features of tonal "grunt" calls in baboons, *J. Acoust. Soc. Am.* 94:1823.
Ploog, D., 1988, Neurobiology and pathology of subhuman vocal communication and human speech, *in:* "Primate Vocal Communication," D.Todt, P. Goedeking, and D. Symmes, eds., Springer-Verlag, New York.
Rabiner, L.R. and Gold, B., 1975, "Theory and Application of Digital Signal Processing," Prentice-Hall, Englewood Cliffs, New Jersey.
Rabiner, L.R. and Shafer, R.W., 1978, "Digital Processing of Speech Signals," Prentice-Hall, Englewood Cliffs, New Jersey.
Richman, B., 1976, Some vocal distinctive features used by gelada monkeys, *J. Acoust. Soc. Am.* 60:718-724.
Robinson, E.A., 1967, Predictive decomposition of the time series with application to seismic exploration, *Geophys.* 32:418-484.
Rowell, T.E., 1962, Agonistic noises of the rhesus monkey (*Macaca mulatta*), *Symp. Zool. Soc. Lond.* 8:91-96.
Rowell, T.E. and Hinde, R.A., 1962, Vocal communication by the rhesus monkey (*Macaca mulatta*), *Proc. Zool. Soc. Lond.* 138:279-294.
Scherer, K. and Kappas, A., 1988, Primate Vocal Expression of Affective State, *in:* "Primate Vocal Communication," D.Todt, P. Goedeking, and D. Symmes, eds., Springer-Verlag, New York.

Schön, M.A., 1971, The anatomy of the resonating mechanism in howling monkeys, *Folia Primatol.* 15:117-132.

Seyfarth, R.M. and Cheney, D.L., 1984, The acoustic features of vervet monkey grunts, *J. Acoust. Soc. Am.* 75:1623-1628.

Seyfarth, R.M., Cheney, D.L., Harcourt, A.H., and Stewart, K.,1994, The acoustic features of gorilla double-grunts and their relation to behavior, *Amer. J. Primatol.* 33: 31-50.

Steiglitz, K., 1974, "An Introduction to Discrete Systems," Wiley and Sons, New York.

Stevens, K.N. and House, A.S., 1955, Development of a quantitative description of vowel articulation, *J. Acoust. Soc. Am.* 27:484-493.

Stoddard, P.K., 1990, Audio computers - theory of operation and guidelines for selection of systems and components, *Bioacoust.* 2:217-239.

Wakita, H., 1976, Instrumentation for the study of speech acoustics, *in:* "Contemporary Issues in Experimental Phonetics," N.J. Lass, ed., Academic, New York.

Yumoto, E., Gould, W.J., and Baer, T., 1982, Harmonics-to-noise ratio as an index of the degree of hoarseness, *J. Acoust. Soc. Am.* 71:1544-1550.

Yumoto, E., Sasaki, Y., and Okamura, H., 1984, Harmonics-to-noise ratio and psychophysical measurement of the degree of hoarseness, *J. Speech Hear. Res.* 27: 2-6.

Zahorian, S.A. and Jagharghi, A.J., 1991, Speaker normalization of static and dynamic vowel spectral features, *J. Acoust. Soc. Am.* 90:67-75.

Zahorian, S.A. and Jagharghi, A.J., 1993, Spectral-shape features versus formants as acoustic correlates for vowels, *J. Acoust. Soc. Am.* 94:1966-1092.

Zhinkin, N.I., 1963, An application of the theory of algorithms to the study of animal speech - methods of vocal intercommunication between monkeys, *in:* "Acoustic Behavior of Animals," R.G. Busnel, ed., Elsevier, Amsterdam.

Zimmermann, E., 1981, First record of ultrasound in two prosimian species, *Naturwissen.* 68:531.

ARTIFICIAL NEURAL NETWORKS FOR ANALYSIS AND RECOGNITION OF PRIMATE VOCAL COMMUNICATION

Axel Zimmermann

Universität Stuttgart
Institut für Informatik
Breitwiesenstr. 20-22
D-70565 Stuttgart

INTRODUCTION

A number of simulated neurons form an artificial neural network. They are connected together in a way similar to biological neural systems. It is the high interconnection rate and the build-in parallelism of these networks that allow completely different processing capabilities in comparison to conventional computer systems. The performance of current pattern recognition systems is far below of humans' abilities. Artificial neural networks offer the potential of providing new approaches to such problems.

In 1943, Warren McCulloch and Walter Pitts published the idea of a new processing element for building up large networks to solve complex logical problems (McCulloch and Pitts, 1943). Some years later, Rosenblatt introduced the ability of learning in his neural network model (Rosenblatt, 1958). Today there are many types of artificial neural networks, but all have three things in common. A neural network can be described in terms of its individual neurons, the connections between them and its learning rule. Together, these three things constitute the neural network model.

There are many applications for artificial neural networks, for example pattern recognition, trend prediction, filtering, extrapolation and the analysis of large amounts of data. There are two main streams of developments for acoustical signal processing: The analysis and classification of technical signals (such as motor vibration) and the automatic speech and speaker recognition for human and nonhuman primates. For both, neural network models must compete with highly developed conventional statistical approaches (O'Malley, 1990; Strathmeyer, 1990).

Speech recognition finds frequent application in product inspection, command and control, manufacturing, medicine and in systems for physically handicapped people. Speaker recognition has been applied in security systems and as user identification. The current speech recognition systems performing best use statistic algorithms (Bartkova and Jouvet, 1991; Deng and Erler, 1992). Neural nets for speech recognition have been explored as part of the recent resurgence of interest in this area and are now reaching the accuracy of statistical algorithms (Iso and Watanabe, 1990; Kosko, 1992).

A new application for neural networks is the analysis and classification of animal sounds. Possible tasks are the recognition of the sender of a sound, the recognition of the population from which the sender stemms from, the classification of a sound repertoire of a species into distinct sound patterns and the identification of the pattern development during ontogeny. With real time neural systems realized in hardware, it will be possible to automatically recognize and classify animal sounds in the field or in the laboratory.

This paper gives a brief introduction to the theory of artificial neural networks and outlines current research activities in neural acoustical signal processing with respect to applications in the analysis of primate vocal communication.

NEURAL NETWORK THEORY

Artificial neural networks approximate functions with raw sample data. Topology and dynamics of a neural network define a black-box approximater from input to output. The approximation accuracy tends to increase as the sample size increases. That means, the neural system learns from experience (Rojas, 1993).

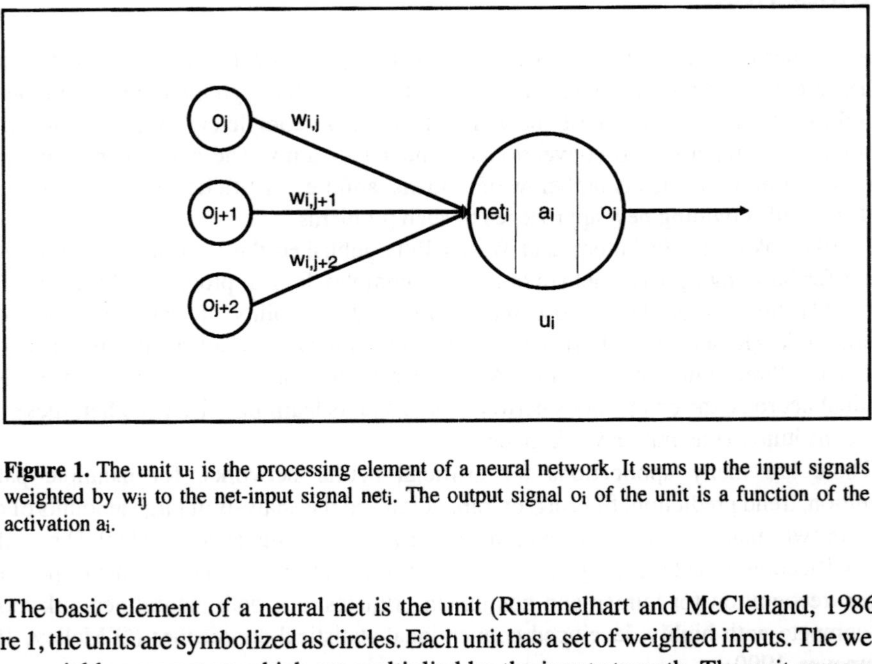

Figure 1. The unit u_i is the processing element of a neural network. It sums up the input signals weighted by w_{ij} to the net-input signal net_i. The output signal o_i of the unit is a function of the activation a_i.

The basic element of a neural net is the unit (Rummelhart and McClelland, 1986). In Figure 1, the units are symbolized as circles. Each unit has a set of weighted inputs. The weights w_{ij} are variable parameters which are multiplied by the input strength. The unit sums up the weighted inputs to the net-input signal. For the unit u_i it is given as:

$$net_i = \sum_j w_{ij} o_j \, .$$

The activation $a_i(t)$ of the unit is a function of previous unit states, the net-input and some global parameters P and is usually described as:

$$a_i(t) = f(\, a_i(t-1)\, ,\, a_i(t-2)\, ,\, \dots\, ,\, net_i(t)\, ,\, P\,)\, .$$

The global parameter could be used to simulate an influence of the environment such as global learning rate or excitation as an analogy to biological nervous systems.

The activation is transferred through the output function to gain the output signal o_i of a unit. Figure 2 shows four simple examples for commonly used output functions in the case that the activation of a unit is equal to the net-input. There is no history information and no global parameter in the activation function therefore, in this example.

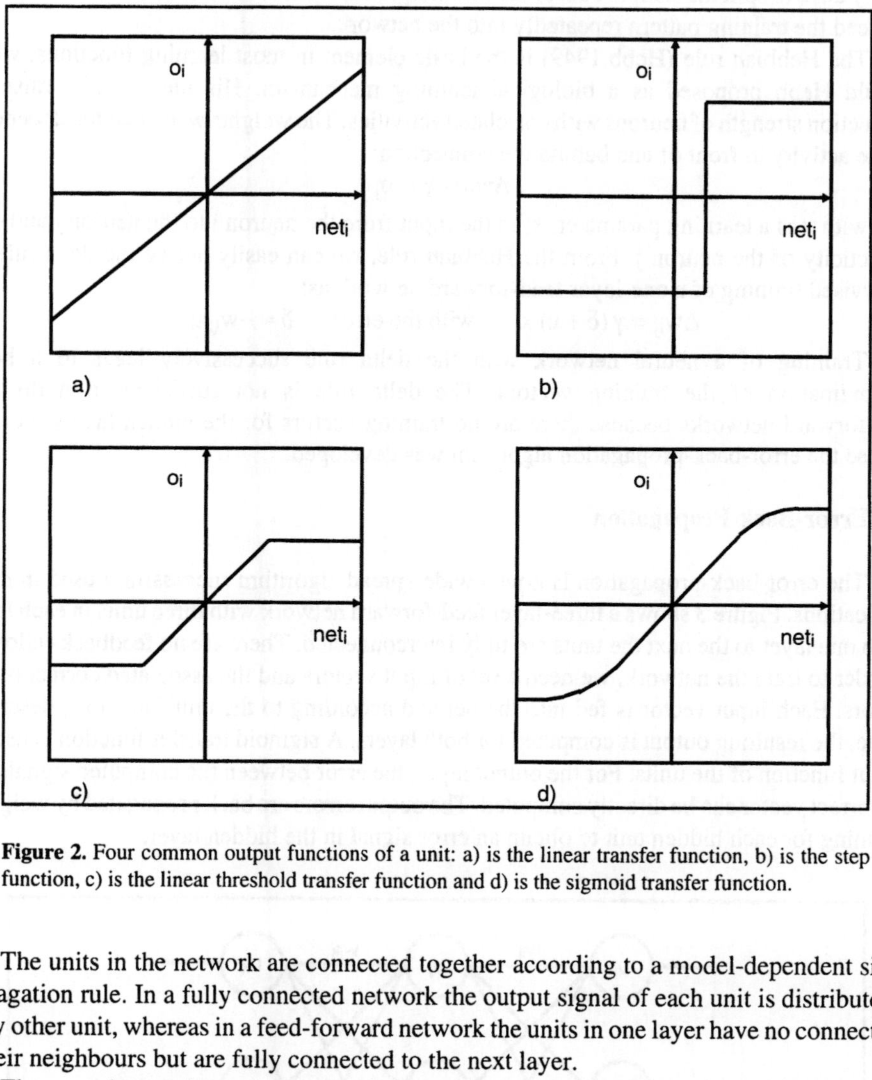

Figure 2. Four common output functions of a unit: a) is the linear transfer function, b) is the step function, c) is the linear threshold transfer function and d) is the sigmoid transfer function.

The units in the network are connected together according to a model-dependent signal propagation rule. In a fully connected network the output signal of each unit is distributed to every other unit, whereas in a feed-forward network the units in one layer have no connections to their neighbours but are fully connected to the next layer.

The most interesting component of a neural net is the learning rule, which determines how the variable parameters (weights, activations, connections) have to be adapted to reach the desired performance. In principle, there are three ways to learn:

1. Development of new connections.
2. Cutting off of existing connections.
3. Adaptation of connection weights and sensitivity of neurons.

The knowledge is stored in the connection strength in most models . Therefore, learning means to train these connections. There is no training of explicit rules. The rules are implicitly contained in the raw sample data. Depending on the network model, two different learning methods are used: the supervised and the unsupervised learning.

If the neural net should classify patterns of known classes, we can use the supervised training method to obtain the desired classification. The training data must have class labels. Two members of this training method are pattern association and error-back- propagation.

On the other hand, if there are given data samples without classification, the unsupervised learning methods (in particular the self-organizing feature-map) could be successful. We simply have to split the sample data of the actual signal space into a training and a test sequence and feed the training pattern repeatedly into the network.

The Hebbian rule (Hebb,1949) is the basic element in most learning functions, which Donald Hebb proposed as a biological learning mechanism. His idea was to raise the connection strength of neurons with correlated activities. The weights were adapted depending on the activity in front of and behind the connection:

$$\Delta w_{ij} = \gamma \, x_i \, o_j$$

with γ as a learning parameter, x_i as the input from the neuron i to the neuron j and o_j as the acticity of the neuron j. From the Hebbian rule, we can easily derive the delta rule for supervised training of a one-layer feed-forward network as:

$$\Delta w_{ij} = \gamma \, (\delta + \varepsilon) \, x_i \qquad \text{with the error} \qquad \delta = -w_{ij} x_i \, .$$

Training of a neural network with the delta rule successively leads to a better approximation of the training vectors. The delta rule is not sufficient in multi-layer feed-forward networks because there are no training vectors for the hidden layers. For this pupose the error-back-propagation algorithm was developed.

The Error-Back-Propagation

The error-back-propagation is now a wide-spread algorithm successfully used in many applications. Figure 3 shows a three-layer feed-forward network with three units in each layer. From one layer to the next the units are fully interconnected. There are no feedbacks allowed. In order to train the network, we need a set of input vectors and the associated correct output vectors. Each input vector is fed into the net and according to the unit functions, described above, the resulting output is computed for both layers. A sigmoid transfer function is used as output function of the units. For the output layer, the error between the computed signals and the correct vector can be directly computed. The output errors are back propagated by weighted summing for each hidden unit to obtain an error signal in the hidden layer.

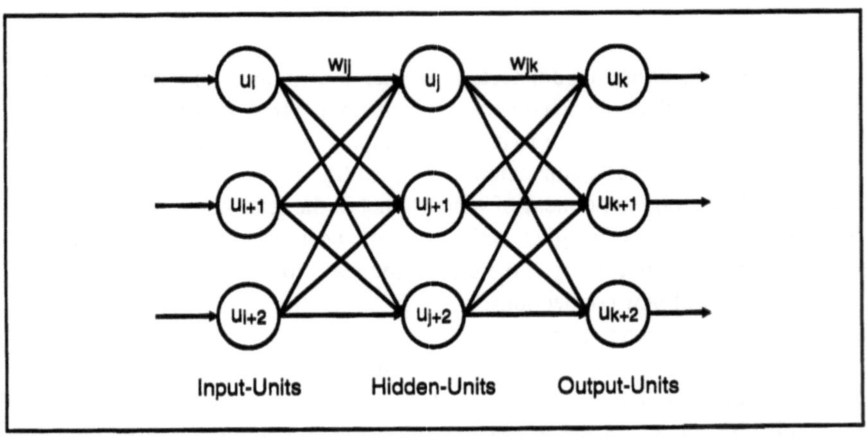

Figure 3. Three layer feed forward network.

The error signal of a unit u_j in the hidden layer is given as:

$$\delta_j = o_j(1-o_j) \sum_{l=1}^{k} w_{jl}\delta_l$$

with $o_{i,j,k}$ as the output signal of the units $u_{i,j,k}$.
Now we can adapt the weights in the hidden layer according to the delta rule:

$$\Delta w_{jk} = \gamma\, o_j\, \delta_k\,.$$

The error-back-propagation algorithm is a gradient descend method which finds a minimum in the error space of the network function. Because that minimum is only local, it can not be guaranteed to find the absolute minimum. Possible changes in the algorithm increase the performance in this respect.

Kohonen's Self-Organizing Feature-Map

Kohonen's self-organizing feature-map procedure (Kohonen, 1984) is highly physiologically motivated. The model proposes one layer of fully interconnected units in the two-dimensional case (Figure 4). The weights of all connections are inhibiting laterally. The input vector is fed through the net to every unit. Associated to every unit is an activation vector with the same dimension as the input vector. The units determine the distance between the input vector and the activation vector and the one with the smallest distance is the maximally excited unit, the activation centre. It could be proved that there exists an activation centre at any time.

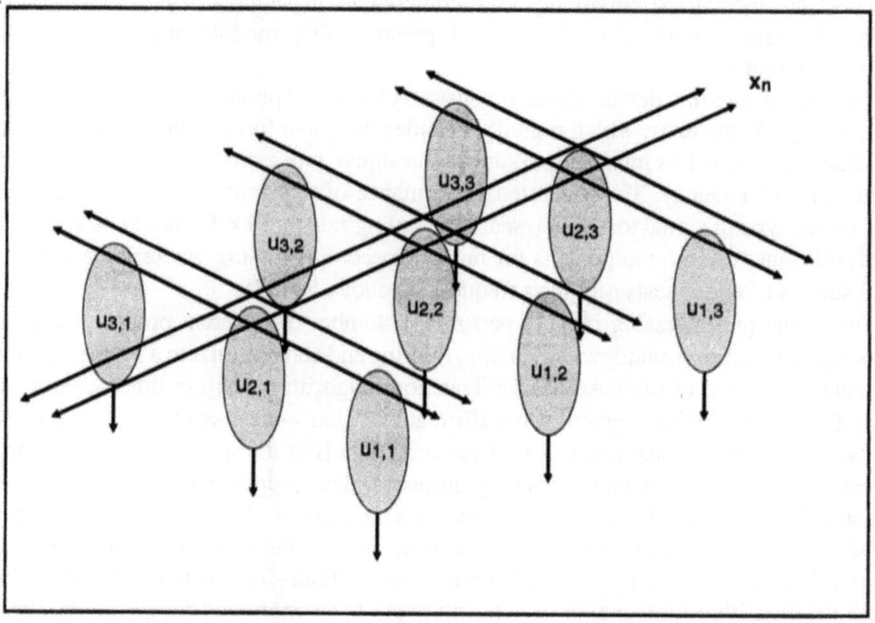

Figure 4. A feature-map is a two-dimensional array of units $u_{i,k}$ which are fully interconnected. The input signal vector x_n feeds to each unit.

Learning is done by successively presenting random input vectors of the input space to the network, computing the centre of activation and adapting the units in a predefined region around the activation centre. The adaptation alters the activation vector of the unit in the direction of the input vector. That means that the distance between the input and the activation vector becomes smaller. The algorithm could be described in four steps (Ritter et al., 1990):

1.Random search for an input vector out of the n-dimensional input space.

2.The vector is fed into each unit. The unit i with a maximum excitation is determined. The associated weight vector has the minimal Euclidian distance to the input vector. This is the activation centre.

3.All units in the neighbourhood of the activation center are adapted according to:

$$\Delta w_{ij} = \gamma\, h(c,k)\, (x_i - w_{jk})$$

with γ as the learning rate and h as a neighbourhood function (for example a Gaussian function).

4.The adaptation radius and the learning rate are adjusted according to a special rule and the procedure is repeated until the predefined maximum number of training steps is reached.

The algorithm tends to depict the input space on the feature-map during the training procedure. It produces classes with similar input vectors as prototypes. In order to reach a good performance of this algorithm it is important to choose adequate parameters, which is often an empiric procedure.

Both, supervised and unsupervised artificial neural network models are used in acoustical signal classification.

SIGNAL ANALYSIS AND SPEECH RECOGNITION

There are three most commonly used components in signal analysis and recognition systems: A signal capture device, a signal preprocessing module and an analysing or recognition module.

The signal capture device usually consists of a microphone or tape-recorder and a analog-to-digital converter, which digitally encodes the wave form of the signal. In order to avoid aliasing effects, it is important to sample the signal with a rate at least twice the highest possible signal frequency. The requested performance of the signal capture device depends highly on the type of signal to be analysed. A sampling rate of 20 kHz should be sufficient to gain all relevant spectral components for human speech processing whereas the analysis of animal sounds often requests sampling frequencies above 50 kHz.

The signal preprocessing module performs a number of different preprocessing steps, such as signal separation, windowing, scaling, filtering and convertion into a frequency domain representation by use of the Fast-Fourier-Transform algorithm. Further preprocessing steps (Linear Predictive Coding, cepstrum transformation, filter banks; see Owren, this volume) have the aim to enhance and retain only those components of the spectral representation that are useful for recognition and analysing purposes. The preprocessor extracts important parameters from the speech wave form to compress the amount of data that must be processed at higher levels. The output of this module is a set of signal parameters for one interval of time, the signal frame. This module is usually realized as a Digital-Signal-Processor (DSP) board because the algorithm for signal preprocessing requests an enormous computing power. The DSP is able to work in real time depending on the amount of processing steps.

The analysing and recognition module has to compare the preprocessed signal input with a number of stored reference signals. It has to decide which output class the signal belongs to. This is the most important module of a signal analysis system and there exists a great number of different processing models. Generally, the algorithms for signal recognition uses highly parallel processing elements, although conventional computers work sequentially. Parallel processing systems are necessary to gain a maximum performance. Universal supercomputers such as the CRAYs or MASSPARs are usually too expensive for frequent use in many applications. A cluster of DSPs could speed up the operation in a satisfactory way. The best result is given by special designed hardware (Mead, 1989), which is adapted to the calculation problem and could be manufactured at low cost. It is possible to recognize speech or other signals in real time with these systems.

Human and nonhuman primates do not have to make a great effort in parallel recognition of different signals. For machines, this is one of the most difficult tasks. We are far away from reaching a performance in automatic speech recognition as good as in biological systems.

Current speech recognition systems can be categorized according to the types of constraints they place on the speech. At one end of the spectrum fall speaker-independent, continuous, unconstrained-grammar, large-vocabulary systems, which are still in the research state (Cole et al., 1991). At the other end of the spectrum are speaker-dependent, single-word, small-vocabulary recognition systems which can achieve accuracies above 99 percent. The best existing speech recognizers perform well only in artificially constrained tasks (Lenning, 1990).

The following description outlines four commonly used techniques in automatic speech recognition: the statistical approaches dynamic-time-warping, the hidden-Markov model, the neural applications time-delay neural network, and self-organizing feature-map systems. The layout and performance of a continuous word recognition system is described in detail as an example.

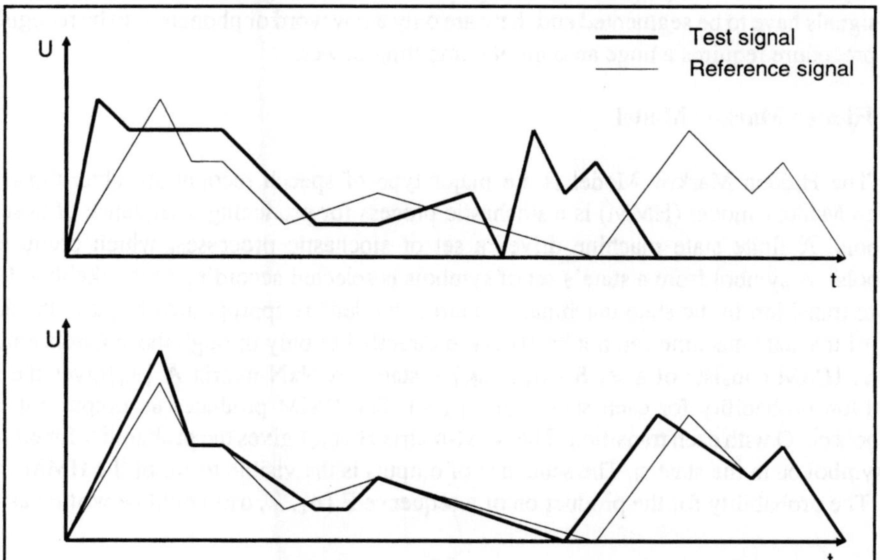

Figure 5. Example of a time warping process (with U as the signal amplitude and t as the time).

Dynamic Time Warping

Dynamic Time Warping is an early used procedure to compare sound patterns with a variable time structure. It performs a warp in the time domain to reach an optimal fitting of a signal against a reference pattern.

If $A\{a_1 \dots a_N\}$ is the signal pattern and $B\ \{b_1 \dots b_N\}$ the reference pattern the algorithm searches for an assignment $W\ \{w_1 \dots w_L\}$ and $w_i=(w_{ia}, w_{ib})$ with:

I) $w_1 = (1,1)$ and $w_L = (N,M)$

II) $w_{i(a-1)} \leq w_{(i-1)a} \leq w_{ia}$

III) $w_{i(b-1)} \leq w_{(i-1)b} \leq w_{ib}$

With II) and III) we can derive:

If $w_i = (a\ , b)$ then

$w_{i+1} = (a\ , b+1)$ or

$w_{i+1} = (a+1\ , b)$ or

$w_{i+1} = (a+1\ , b+1)$.

In order to compare the time warping paths, we have to define a distance measure which is summed up over the complete path. With a special dynamic programming method an optimal warping path could be computed.

In Figure 5 the functionality of the Dynamic Time Warping algorithm is demonstrated. The speech frames that make up the test and the reference templates are shown as amplitude values plotted on a graph with time as the x axis. They are multidimensional vectors in real applications. The frame distance between the signal frame and the reference template is computed. Then a "time warp" is done on the utterance (and a scaling in the length) so that similar frames in the speech signal line up better with the reference frames. A dynamic procedure finds a warp that minimizes the sum of frame distances in the template comparison. The distance produced by this warp is chosen as the similarity measure. The graphs show how warping one of the templates improves the match between them (O'Shaunessy, 1987).

Dynamic time warping is used in highly constrained single word recognition systems. The signals have to be segmented and there are only a few word or phonemes to be recognized. The procedure requires a huge amount of computing power.

The Hidden Markov Model

The Hidden Markov Model is the major type of speech recognition algorithms. The hidden Markov model (HMM) is a stochastic process for producing a sequence of observed symbols. A finite state machine drives a set of stochastic processes, which produce the symbols. A symbol from a state's set of symbols is selected according to its likelihood after a state transition in the state machine. The term "hidden" is appropriated because the actual state of the state machine can not be observed directly but only through the symbols emitted.

A HMM consists of a set $S\ \{s_1, \dots s_N\}$of states. A NxN-matrix $A\ \{a_{ij}\}$gives the state transition probability for each state pair $(s_j\ , s_i)$. The HMM produces an output out of M symbols $o \in O$ with each transition. The NxM-matrix $B\ \{b_{ik}\}$ gives the probability for emitting the symbol o_k in the state s_i. The sequence of outputs is the visible result of the HMM.

The probability for the production of a sequence $F\ \{o_1, \dots, o_T\}$ could be written as:

$$P(F \backslash M) = \sum_{\mu \in S^{T-1}E} P(F\ , \mu \backslash M)$$

$$P(P \backslash M) = \sum_{\mu = s_1, s_2, \dots s_T} \pi_1 b_{s_1 o_1}\ a_{s_1 s_2} b_{s_2 o_2} \dots a_{s_{T-1} s_T} b_{s_T o_T}$$

The example in Figure 6 shows the production of symbols. The finite state machine has five states symbolized as circles. If we take the state number as the output of the HMM, then we can, for example, produce the following sequences:

1. 1-3-5
2. 1-2-3-4-5
3. 1-1-1-2-4-4-4-4-5
4. 1-3-5-5-5-5-5-5-5 ...

Probabilities are attached to the state transitions in real speech recognition applications, and discrete probability distributions control the symbol output for each state. Each word in the vocabulary has a corresponding HMM. These HMMs might actually consist of other HMMs that model phonemes connected together to form a single word. The observable symbols correspond to speech frame measurements. An algorithm finds a set of state transition probabilities and symbol output distributions for each HMM. The probability for each HMM produced the observed sequence of symbols is computed, and the model with the highest probability represents the recognized word in speech recognition (Rabiner et al., 1988).

As well as representing low-level speech segments and transitions, HMMs provide a framework to model higher level structures in continuous speech signals and incorporate other knowledge about the communication.

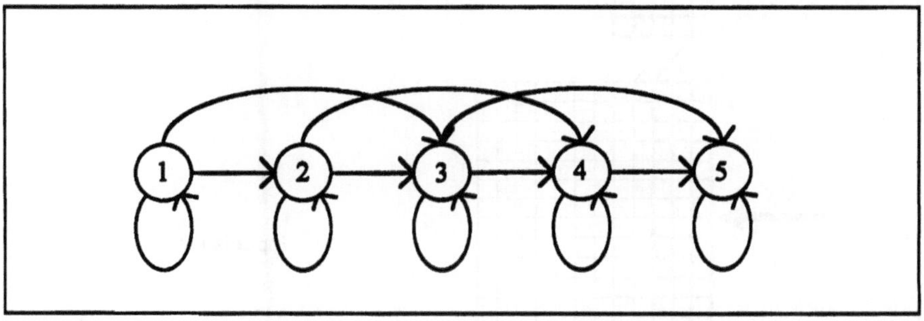

Figure 6. Hidden Markov model with 5 units.

Time Delay Neural Networks

Good results in speech recognition have been obtained using multilayer artificial neural networks with delays and some form of temporal integration in output nodes (Hampshire and Waibel, 1990, Unnikrishnan et al., 1991).

Figure 7 shows a multilayer perceptron with time delays. The boxes on the x axis in this figure represent fixed delays. On the lower left, 16 spectral coefficients form the input in a speech frame of 10 msec. The three columns on the bottom thus represent an input buffer containing a context of three frames. The outputs of these boxes are connected to 8 hidden neural units in the first layer and are buffered across the five boxes in this layer. Outputs from these boxes feed three hidden units in the second hidden layer. Finally, outputs from these three units are integrated over nine time steps into the output unit.

This time delay net was trained to recognize the voiced stops "b", "d" and "g" (Waibel et al., 1987). The speech signal was sampled with 12 kHz and multiplied each 5 ms with a Hamming window. Then a 256-point Fourier-transformation was performed. Out of the resulting spectrum 16 logarithmized Mel-scale coefficients were computed. The average of two such vectors forms the input to the neural network.

The neural net classifier provided an error rate of 1.5% compared to an error rate of 6.5% provided by a simple HMM recognizer. A more recent experiment (Waibel et al., 1989, Lang et al., 1990) resulted in low error rates of 1.4% for the consonants "b,d,g,p,t,k" and 1.4% for the vowels "i,a,u,e,o". The largest net designed from smaller subnets provided a talker-dependent error rate of 4.1% for 18 consonants. An advanced HMM recognizer provided an error rate of 7.3% on this task. This demonstrates that good performance can be provided by time-delay neural networks for specific problems.

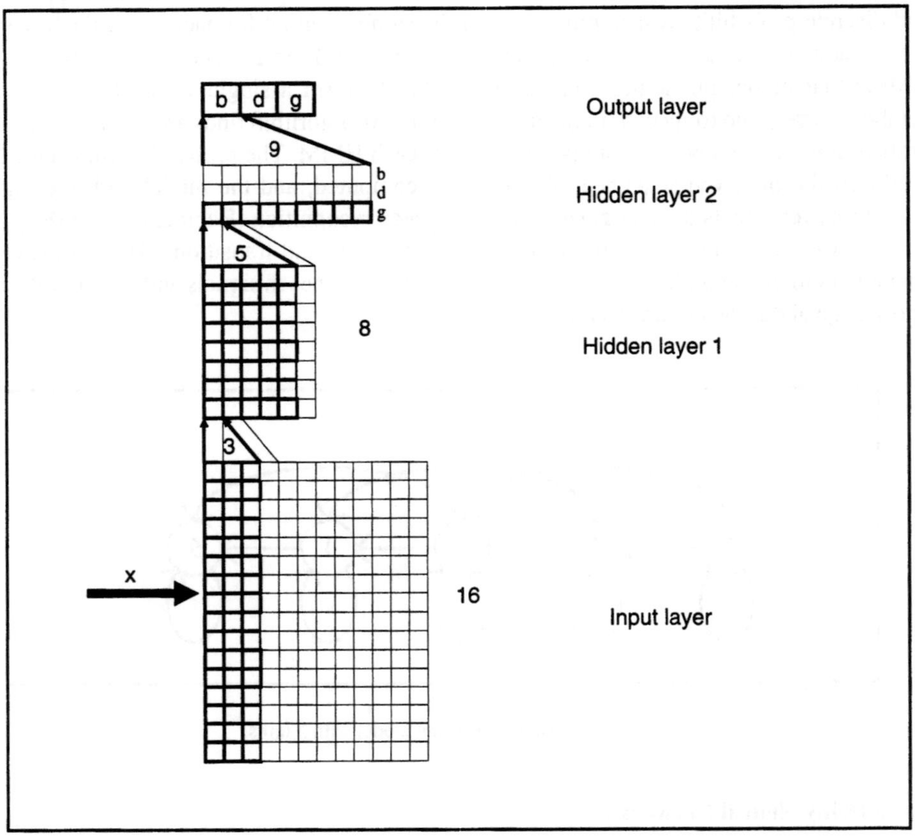

Figure 7. Speech recognition system with a time delay neural network. Each box represents one neuron (Waibel et al., 1989). The system was used to recognize the voiced stops "b", "d" and "g" out of the input-signal X.

Self-Organizing Feature-Maps

In 1984, Teuvo Kohonen proposed a network model with self-organizing neurons which can be used for unsupervised classification tasks. In 1988, he describes a feature-map for speech recognition purpose, the "Neural Phonetic Typewriter" (Kohonen, 1988). It was able to understand Finnish and Japanese phonemes and was realized as an add-on board for IBM-PCs. After some adaptation to the speaker, the system has a performance of 80-90% correctness. With a special postprocessing procedure the performance reaches 92-97% correctness.

The speech signal is sampled with 13020 Hz through a 5300 Hz low-pass filter, then multiplied each 9.82 msec with a Hamming window and Fourier-transformed with 256 points. The spectral lines were logarithmized and filtered to get 15 channel vectors. To receive some dynamic information, the feature-map input vector is formed of two channel vectors with

30 msec delay. Figure 8 shows the feature-map and the association of each unit to a Finnish phoneme.

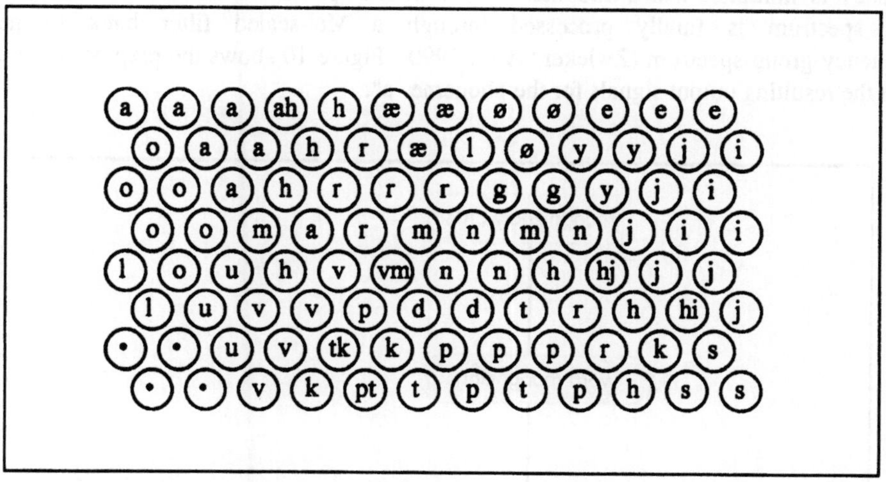

Figure 8. Layout of the phonetic typewriter feature-map (Kohonen 1988).

Figure 8a shows the activation sequence in the feature map when the Finnish word "humppila" was uttered. Each arrow is associated with a time interval of 9.83 msec.

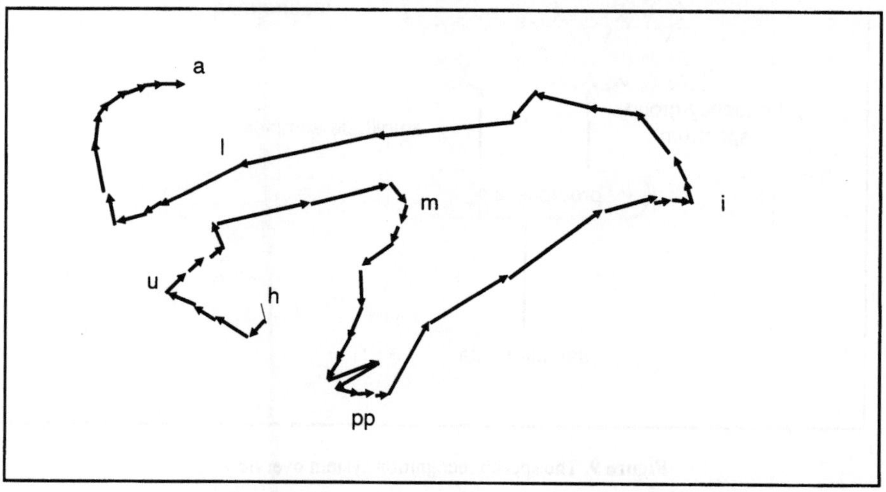

Figure 8a. Example of an activation sequence in the feature-map (Kohonen, 1988).

A Speech Recognition System with Self-Organizing Neural Nets

At the Department of Computer Science at Stuttgart University, we developed a high performance system for continuous speech recognition. The system is dedicated to be implemented in special parallel hardware.

The speech recognizer consists of three processing modules, the preprocessing unit, the phoneme-classification unit and the word recognition unit (Figure 9).

The speech signal is sampled with 16-bit resolution and 22.05kHz sampling rate. Every 5 msec it is multiplied with a Gaussian-window of 1024 points and then Fourier-transformed. The spectrum is finally processed through a Mel-scaled filter bank to get a frequency-group-spectrum (Zwicker, 1982, 1990). Figure 10 shows the preprocessing steps with the resulting output signals for the phoneme "a".

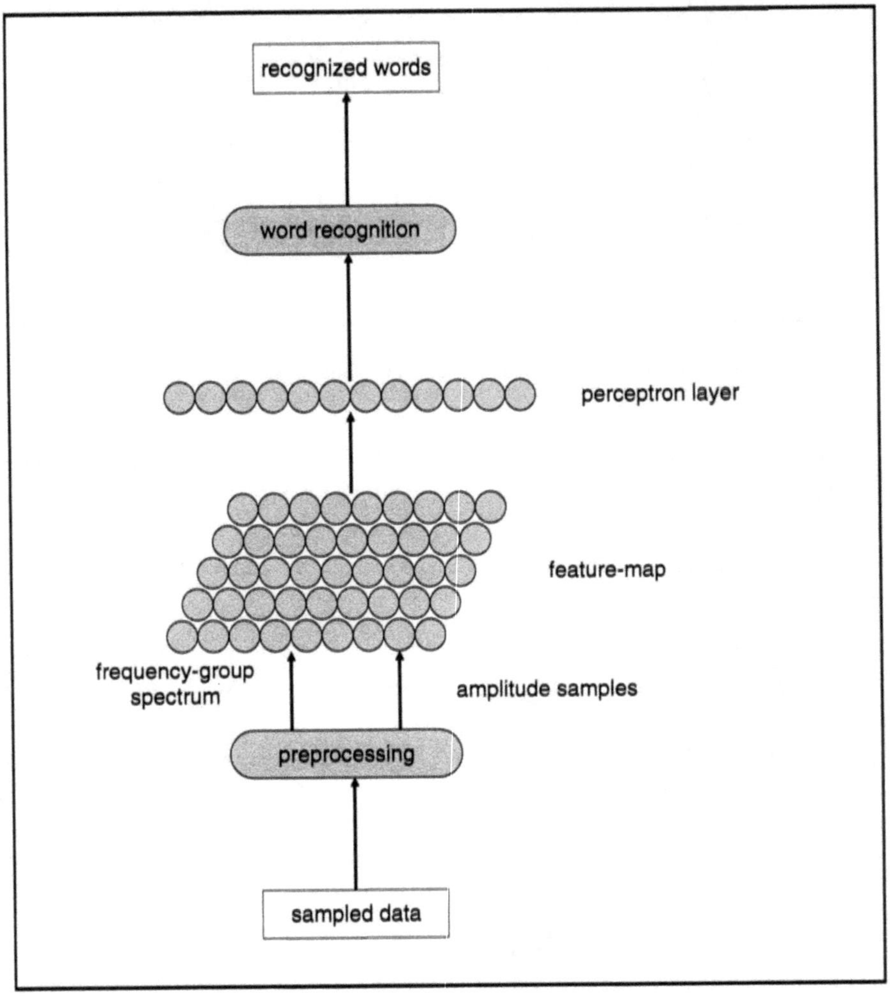

Figure 9. The speech recognition system overview.

The normalized frequency-group-spectrum and the most recent 20 amplitude-samples form the input signal to the feature-map. Therefore, the neural net has a view of a 100 msec time window of the signal.

The phoneme-classification unit consists of a feature-map and a one layer perceptron. Each unit in the feature-map is connected to the perceptron units. The feature-map is trained according to the Kohonen adaptation rules and there is only one unit active at the time. The units in the perceptron layer represent the phonemes to be classified. The activation function of the perceptron units is realized as a time-delay term to get some integration over time. If the unit j in the feature-map is active the function for the unit i is given as:

$$a_i(t) = \max(1, \alpha a_i(t-1) + \beta(1-\alpha) w_{ij})$$

with α as the delay-term and β as a limit for the activation.

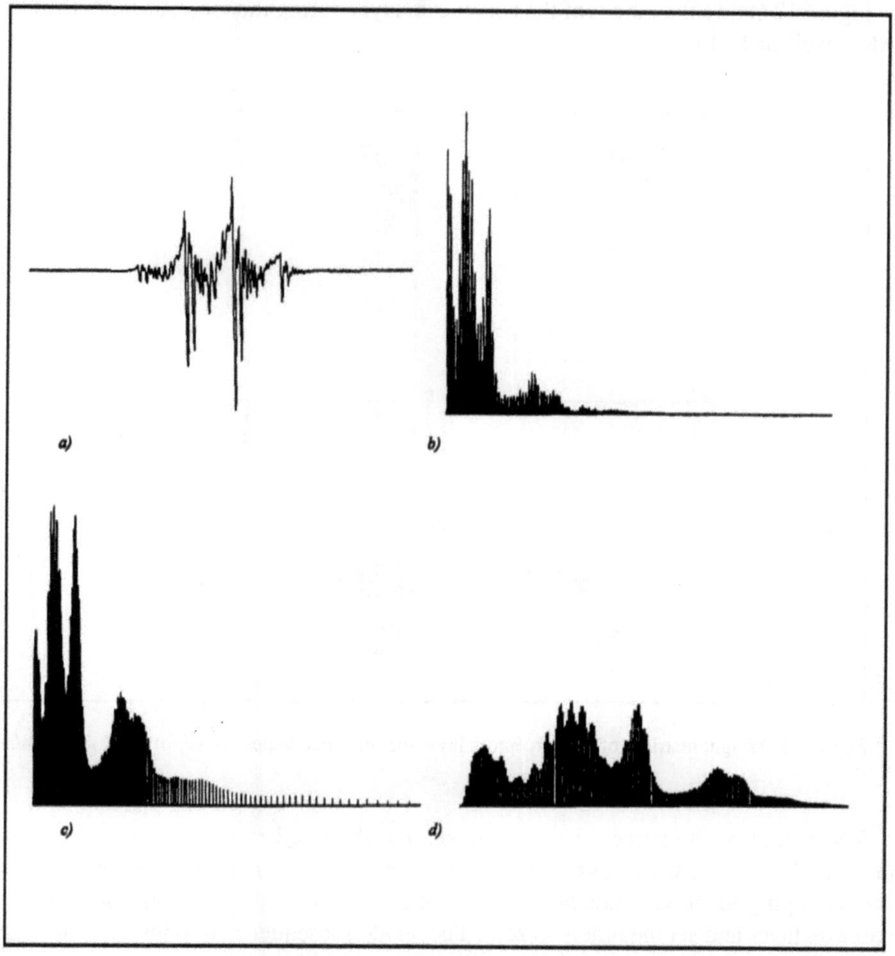

Figure 10. Preprocessing step of the phoneme /a/ out of the German word "eins". a) Sampled signal multiplied with the time window. b) Fourier spectrum. c) Frequency group spectrum in a Hz scale. d) Frequency group spectrum in a Mel scale.

The training of the perceptron layer needs some target values for each unit. The beginning and the end of a phoneme has to be marked in the training data sets with the associated phoneme symbol. The training target is to get the maximum activation of the correct unit and a minimum activation of the other units during one phoneme. A special learning rule is implemented for this purpose. The difference between a one-activation and the maximum activity of the unit in a phoneme time-slot is calculated for the target unit which represents the actual phoneme. The result is multiplied with the learning rate and added to the net-input signal of the unit. The difference between a zero-activation and the actual activation is taken for the other units. The connection weights are all zero at the beginning of the training. During the training the activation pattern on the feature-map is associated to the perceptron unit with a one-activation.

The learning rule is:

$$\Delta w_{ij}(t) = q_{ij}(t) b_i(t)$$

with q_{ij} as a phoneme indicator and b_{ij} the winner-neuron indicator.

Figure 12 shows the output of the perceptron layer during the recognition of the German words "zwei" and "drei".

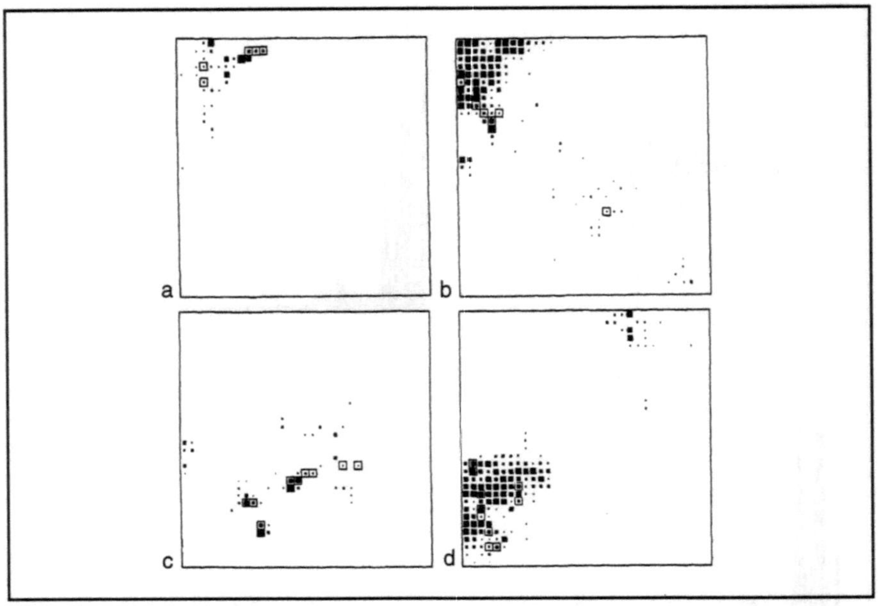

Figure 11. Weight matrices of the perceptron layer for some phonemes: a) /ts/, b) /s/, c) /v/, d) /a/.

The output of the perceptron layer feeds into the word recognition unit which has to determine the correct words associated with the phoneme activation in the time course. In order to simplify the recognition task it is possible to take only the maximum activation of the perceptron layer and set the others to zero. The result is a sequence of phonemes in which we can eliminate equal neighbours to form a word. Because this is a very faulty procedure a new algorithm was developed which includes the use of lexical knowledge.

Each word has a proprietary recognizer which tries to find the word in the input stream. The recognizer determines how good its word fits the input. All units work in parallel and stop if their actual error becomes too large. They also stop after some steps, if they did not reach a given minimal correctness. It is therefore possible that all recognizers stop at a difficult point. In that case, they will be restarted behind the last detected word. A small part of the signal is discarded to avoid an infinite loop. This procedure is repeated until the difficult part is left and the next word could be recognized.

As soon as one recognizer has found the last phoneme of its word, it signals the end of the actual search and all units start again. It is possible that a different recognizer is still searching concurrently at this time. One disadvantage of this implementation is that a word cannot be recognized if only one single phoneme is missing. A different word recognizing unit based on hidden Markov models is being developed to overcome this problem.

The performance of the speech recognizing system was tested in two different ways. The first task was to recognize and distinguish the syllables "be","de","ge" and "e". We used 33 data sets of one speaker, each containing the four utterances for training and 17 data sets for testing purposes. There was a short pause between two syllables to eliminate distortions from the environment.

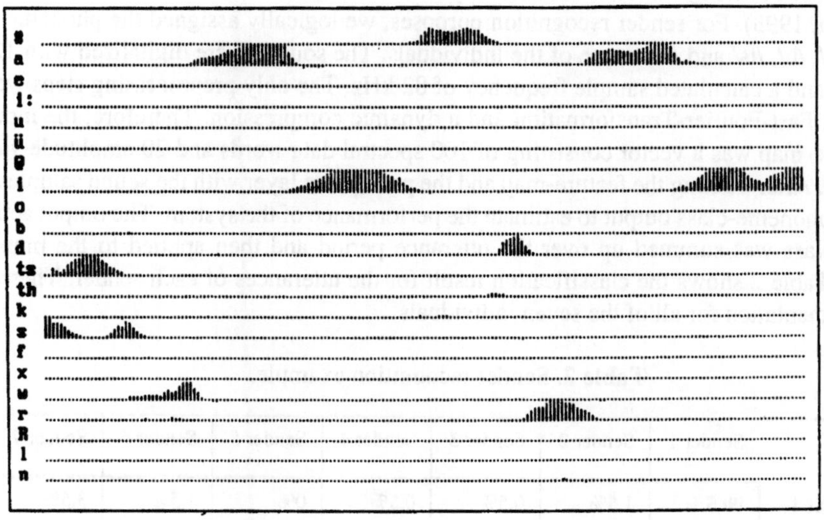

Figure 12. Output of the phoneme recognition unit.

The second test was to recognize the digits "zero" to "nine". We sampled 30 sequences of continuous speech without pauses between the words, 20 for training and 10 for testing. The recognition of syllables is more difficult than the recognition of digits because the acoustic signals of the syllables differ not very much. The result of the digit task was a 100% recognition rate for both test and training data sets. For the syllable task the result was 98.5% recognition rate (one syllable out of 68 was misclassed). Table 1 shows the recognition rate as a function of the frequency resolution of the input signal (line count in spectral data).

Table 1. Performance of the speech recognition system.

Number of frequency groups	212	150	106	75	53
Error number (syllable task)	1	2	1	2	2
Recognition rate (syllable task)	98.5%	97.1%	98.5%	97.1%	97.1%
Error number (digit task)	0	0	1	2	5
Recognition rate (digit task)	100%	100%	99%	98%	95%

A new application of the speech recognition system is the analysis of nonhuman primate vocal signals. For example, it should be possible to automatically distinguish sounds of different individuals of the same species, which is a sort of speaker recognition. Another interesting question is to classify different groups of the same animals with acoustical characteristics.

In both cases, we can use the unchanged speech recognition system for a first approach. Every class is associated with a phoneme symbol of the human speech. The sound signals are segmented and labeled with the phoneme numbers and then fed into the system.

In our experiment, we used sound recordings of seven mouse lemur males (*Microcebus murinus*) each uttering ten times the same type of vocal signal ("Trill" call, see Zimmermann and Lerch, 1993). For sender recognition purposes, we logically assigned the phonemes /a/, /i:/, /ü/, /i/, /b/, /ts/ and /k/ to one of the individuals. The sounds were digitalized with 12 bit accuracy and a calculated sample frequency of 82 kHz. The only preprocessing steps were a 256-point Fast-Fourier-Transformation and a dynamic compression. Therefore, the input to the feature-map was a vector consisting of 100 spectral data words and 20 amplitude values per frame. After training the feature-map and the perceptron layer with the sound samples, we used the phoneme-class output to estimate the performance of the system. The output activity of each class was summed up over the utterance period and then applied to the maximal activity. Table 2 shows the classification result for the utterances of each sender. There is a good segmentation for all of the seven individuals.

Table 2. Sender recognition example.

	Sender 1	Sender 2	Sender 3	Sender 4	Sender 5	Sender 6	Sender 7
Class 1	**90,5%**	1,5%	0,5%	0,5%	0%	1,5%	3,5%
Class 2	0%	**75,5%**	1,0%	0%	0%	0%	0%
Class 3	2,0%	3,5%	**87,0%**	0,5%	0,5%	0%	2,5%
Class 4	0,5%	4,5%	1,5%	**86,5%**	4,0%	0%	0%
Class 5	0%	3,0%	0%	0%	**92,5%**	0%	0%
Class 6	1,5%	4,5%	2,0%	0%	1,5%	**79,0%**	1,0%
Class 7	3%	2,5%	1,5%	0%	0,5%	8,5%	**71,5%**

Hardware Implementation of Artificial Neural Networks.

Most research on artificial neural networks concerns software simulators on sequential computers. The main advantage of these neurosimulators is their flexibility, because they can be used for several different network models and topologies. They are easy to use in training and testing applications, and in many cases the software is in the public domain. On the other hand, neurosimulators work very slowly, even if they are installed on high performance computers. The speech recognition system described in the previous chapter is written in C and runs, for example, on a 486-PC running at 33MHz. The training and testing procedure in the animal speaker-recognition task requires approximately 55 minutes with a 32x32 feature-map. Automatic analysis of speech or other vocal communication is therefore not possible with conventional systems.

We are currently developing a hardware solution of the speech recognition system at our institute, which should be capable of real-time analysing and recognition. The prototype of the feature-map processor has a performance of six million adaptation and feed-forward cycles per second. This leads to a proposed real-time frequency bandwidth of approximately 100kHz with a 32x32 map and a 200 point spectral resolution. The preprocessing is done with a signal-processor board, and a special perceptron hardware completes the system (Zimmermann, 1992).

The feature-map processor is a vector parallel machine and the existing prototype has eight vector-nodes. The above described computing performance requires one vector-node per spectral point. The hardware and the accompanying software are still in the research state. The next step is to build a complete system with a greater number of vector-node processors.

CONCLUSION

In this paper, the principles of the artificial neural network theory and its applications in speech recognition and analysing systems have been discussed. Two major neural techniques, the time-delay network model and the self-organizing feature-map has been described and compared with two statistical procedures, the dynamic-time-warping and the hidden-Markov model.

Artificial neural networks were recently used in many successful applications. They are easy to use and highly adaptive. It is possible to build systems for continuous and speaker independent speech recognition as well as for analyzing nonhuman primate vocal communication.

In a sender recognition experiment with sounds of mouse lemurs it has been shown that the neural speech recognition system designed for human phonemes performs well in the recognition of nonhuman utterances. Further investigations with respect to sound classification can lead to a sound recognition system for animal vocal communication.

Artificial neural networks require an enormous computing power. Unlike simulation programs on conventional computers, real time systems realized in hardware will be capable of automatically recognize and classify human and animal sounds in the field or in the laboratory.

ACKNOWLEDGEMENT

I would like to thank W.H. Burkhardt, G. Palm and M.-T. Schneider for critically reading and correcting the manuscript and E. Zimmermann for the sound material and many helpful discussions.

REFERENCES

Bartkova, K., and Jouvet, D., 1991, Modelization of allophones in a speech recognition system, *Proc. XIIth International Conference of Phonetic Science.* Vol. 4: 474-477.

Cole, R. A., Fanty,M., Gopalakrishnan, M., and Janssen, D. T., 1991, Speaker-independent name retrieval from spelling using a database of 50,000 names, *IEEE Proc ICASSP*: 14-17.

Deng, L., and Erler, K., 1992, Structural design of a hidden Markov-model-based recognizer using multivalued phonetic features: Comparison with segmental speech units, *J. Acoust. Soc. Am.*, 92:3058-3067.

Hampshire, J. B., and Waibel, A. H., 1990, A novel objective function for improved phoneme recognition using time-delay neural networks, *IEEE Trans. Neural Networks.* Vol. 1, 2:443-446.

Hebb, D., 1949, "The Organization of Behavior", John Wiley, New York.

Iso, K., and Watanabe, T., 1990, Speaker-independent word recognition using a neural prediction model, *IEEE Proc. ICASS. Vol. 4: 443-446.*

Kohonen, T., 1984, "Series in Information Sciences: Vol. 8. Self-Organization and Associative Memory", Springer-Verlag, Berlin-Heidelberg-New York-Tokyo.

Kohonen, T., 1988, The neural phonetic typewriter, *IEEE Computer*, 3:11-22.

Kosko, B. (ed.), 1992, "Neural Networks for Signal Processing", Prentice-Hall, New Jersey.

Lang, K. J., Waibel, A., and Hinton, G., 1990, A time-delay neural network for isolated word recognition, *Neural Networks,* 3: 23-29.

Lenning, M., 1990, Putting speech recognition to work in the telephone network, *IEEE Computer*, 8: 35-41.

McCulloch, W., and Pitts, W., 1943, A logical calculus of the ideas immanent in nervous activity, *Bull. Math. Biophysics*, 5:115-121.

Mead, C., 1989, "Analog VLSI and Neural Systems", Addison-Wesley.

O'Malley, M.H., 1990, Text-to-speech conversion technology, *IEEE Computer*, 8:17-23.

O'Shaunessy, D., 1987, "Speech Communication: Human and Machine", Addison-Wesley.

Rabiner, L. R., Wilpon, J. G., and Soong, F. K., 1988, High-performance connected-digit recognition using hidden Markov models, *IEEE Proc. ICASSP*: 119-125.

Ritter, H., Martinetz, T., and Schulten, K., 1990, "Neuronale Netze", Addison-Wesley, Bonn.

Rosenblatt, F., 1958, The perceptron: a probabilistic model for information storage and organization in the brain, *Psychol. Review*, 65:386-408.

Rojas, R., 1993, "Theorie der neuronalen Netze", Springer, Berlin, Heidelberg, New York.

Rummelhart, D., and McClelland, J., 1986, "Parallel Distributed Processing", *MIT Press,* Cambridge-Massachusetts.

Strathmeyer, C. R., 1990, Voice in computing: an overview of available technologies, *IEEE Computer*, 8:10-15.

Unnikrishnan, K. P., Hopfield, J. J., and Tank, D. W., 1991, Connected-digit speaker-dependent speech recognition using a neural network with time-delayed connections, *IEEE Trans. Sig. Proc.* Vol. 39, 3:698-713.

Waibel, A., Hanazawa, T., Hinton, G., Shikano, K., and Lang, K., 1987, Phonem recognition using time-delay neural networks, *IEEE Trans. ASSP.* Vol. 3: 393-404.

Waibel, A., Sawai, H., and Shikano, K., 1989, Consonant and phonem recognition by modular construction of large phonemic time-delay neural networks, *Proc. ICASSP*: 405-408.

Zimmermann, A., 1992, Mikroelektronische Schaltung zum Aufbau selbstorganisierender Karten aus digitalen Neuronen, Offenlegungsschrift, Deutsches Patentamt DE 4227707 A1, G 06 F 15/18.

Zimmermann, E., and Lerch, C., 1993, The complex acoustic design of an advertisement call in male mouse lemurs (*Microcebus murinus*) and sources of its variation, *Ethology* 93:211-224.

Zwicker, E., 1982, "Psychoakustik", Springer, Berlin,Heidelberg,New York,Tokyo.

Zwicker, E., and Fastl, H., 1990, "Psychoacoustics - Facts and Models", Springer, Berlin, Heidelberg.

LOUD CALLS IN NOCTURNAL PROSIMIANS: STRUCTURE, EVOLUTION AND ONTOGENY

Elke Zimmermann

German Primate Center
Laboratory of Biocommunication
37077 Göttingen, Germany

INTRODUCTION

Loud, long-distance or advertisement calls belong to the most distinctive primate vocalizations. They travel over long distances and have been described in virtually every genus. Thus, they are found in almost all families of strepsirhines (e.g., Petter and Charles-Dominique, 1979; Zimmermann, 1990, in press, for reviews) and anthropoids (tarsiers: e.g., Niemitz, 1984; marmosets and tamarins: e.g., Snowdon, 1989; cebids: e.g., Oppenheimer, 1977; Robinson, 1981; Sekulic, 1982; cercopithecids: e.g., Byrne, 1982; Gautier and Gautier, 1977; Hohmann, 1991; Waser, 1982; gibbons: e.g., Deputte, 1982; Marshall and Marshall, 1976; Mitani, 1992; apes: e.g., Boehm, 1989; Hohmann and Fruth, this volume; Marler and Tenaza, 1977; Mitani and Nishida, 1993). Since loud calls have a highly stereotypic and species-specific vocal pattern across all primate taxa, they are often used in combination with reproductive or biochemical characters to identify species within monomorphic species groups, to clarify the status in polymorphic species or to develop inferences about phylogeny (e.g., Struhsaker, 1970, for guenons; Oates and Trocco, 1983, and Struhsaker, 1981, for Colobus monkeys; Haimoff et al., 1982, Marshall and Marshall, 1976, Marshall and Sugardjito, 1986, for gibbons; Snowdon et al., 1986, for golden lion tamarins; Gautier, 1988, for guenons; Zimmermann et al., 1988, Zimmermann, 1990, for bushbabies; Niemitz et al., 1991, for tarsiers; Macedonia and Stanger, submitted, for diurnal lemurs).

Generally, loud calls are produced either by sexually mature males and females or by sexually mature males only (see Hohmann and Fruth, this volume). Loud calls are used in a broad variety of contexts regulating social encounters among individuals or groups over longer distances such as mate attraction, defense of restricted resources (e.g., territory, mating partners) or rallying of group members. Their assumed functions are in close accordance with those assumed for bird songs.

During the last decade, research on bird song acquisition has had an enormous impact on our knowledge of behavioral development and its neurological basis. Acquisition mechanisms revealed parallels to language development in human infants (e.g., Konishi,

1985; Marler, 1975, 1990; Snowdon et al., 1986). Loud or long calls of marmosets and tamarins, squirrel monkeys and gibbons showed an impressive similarity to bird song in their usage. Consequently, research interest in the past has focused on these groups in order to find possible suitable primate models for the investigation of human speech acquisition. It appeared, however, that the song of birds was ontogenetically much more flexible and more influenced by socio-acoustical experience than were vocalizations of primates (see Biben and Bernhards,this volume; Newman, this volume).

In the past few years, my laboratory has been involved in a comparative study of the organization and ontogeny of vocal communication of a scarcely known primate group, the nocturnal strepsirhines. I was impressed by the relatively elaborate vocal signalling system these "primitive", "solitary foraging" primates (Bearder, 1987; Richard, 1987) displayed and surprised to find, in one thoroughly studied taxon, an unexpected flexibility in the acoustic properties of a particular social call, the loud or advertisement call .

In this chapter, I will review what is currently known about loud or advertisement calls in nocturnal strepsirhines. I will illustrate and discuss major patterns as well as similarities and divergences in loud call structure among different species. Based upon this, I will examine potential causes for these differences and outline a hypothesis for the evolution of loud call structure. I will focus on the ontogeny of vocal communication in nocturnal strepsirhines with a special emphasis on loud calls, demonstrate how flexible the loud call structure is within a homogenous behavioral situation in a phenotypically homogenous population and emphasize possible underlying mechanisms. Finally, I will discuss these findings with those obtained for other primates and discuss prospects for future research.

STRUCTURAL ASPECTS OF LOUD CALLS

Representative loud calls of 17 nocturnal strepsirhine species are presented in Figs. 1, 2 and 3. Additional information on the relationship between taxonomic group and mean fundamental frequency, mean body weight, habitat, locomotion and home range size is shown in Table 1.

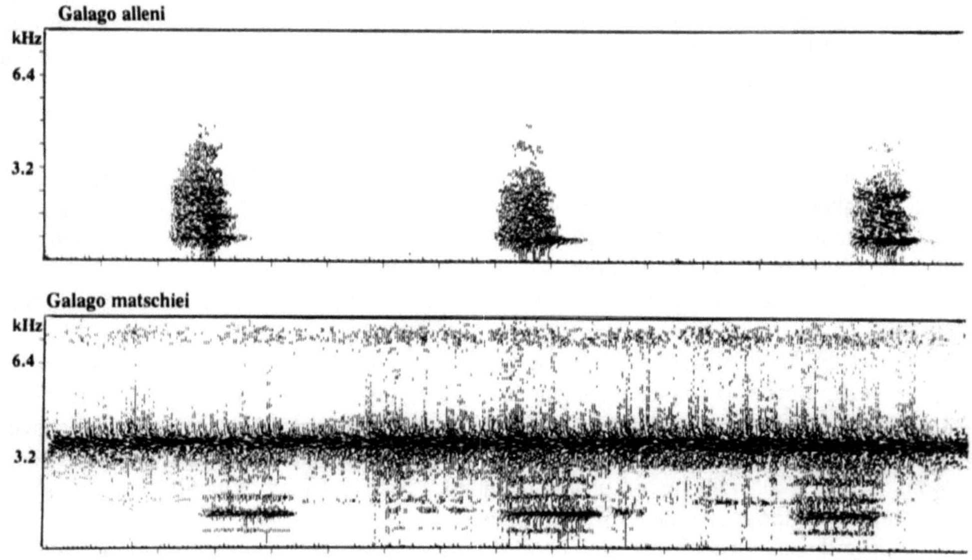

Figure 1a. Representative oscillograms, spectrograms and power spectra of the loud call in lesser bushbabies.

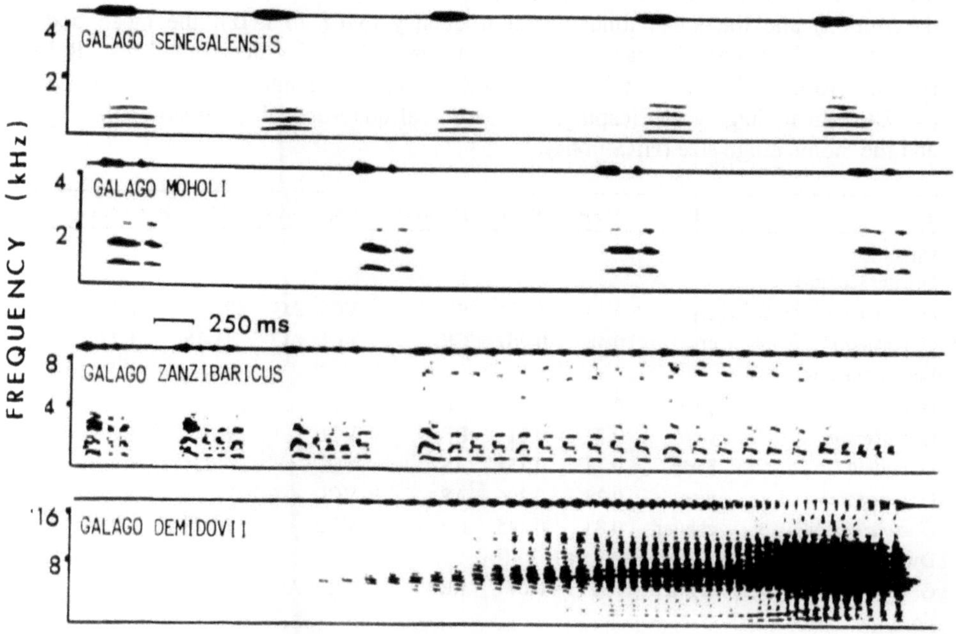

Figure 1b. Representative oscillograms and spectrograms of the loud call in three lesser and a dwarf bushbaby. The three lesser bushbabies were treated as the same species until recently. Differences in the structure of the loud call together with additional differences in reproductive behavior reveal three different species (Zimmermann, 1990). The valid species name of the dwarf bushbaby is G. demidoff.

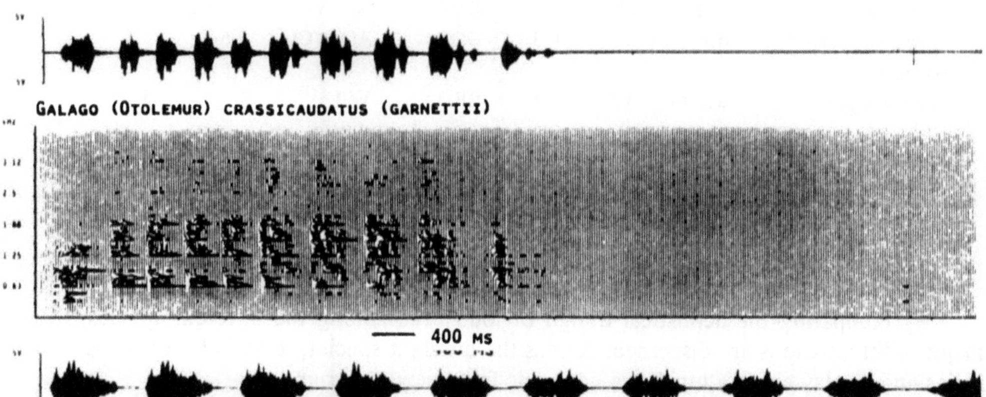

Figure 1c. Representative oscillograms and spectrograms of the loud call in two greater bushbabies.

Table 1. Data set for the relationships between taxa (OD= onomatopoetic description) and the mean fundamental frequency (MFF in kHz), the mean body weight (MBW in kg), the habitat (RF=rain forest; DDF= dry decidous forest; RV= riverine forest; CF= coastal forest; AS= Acacia savanna), the locomotion (VCL=vertical clinging and leaping; AQ= arboreal quadrupedal; SL= slow climbing) and the home range size (HRS in ha).

Taxa	OD	MFF	MBW	Habitat	Locomotion	HRS	Reference
Galagidae							
Greater galagos							
Galago crassicaudatus	cry	0.45	1.131	RV	VCL, AQ	7-10	1,2,6
G. garnettii	cry	0.48	0.767	CF	VCL, AQ	11-18	1,2,6
Lesser galagos							
G. alleni	croak	0.81	0.314	RF	VCL	8-50	1,2,6
G. matschiei	woo	0.7	0.21	RF	VCL	?	1,2,6
G. moholi	bark	0.73	0.158	AS	VCL	4-23	1,2,6
G. senegalensis	woo	0.38	0.206	AS	VCL	?	1,2,6
G. zanzibaricus	chatter	0.83	0.145	CF	VCL	16-29	1,2,6
Dwarf galago							
G. demidoff	chatter	0.8	0.07	RF	VCL, AQ	0.6-2.7	1,2,6
Lorisidae							
Loris tardigradus	whistle	2.5	0.275	RF	SC	?	1,8
Nycticebus coucang	whistle	6	1.25	RF	SC	?	1,5,9
N. pygmaeus	whistle	7.5	0.6	RF	SC	?	9
Perodicticus potto	whistle	11	1.05	RF	SC	?	1,3,9
Cheirogaleidae							
Cheirogaleus major	whistle	13	0.4	RF	AQ	?	1,3,9
Ch. medius	whistle	16	0.18	DDF	AQ	3-4	1,3,9
Microcebus murinus	trill	21	0.08	DDF	AQ, VCL	0.07-2	1,7,8
Mirza coquereli	trill	6.28	0.3	DDF	AQ, VCL	2.5-3	1,4
Indridae							
Avahi laniger	whistle	3.5	1.3	RF	VCL	?	1,3

References: (1) Harvey et al., 1987; (2) Nash et al., 1989; (3) Petter and Charles-Dominique, 1979; (4) Stanger, 1993; (5) Zimmermann, 1985 a; (6) Zimmermann, 1990; (7) Zimmermann and Lerch , 1993; (8) Zimmermann, in press; (9) Zimmermann, unpubl.

By comparing the acoustical design of loud calls among the different species, three major differentiations are discerned. Across the different species, loud calls are produced in series and display either a tonal or a harmonic frequency spectrum:

1.) Mean fundamental frequency (MFF= [Fpeak-Fmin]/2) of the loud call in all bushbabies lies in the lower frequency range (MFF= 0.647 kHz, Tab. 1, Fig. 1) compared to the other investigated taxa. In four species of the lesser bushbabies (*Galago alleni*, *Galago matschiei*, *Galago moholi*, *Galago senegalensis*), the loud call consists of relatively long series (> 10 s) of rhythmically repeated units with a simple, almost constant fundamental frequency band and an almost equal intonation pattern throughout the whole series. In one species of the dwarf bushbabies (*Galago demidoff*), one species of the lesser bushbabies (*Galago zanzibaricus*) and both species of the greater bushbabies (*Galago garnetti* and *Galago crassicaudatus*), the loud call is of shorter series duration (< 10 s) and shows a more complex temporal and intonation pattern (see Zimmermann, 1990, for details). 2.)

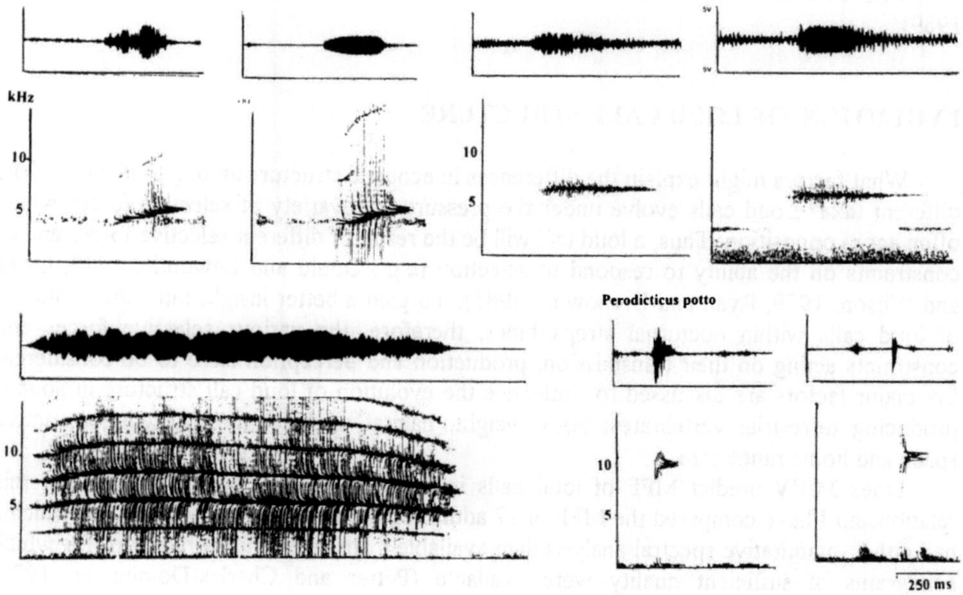

Figure 2. Representative oscillograms and spectrograms of the loud call in lorisids.

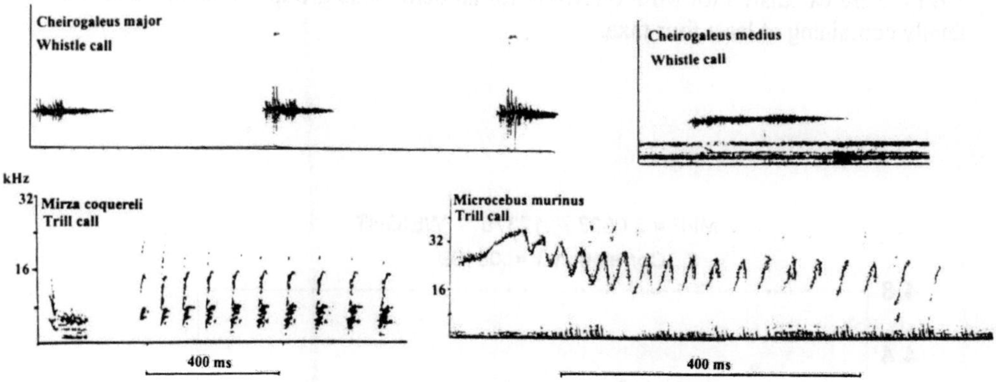

Figure 3. Representative spectrograms of the loud call in cheirogaleids. Frequency and time axis of sonagrams of Cheirogaleus correspond to Microcebus.

MFF of the loud call in African and South East Asian lorisid taxa lies in a medium frequency range (MFF=9 kHz, Tab.1, Figs. 2 and 3) in all Malagasy cheirogaleids in a high range (MFF=14.07 kHz, Tab. 1, Fig. 3). Lorisids and the Malagasy genera *Cheirogaleus* and *Avahi* use series of irregularly repeated, narrowband and almost constant frequency whistle units with relatively equal intonation. The Malagasy genera *Microcebus* and *Mirza* utter short, broadband trill series with complex frequency, temporal and intonation pattern.

Interestingly, the major acoustic characteristics of the loud call of the African bushbabies correspond to that of Old World monkeys (Gautier and Gautier-Hion, 1977, for review), that of the African and South-East Asian lorisids, the Malagasy genera

Cheirogaleus and Avahi to tamarins and marmosets in the New World (e.g., Snowdon, 1989).

EVOLUTION OF LOUD CALL STRUCTURE

What factors might explain the differences in acoustic structure among loud calls of the different taxa? Loud calls evolve under the pressure of a variety of selective forces which often act in opposition. Thus, a loud call will be the result of different selective forces and of constraints on the ability to respond to selection (e.g., Gould and Lewontin, 1979; Oster and Wilson, 1979; Ryan and Brenowitz, 1985). To gain a better insight into the evolution of loud calls within nocturnal strepsirhines, therefore, the various selective forces and constraints acting on their transmission, production and perception have to be considered. Six major factors are discussed to influence the evolution of loud call structure in sound-producing terrestrial vertebrates: body weight, habitat, ambient noise, predation, active space and home range size.

Does MBW predict MFF of loud calls in nocturnal strepsirhines? To evaluate this relationship I have compared the MFF in 17 adult nocturnal strepsirhine species for which I had either quantitative spectral analysis data available from my own laboratory or for which sonagrams of sufficient quality were available (Petter and Charles-Dominique, 1979; Stanger, 1993). The data set consisted of 17 species from 9 different genera and four families and a total of 456 vocalizations (Tab. 1). Data on body weight were taken from Harvey et al.'s (1987) life history analysis and from Nash et al.'s (1989) revision of the family Galagidae. Regressions (Pearsons product moment correlation, Rohlf and Sokal, 1991) were calculated for MFF on MBW for all taxonomic groups combined and for each family containing at least four taxa.

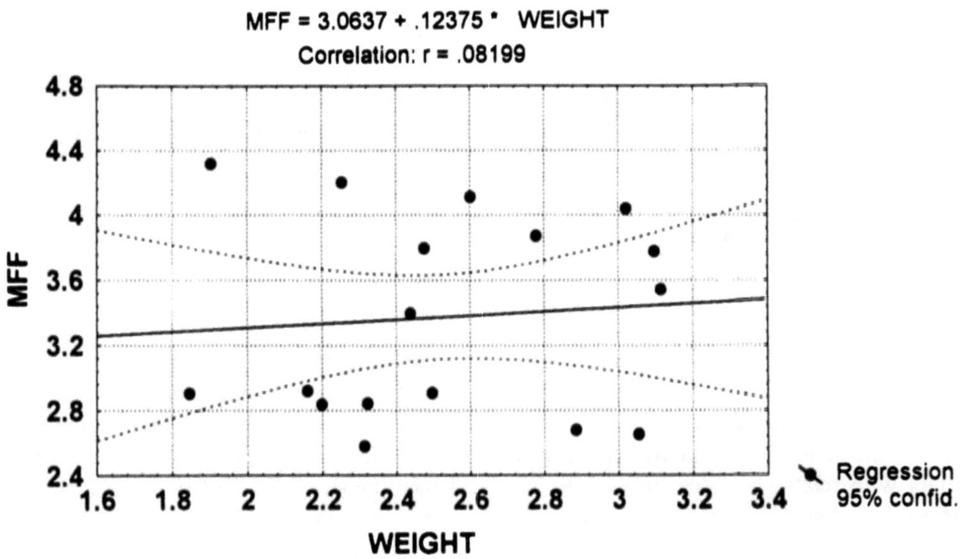

Figure 4. Regression of mean body weight (kg) on mean fundamental frequency (kHz) for 17 nocturnal strepsirhines combined ; both axes log transformed . For species see table 1.

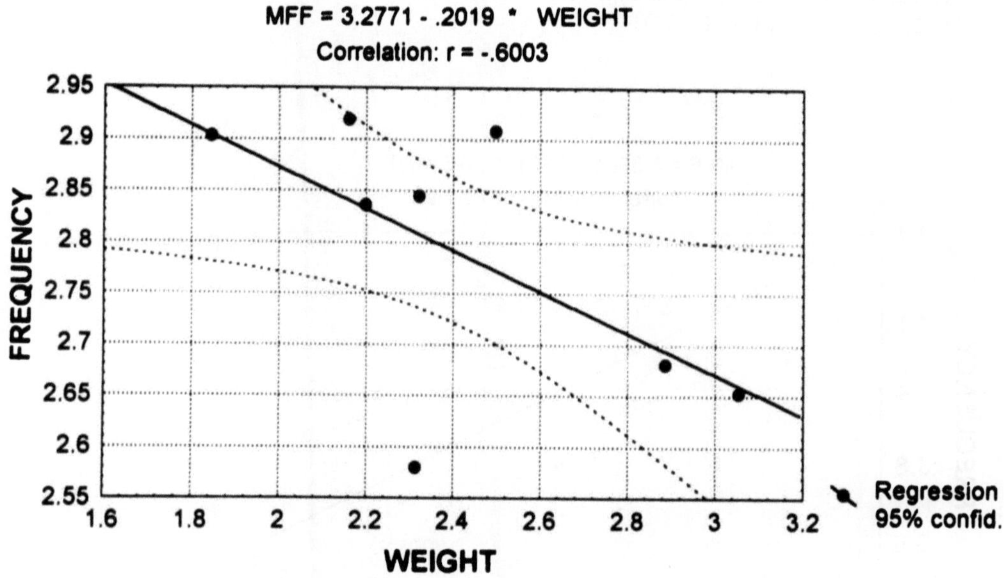

Figure 5. Regression of mean body weight (kg) on mean fundamental frequency (kHz) in busbabies. Axes are both log transformed.

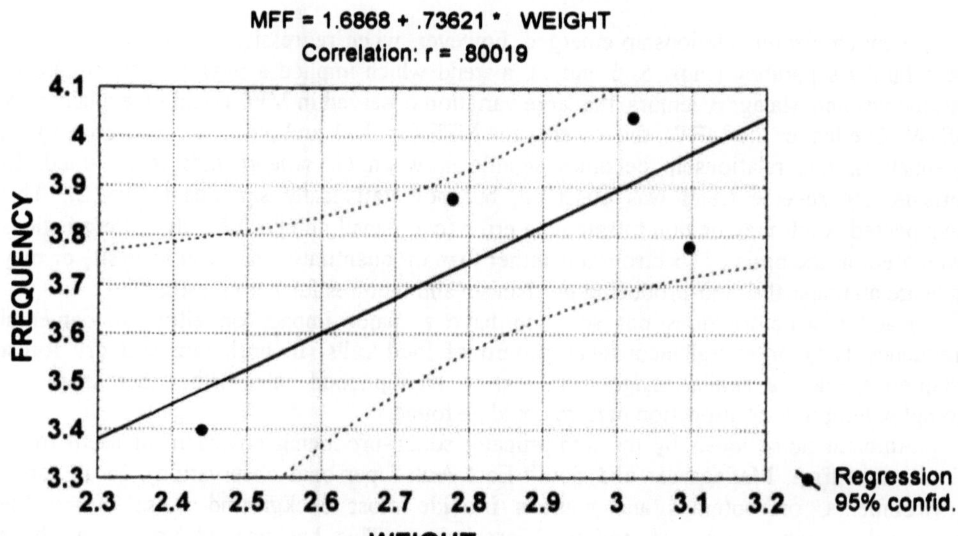

Figure 6. Regression of mean body weight (kg) on mean fundamental frequency (kHz) in lorisids. Axes are log transformed.

Fig. 4 shows the results for the data on log MBW versus log MFF for all species combined. No statistically significant relationship could be found (r(x/y)= 0.081989, r^2= 0.006722, t= 0.318613, p=0.754).

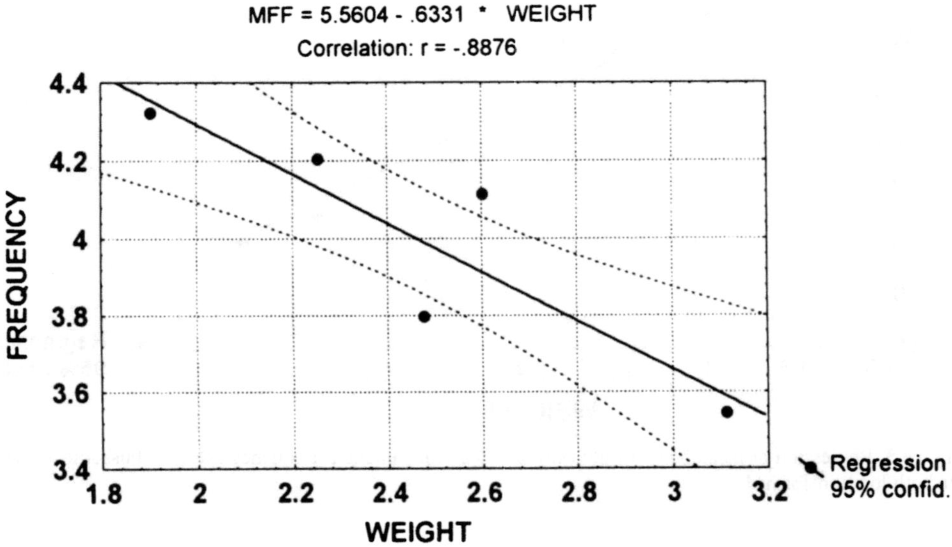

Figure 7. Regression of mean body weight (kg) on mean fundamental frequency (kHz) in nocturnal Malagasy lemurs. Axes are log transformed.

A much clearer relationship emerged, however, when regressions were calculated for each family separately (Figs. 5, 6 and 7), a trend which implied a phylogenetic origin. In bushbabies and Malagasy lemurs, the large variation observed in MFF could be explained by MBW. The higher the MBW, the lower is the MFF (p= 0.11 and p= 0.04 , respectively). In bushbabies, this relationship becomes significant when *G. senegalensis* is excluded. In lorisids, the reverse trend was observed, but not statistically significant (p=0.2). This unexpected result may be due to sampling error (e.g., small number of calls in Perodicticus evaluated on the basis of spectrograms rather than on quantitative spectral analysis) or may indicate an unusual sound-producing mechanism and requires further investigation.

Habitat structure does not seem to have a major impact on either fundamental frequency (Fo) or overall acoustical pattern of loud calls. In both rain and dry forest adapted species, a similar range from low to high pitched calls with either simple or complex temporal or intonation pattern could be found.

Ambient noise levels by the non-primate, sound-producing environment seem to be similar in Africa, Madagascar and South East Asia (personal observation). In all forest environments, orthopterans and anurans provide most background noise during the nocturnal activity period of nocturnal prosimians. This background noise may have prominent masking effects in the range between 1.5 and 8 kHz. Loud calls in nocturnal strepsirhines seem to be designed to by-pass background noise by producing loud calls with F_0 and dominant frequencies either below (galagids) or above (lorisids, cheirogaleids) this range (an example is given in Fig. 8).

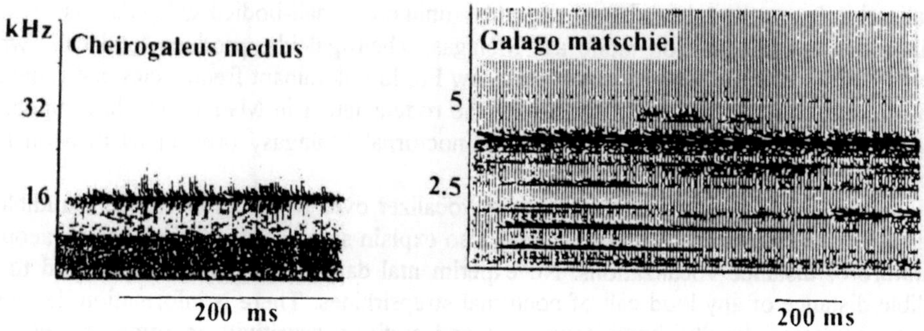

Figure 8. Unfiltered sound spectrograms of the loud call in a Malagasy lemur (*Cheirogaleus medius*, Morondava) and an African bushbaby (*Galago matschiei*, Uganda) showing the amount of ambient noise in the respective habitats.

Locomotion and body size should influence the probability of being captured by predators. Thus, the vulnerability of slow-climbing primates is expected to be higher than that of leapers, and small-bodied primates are expected to be more vulnerable than larger-bodied ones. Nocturnal prosimians are confronted with similar types of predators, irrespective of geographical distribution and habitat. Nocturnal snakes do not show any responsiveness to sounds but localize their prey by temperature or vibrational cues and can therefore be excluded from this consideration. Nocturnal predators such as carnivores and owls are specialized to locate prey by auditory cues. They have a highly elaborated auditory system (e.g., Fay, 1990) and are known to prey on nocturnal strepsirhines (Barrett, 1984; Cheney and Wrangham, 1987; Goodman et al., 1993). Their best ability to localize, however, is restricted to the frequency range of best hearing. Thus in barn owls (*Tyto alba*), the species most thoroughly studied, localization ability decreases quite rapidly for frequencies above 9 kHz and below 6 kHz . Wide-band noises in this range are better located than pure tones (Konishi, 1973). Owls are known to be the major threat to small-bodied nocturnal Malagasy lemurs such as mouse and dwarf lemurs (Goodman et al., 1993) and to African dwarf and lesser galagos (Bearder, personal communication). For tropical viverrids comparable psychoacoustical data are not yet available. Audiograms of the least weasel (*Mustela nivalis*) and the ferret (*Mustela putoris*) display a best frequency range from 1 to 16 kHz and 10 to 16 kHz, respectively (Fay, 1990). Most probably, tropical viverrids show similar hearing sensitivities. It is to be expected, however, that viverrids impose a major threat mainly to species living outside the dense and bushy undergrowth of tropical forests such as larger-bodied nocturnal lemurs and lesser and greater galagos. Furthermore, predation pressure on small Malagasy lemurs seems to be higher than for their ecological counterparts in Africa, the bushbabies. Thus in South-African bushbabies (*Galago moholi*), the total predation rate was estimated to be about 15 % of the population per year (Cheney and Wrangham, 1987) whereas in the Malagasy mouse lemur predation rate by barn owls was found to be greater than 25 % of the population per year (Goodman et al., 1993). Differences in predation risk are expected to be correlated with differences in the structure of distance calls. Marler (1955) predicted that sounds with wide frequency spectra should be more easily detected than pure tones or narrow-band tones. The high pitched narrow-band loud calls ("Whistle", "Trill") in all small-bodied cheirogaleids and in slow-climbing lorisids have, indeed, an acoustic structure which seems to make sound localization by nocturnal predators extremely unlikely. Small-bodied galagids, displaying a similar size and locomotion mode as Malagasy cheirogaleids, produce loud calls which should be easily located as a result of their low Fo, low dominant frequencies and repetitive acoustic structure. Perhaps the paucity of the rodent fauna in Madagascar has imposed a much stronger predation pressure on small nocturnal Malagasy prosimians than on their ecological counterparts in Africa.

Active space, the area surrounding the vocalizer over which an utterance is audible to conspecific recipients (Brown, 1989), may also explain some of the differences in acoustic structure of distance vocalizations. No experimental data are available with regard to the audible distance of any loud call of nocturnal strepsirhines. There is information, however, about population density, home range size and auditory sensitivity in some species. This allows an assessment of the importance of audible space for the evolution of loud call structure. Best frequencies of hearing were found in the range of 8 to 32 kHz in *Galago moholi* , in the range of 8 to 16 kHz in *Nycticebus coucang* and *Perodicticus potto* (all Fay, 1990) and in the range of 2 to 30 kHz in *Microcebus murinus* (Niaussat and Petter, 1980). This means that the loud call in *Galago moholi* lies outside the range of its best hearing sensitivity, whereas it lies almost or totally within this range in the other three species. Population density is lower and home range size strikingly larger in African galagos than in Malagasy cheirogaleids of similar size and mode of locomotion. Thus the home range of the African dwarf bushbaby, *Galago demidoff*, is described to be about twice the size of its Malagasy counterpart, the grey mouse lemur, *Microcebus murinus*, and that of the African lesser bushbaby, *G. moholi*, is about 6 times the size of the Malagasy Coquerel's dwarf lemur, *Mirza coquereli* (see Tab. 1). Habitat effects are known to affect transmission of loud calls and increase with increasing distance between the sound-producing animal and its receiver (Brown and Gomez, 1992). It should therefore impose a much stronger effect on the evolution of loud call structure in African bushbabies than in Malagasy cheirogaleids. Indeed, loud calls in all bushbabies closely fit predictions for optimal adaptation of longer distance propagation in primates (e.g., Brown and Gomez, 1992; Krebs and Davis, 1993; Waser and Waser, 1977; Wiley and Richards, 1978). Irrespective of species and size, the loud call of bushbabies consists of either long series of rhythmically repeated, species-specific units or of irregulary repeated, distinct, short series with a species-specific dynamic pattern for which the main energy lies in the lower frequency range, between 0.3 and 1.5 kHz. In contrast, the high-pitched loud call of all the remaining nocturnal strepsirhines appears to be insufficient for long distance transmission. At least three of these species for which data are available, have their highest auditory sensitivity in the range of their loud calls. In these species, it might also be the case that there was no pressure to evolve energetically expensive loud calls suited for long distance transmission because of their much smaller home range size and increased predation risk. Our field studies in a dry forest habitat in the vicinity of Morondava, Madagascar, suggest that at least the loud call of the grey mouse lemur, with its high pitched fundamental frequency, induces strong antiphonal and phonotactic responses in territorial animals over distances exceeding the home range size of large-bodied males (Zimmermann, unpublished).

Based on these findings, the following predictions for the evolution of loud call structure in nocturnal primates can be made:

The structure of the loud call should be optimally adapted to the active space the respective species needs for acoustically evoked social responses of conspecifics over longer distances.

Species with large home range size (probably induced by dispersed resources) or less vulnerable species may evolve low-pitched loud calls with regular unit repetition rate, whereby the main energy spreads over several harmonics. Such acoustic characteristics will be optimally adapted for long distance transmission. However, they also bear the risk of being easily located by predators.

Species with small home range size or highly vulnerable species, e.g., small-sized species with a slow-climbing locomotion mode, may evolve the less risky high-pitched, narrow-band loud calls with irregular repetition rate, together with best auditory sensitivity just within that range. This would minimize predation risk and may outweigh the disadvantage of poor signal transmission.

ONTOGENETIC AND DEME-SPECIFIC FLEXIBILITY OF VOCAL STRUCTURES WITH A SPECIAL EMPHASIS ON LOUD CALLS

One of the most obvious features of human speech acquisition is learning by perception and vocal imitation (e.g., Crystal, 1992; Oller et al., 1976). Evidence for vocal learning in non-human primates is, to date, ambiguous (see Biben and Bernhards, this volume; Newman, this volume). Snowdon and Elowson (1992) suggested four major mechanisms of vocal development in primates which might interact in complex ways to produce the species-specific vocal behavior: (1) Acoustic structure fixed at birth. (2) Physical maturation. (3) Learning from tutors in closed vocal systems. (4) Learning from tutors in open vocal systems.

These mechanisms will be considered by reviewing current knowledge of vocal development in nocturnal strepsirhines with a special emphasis on loud calls. Tab. 2 presents an overview of the descriptive information available for vocal ontogeny in this group.

Evidence for mechanisms 1, 2 and 4 can be found by comparing the different species from birth until adulthood.

Table 2. Descriptive information available for the ontogeny of vocal communication in nocturnal strepsirhines.

Genus	Species	Documentation Source
Galagidae		
Galago alleni	limited	Charles-Dominique, 1977; Petter and Charles-Dominique, 1979;
Galago crassicaudatus	limited	Bearder, 1974; Zimmermann, unpubl.
Galago demidoff	limited	Charles-Dominique ,1977; Petter and Charles-Dominique, 1979
Galago moholi	limited	Mascagni and Doyle ,1993; Zimmermann, unpubl.
Galago senegalensis	detailed	Zimmermann ,1989 a
Lorisidae		
Nycticebus coucang	detailed	Andrew, 1963; Zimmermann, 1989 b
Perodicticus potto	limited	Petter&Charles-Dominique, 1979; Zimmermann, unpubl.
Cheirogaleidae		
Cheirogaleus medius	detailed	Stanger, 1988
Microcebus murinus	detailed	Stanger, 1988; Kuhn, 1989; Zimmermann, 1991, unpubl.
Mirza coquereli	detailed	Stanger ,1993

Comparison of Vocal Development in Nocturnal Strepsirhines

There are vocalizations in infant nocturnal strepsirhines which occur fully developed at birth and undergo neither major modification in vocal structure nor in vocal usage. This pattern closely fits mechanism 1 of Snowdon and Elowson (1992) and can be found in a) "Purr" calls of the genus *Microcebus*, *Mirza* and *Cheirogaleus*, which are given during affiliative social interactions with tactile contact (Kuhn, 1989; Stanger, 1988, 1993; Zimmermann, 1991) and signalling comfort, b) in "Scream" calls of all cheirogaleids, galagids and lorisids, signalling discomfort, when extremely hungry, injured or transported in an awkward way by the mother (see Tab. 2) and in c) "Grunt" calls of these groups, signalling anger, when the animals are cornered. Vocalizations are combined with a particular attack display in cheirogaleids and galagids. Calls are elicited involuntarily as soon as the correct stimulus is given. They most probably reflect changes in the caller´s emotional state and may be treated as fixed action patterns with fully developed neuronal pathways in neonates from tactile or visual perception to vocal production.

All other vocalizations in the adult repertoire emerge either in adults or undergo more or less pronounced modification in vocal usage and/or structure during maturation. Learned and maturational effects (mechanisms 2-4 of Snowdon and Elowson, 1992) are thus difficult to distinguish in pure descriptional information. In addition, there are vocal patterns restricted to immatures in galagids and cheirogaleids (Zimmermann, 1989a , 1991).

Most interestingly, all cheirogaleids and galagids do show a period of high vocal plasticity during early infancy, as do human infants and some marmosets (Snowdon, 1989). During affiliative social interactions with siblings, mother or other social companions, they produce streams of vocalizations (onomatopoetically described as "Week", "Weak", "Wak" for *Galago senegalensis*; Zimmermann, 1989 a, 1991 ; "Twitter", "Chirp", "Proto-Trill", "Proto-whistle" for *Microcebus murinus*, *Mirza coquereli* and *Cheirogaleus medius*; Kuhn, 1989; Stanger, 1988, 1993; Zimmermann, 1991) displaying such a high degree of structural variability that it is currently impossible to assess them accurately in a quantitative way (see Fig. 9 for an example). This plasticity in vocal morphology regresses around the time of nutritional weaning. In lorisids, such a period of high structural variability in vocal morphology could not be detected (Zimmermann, 1989b).

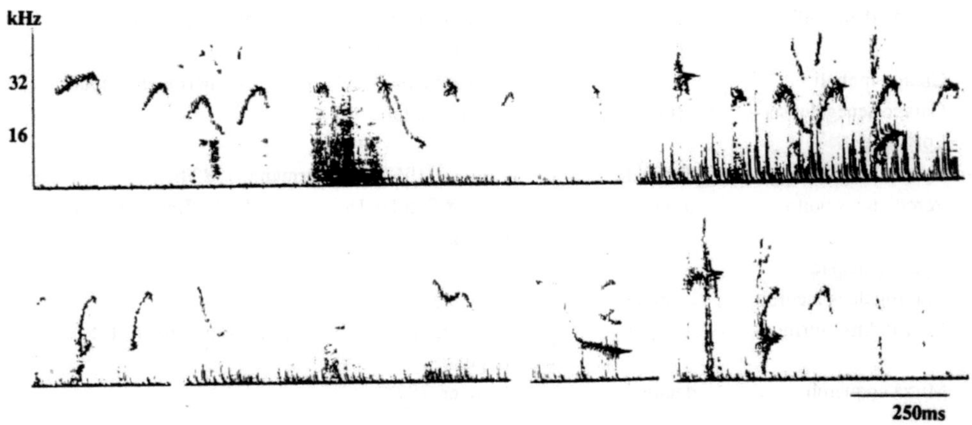

Figure 9. A vocal stream of highly variable vocal patterns in a neonate mouse lemur male during grooming by its mother.

In contrast to the broad variety of distinct vocal patterns produced by some anthropoids during short-term isolation from mother or social companions (see Newman, this volume), short-term isolation in the strepsirhine infants investigated (see Table 2) evokes relatively stereotyped vocal patterns typical for the respective phylogenetic group. In all species, "infant isolation calls" evoke immediate behavioral responses in mothers. Series of short, broadband "click" or "zek"- calls characterize the acoustic pattern of all galagid and lorisid species, whereas series of narrow-band "whistle" calls were found in all cheirogaleids (see Fig. 10 for an example and Tab. 2 for references). In galagids, "clicks" are gradually transformed into the adult, species-typical loud call after a "voice break" (Zimmermann, 1990). In lorisids, they retain their overall acoustic pattern until adulthood, where calls similar in structure to "zeks" can be found in "ak"-calls, which are contact-seeking calls at short distance (Zimmermann, 1990). In cheirogaleids, the "whistle" call structure is found in the "whistle" loud call of adults in the genera *Cheirogaleus*; a part of the "whistle" element can also be found at onset of the "trill" loud call of the genera *Mirza* and *Microcebus*.

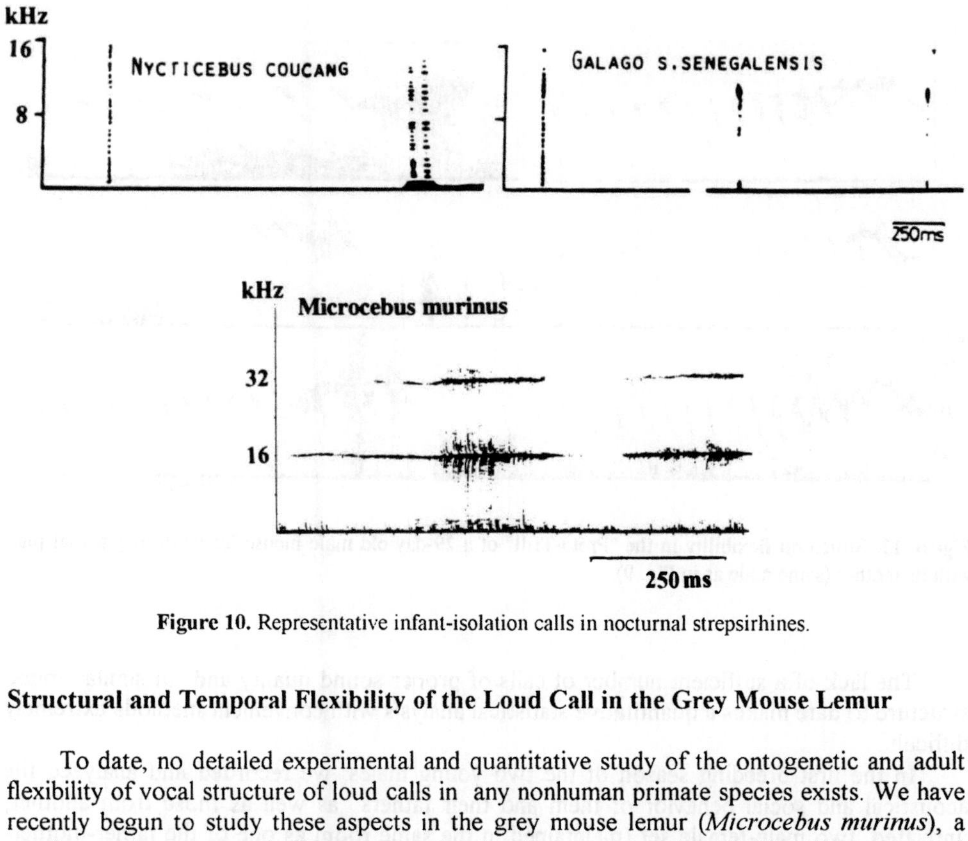

Figure 10. Representative infant-isolation calls in nocturnal strepsirhines.

Structural and Temporal Flexibility of the Loud Call in the Grey Mouse Lemur

To date, no detailed experimental and quantitative study of the ontogenetic and adult flexibility of vocal structure of loud calls in any nonhuman primate species exists. We have recently begun to study these aspects in the grey mouse lemur (*Microcebus murinus*), a species with an acoustically complex loud call structure, the "Trill". The acoustical complexity of this call and its usage (used most frequently during the mating season by dominant males, Zimmermann and Lerch, 1993), as well as the breeding mode of mouse lemurs (territorial males, seasonal reproduction) show interesting parallels to song in oscine birds.

We established two pairs of mouse lemurs and investigated the vocal and social development of two young males born and reared in these groups to obtain information about the pattern and course of the development of the "Trill". The two groups were kept in different rooms. "Trill" call elements occurred on the first day after birth. They were, however, intermingled with streams of young-specific vocalizations uttered during close range affiliative social interactions (e.g., being groomed, social play, see Fig. 9). Neither of these two males nor other infants for which we have additional cross-sectional data, produced the individual-specific, three component "Trill" call (whistle + sweep + train of frequency-modulated units) typical of adults (Kugel and Zimmermann, 1993; Zimmermann and Lerch, 1993). The period of vocal play, characterized by a high plasticity in vocal morphology during affiliative social interactions, extends from just after birth until nutritional weaning. Weanlings started to show a new variation of play behavior, sexual play. They try to catch the tail or parts of the body of their mother, father or (in socially reared infants) siblings and show mounting behavior with rhythmical pelvic thrust movements. At this time, and most obviously during sexual play interactions, the two males produced the three component "Trill" call in a similar order as adults. In contrast to the latter, however, this "Proto-Trill" has a fairly low sound intensity and a strikingly high structural flexibility in frequency and amplitude modulation and in intonation (Fig. 11).

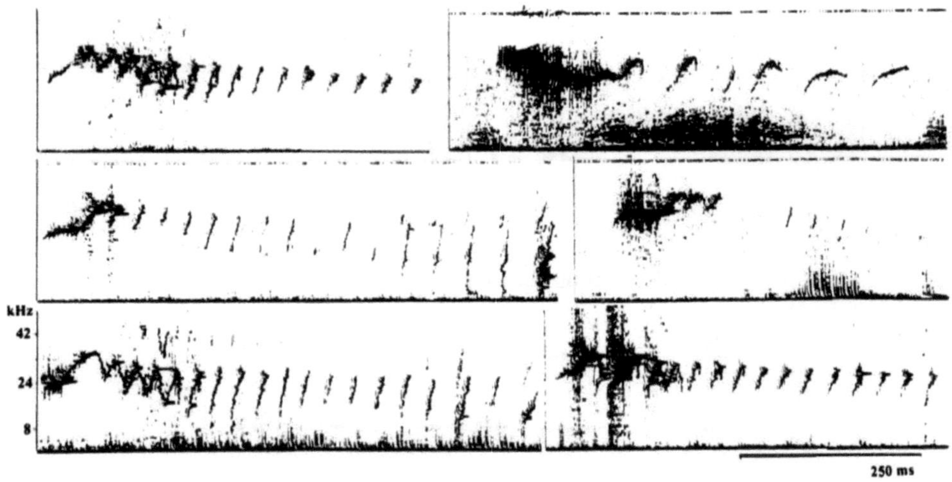

Figure 11. Structural flexibility in the "Proto-Trill" of a 29-day old male mouse lemur during sexual play with its mother (same male as in Fig. 9).

The lack of a sufficient number of calls of proper sound quality and of similar sound structure to date makes a quantitative statistical analysis with convenient methods extremely difficult.

In the first breeding season of the two young males, we recorded and analyzed the acoustical and social behavior of them and their fathers as well as those from another, unrelated, two male-female set (maintained in the same room as one of the father-mother-son sets) with respect to the acoustical flexibility of the "Trill" call during pre-estrous and estrous of the females. We characterized 138 "Trill" calls of these six males, using 33 acoustical characters and performed a multivariate statistical analysis to point out possible acoustical similarities and differences among the different males and male groups (see Zimmermann and Lerch, 1993 for a detailed explanation of the procedure). The result of an average linkage cluster analysis is presented in Fig. 12.

The degree of acoustical similarity in the structure of the "Trill"call is reflected in the average Euclidean distance. The more similar the calls, the shorter the respective distance values. Individual males could be reliably identified by the structure of their loud call as shown by an almost homogenous clustering of individuals by calls. The acoustical design of the "Trill" call in the two mature males is therefore as stable as that of the other adult males. "Trill" calls of individual males of each male group, however, show closer similarity to each other than to any other male group. Thus, "Trill" calls of the unrelated male group were combined to form one cluster at an average distance level of about 1.0, one father-son group at about 0.8 and the other, acoustically isolated one at about 0.65. The acoustical variability in this latter group is higher than in both of the other groups.

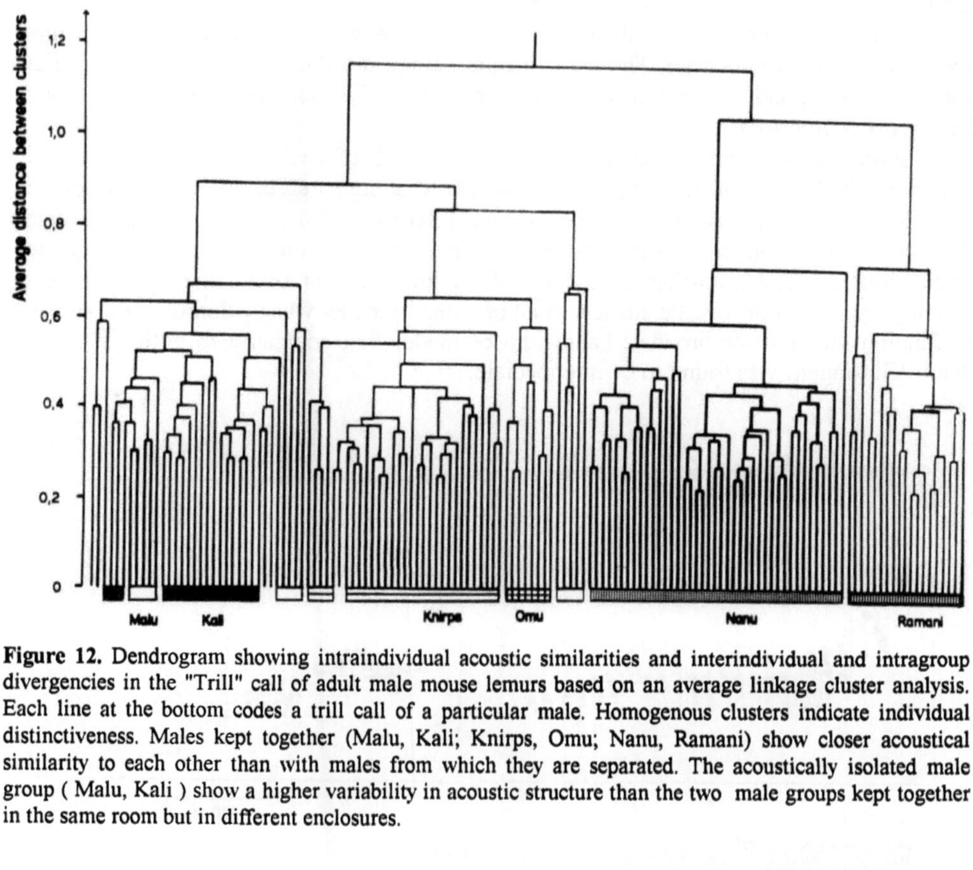

Figure 12. Dendrogram showing intraindividual acoustic similarities and interindividual and intragroup divergencies in the "Trill" call of adult male mouse lemurs based on an average linkage cluster analysis. Each line at the bottom codes a trill call of a particular male. Homogenous clusters indicate individual distinctiveness. Males kept together (Malu, Kali; Knirps, Omu; Nanu, Ramani) show closer acoustical similarity to each other than with males from which they are separated. The acoustically isolated male group (Malu, Kali) show a higher variability in acoustic structure than the two male groups kept together in the same room but in different enclosures.

What might be the reason for this close accordance in vocal structure within and the higher divergence among the different male groups? A high degree of similarity in a contextually homologous vocal pattern would be expected when paternal traits are inherited from one generation to the other or when an animal adapts its vocal output toward that of another animal by perception of acoustically similar vocal patterns and by copying them vocally. Neither paternal effects nor vocal learning have been suggested to influence the acquisition of vocal morphology in any non-human primate species (Lieblich et al., 1980; Ploog, 1992). All mouse lemur males from this experiment originated from one founder pair and were therefore assumed to possess a high degree of genetic concordance. Their loud calls, however, possess an individual-specific acoustical design. The acoustic similarity of the two father-son pairs raised together from birth until the first breeding period corresponds to that of the set of two relatively unrelated adult males, raised apart from each other until their first breeding period. Although paternal effects cannot be excluded, the fact that the "Trill" call originated from highly variable infant calls and showed a close acoustical resemblance among individuals both within related male groups and within a relatively unrelated male group provides evidence for a social influence on vocal production.

How flexible the acoustical design of the "Trill" call of a phenotypically homogenous population within a homogenous behavioral situation may be, we discerned from a pilot study we conducted recently using the same methodology. We recorded, analysed and compared the "Trill" call of males during the pre-estrous and estrous period of a female in a phenotypically homogenous mouse lemur population kept at three different primate facilities in Europe (Brunoy, F; Moskau(Kassel), R; Göttingen, G). Most interestingly, dialect-like differentiations in "Trill" call structure emerged.

I will outline here results of this ongoing study, based on seven males and a sample size of 10 vocalizations/male. The experimental set-up and the procedure of multivariate statistics correspond to methods already described (Zimmermann and Lerch, 1993; Zimmermann, in press).

Sound spectrograms of representative "Trill" calls of each deme are presented in Figure 13. Trill calls of each deme seem to show a high degree of intrademe-specific similarity and a pronounced interdeme-specific difference. Thus, the "Trill" call in deme "Moskau" show a longer whistle duration, a shorter call duration, a more rapid frequency modulation rate and a smaller bandwidth of modulation than both other demes. Deme "Brunoy" has either no whistle element at all or a much smaller whistle duration, the slowest modulation rate and the broadest bandwidth of modulation compared to both the others. Deme "Göttingen" was found to be intermediate.

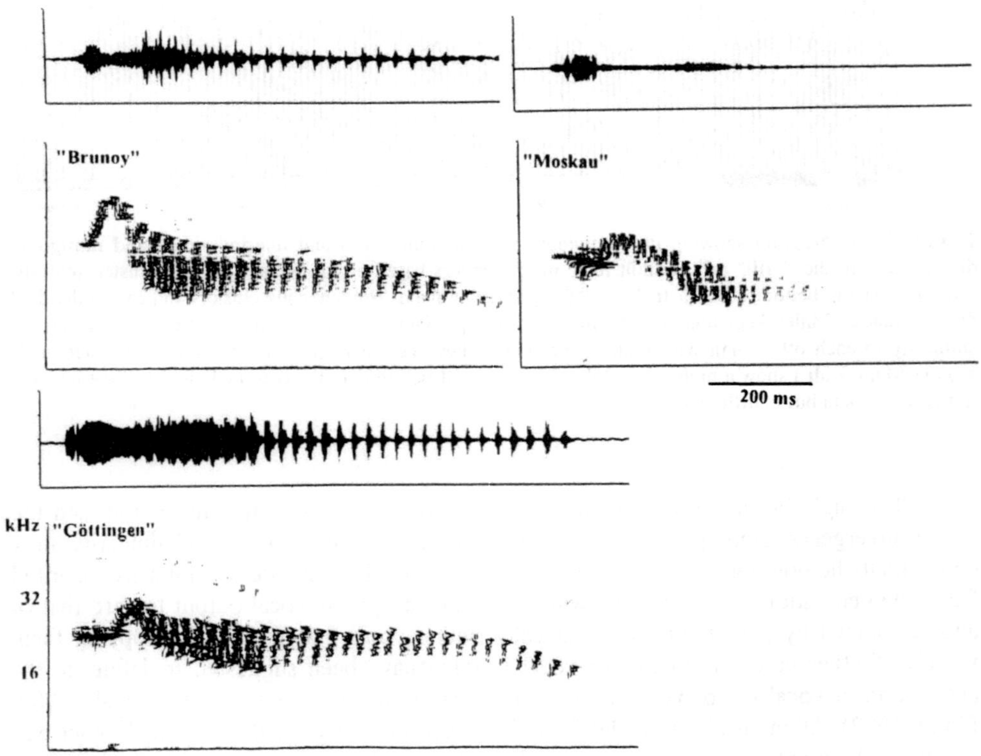

Figure 13. Dialect-like vocal patterns in "Trill" calls of males in three different mouse lemur demes as shown by representative oscillograms and spectrograms.

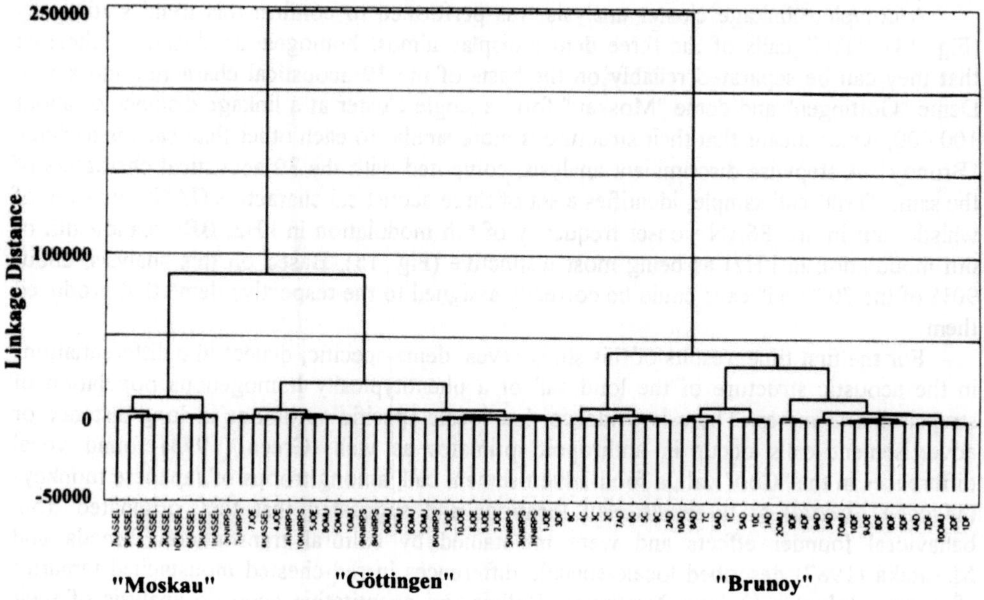

Figure 14 . Dendrogram of the acoustic similarities and divergencies in 70 trill calls in seven mouse lemur males of three different demes, as revealed by a complete linkage cluster analysis and squared Euclidean distances.

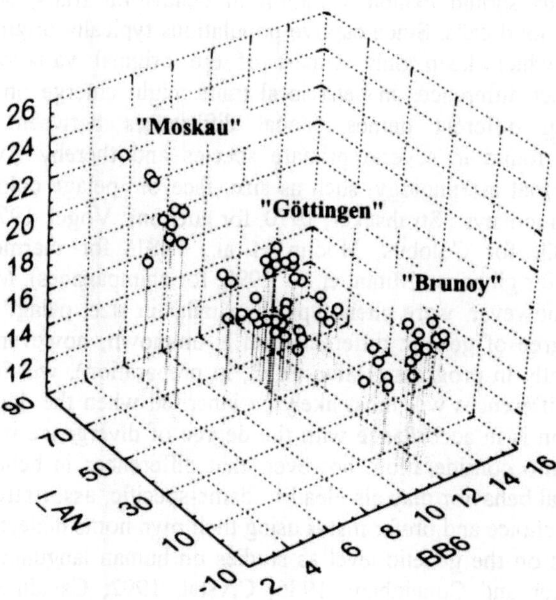

Figure 15. Acoustic character set separating three different demes of mouse lemurs based on a stepwise discriminant analysis.

A complete linkage cluster analysis was performed to confirm this trend statistically (Fig. 14). "Trill" calls of the three demes display almost homogenous clusters, indicating that they can be separated reliably on the basis of the 39 acoustical characters measured. Deme "Göttingen" and deme "Moskau" form a single cluster at a linkage distance of about 100 000, which means that their structure is more similar to each other than each is to deme "Brunoy". A stepwise discriminant analysis, computed with the 39 acoustical characters of the same "Trill" call sample, identifies a set of three acoustical characters (IAN= duration of whistle unit in ms; F6AN= onset frequency of 6th modulation in kHz; BB6=bandwidth of 6th modulation in kHz) as being most distinctive (Fig. 15). Based on this analysis, about 90% of the 70 "Trill" calls could be correctly assigned to the respective deme that produced them.

For the first time, results of this study reveal deme-specific, dialect-like differentiations in the acoustic structure of the loud call of a phenotypically homogenous population of strepsirhine primates. There is evidence that locale-specific variations in long-distance or advertisement calls occur in anthropoid primates as well. Green (1975) found vocal differences in the "Coo" call, a food-advertisement call, among troops of Japanese monkeys (*Macaca fuscata*) at three different locations and suggested that they originated from behavioral founder effects and were maintained by cultural transmission. Maeda and Masataka (1987) described locale-specific differences in red-chested moustached tamarins (*Saguinus labiatus l.*) from Northwest Bolivia, by quantitative acoustic analysis of their long calls. The two populations were found to occur only 27 km apart from each other, without any geographical barrier. Recently, Mitani et al. (1992) discerned geographical variation in the acoustical structure of chimpanzee pant hoot, a loud, species-typical long distance call in common chimpanzees (*Pan troglodytes*). They suggest that these variations might be caused by learning.

Three major factors may be responsible for vocal differences among demes of the same species. I will discuss these with regard to our findings on the ontogenetic and adult variation of loud calls in the mouse lemur:

- Genetic differences among different demes caused by genetic drift and founder effects. Natural populations should exhibit variation in behavioral traits, such as the acoustic characters of their loud calls. Since captive populations typically originate from only a few founder animals which keep only a part of the original variation found in natural populations, distinct differences in behavioral traits might emerge on the basis of genetic differences among different demes. Vocal differences between genetically isolated populations were found in several primate species and thereby correlated with subtle differences in external morphology, such as size, face or pelage coloration (e.g., Winter, 1969, for squirrel monkeys; Struhsaker, 1970, for guenons; Vogel, 1973, for langurs; Oates and Trocco, 1983, for Colobus; Hodun et al., 1981, for marmosets; Marshall and Sugardjito, 1986, for gibbons; Mitani et al., 1992, for chimpanzees). Mouse lemur males of the three demes, however, were phenotypically similar in size, pelage coloration and body weight. Their degree of genetic differentiation is unknown, however. DNA-amplification techniques, currently in progress (Hafen et al., in preparation), will hopefully resolve this question. Vocal differences will most likely be inherited when the degree of divergence in acoustical variation is in accordance with the degree of divergence in genetic variation. It has to be taken into consideration, however, that differences in behavioral traits such as differences in vocal behavior may also lead to deme-specific, assortative matings, if females show active mate choice and prefer mates using their own home dialect. This, in turn, would induce differences on the genetic level as studies on human language and bird song have shown (e.g., Baker and Cunnigham, 1985; Crystal, 1992; Cavalli-Sforza et al., 1988). Monozygotic siblings raised in different acoustical environments may help to illuminate the degree of inheritance of acoustic traits.

- Differences in context or arousal state. Subtle differences in acoustical characters of vocalizations may often encode information about slightly different contexts or arousal states, as known from macaques, guenons and humans (e.g., Cheney and Seyfarth, 1990; Crystal, 1992; Gouzoules et al., 1974; Green, 1975; Hauser, 1992). All mouse lemur males were tested, however, in the same homogenous context (two males with a pre-estrous or estrous female). Furthermore, using discriminant function analysis, we have investigated whether slightly different variations of the sound-correlating behavior of the sender will have an effect on trill call structure. A total of 237 calls characterized by 39 acoustical characters and produced by four different males in five different behavioral situations (approach to female, n=81; retreat from female, n=114; attacked by male, n=3; chased by female, n=18; spontaneously, n=21) were analyzed. As Table 3 outlines , classification accuracy was poor. Thus, consistent acoustical differences between calls used during different sound-correlated behaviors in the same behavioral context could not be detected.

Table 3. Classification of 237 "Trill calls", in relation to the sound-correlated behavior of the sender using discriminant function analysis with 39 acoustical characters.

Context Correctly classified	Retreat	Attacked	Chased	Approach	Spontaneously	%
Retreat	69	2	6	30	7	58.5
Attacked	1	0	0	2	0	0
Chased	11	0	1	2	4	5.3
Approach	49	1	2	23	6	28.4
Spontaneously	4	0	6	6	5	23.8

Summarized error rate: 59.5 %

If genetical and contextual factors do not account for the observed differences in trill call structure among different demes, it may be proposed as a final explanation that this variation is caused by

- Social adaptation as found in learned dialects of various song birds (e.g., Kroodsma, 1982; Marler, 1990; Payne, 1981) and in human languages (Crystal, 1992). The social system of mouse lemurs would create an appropriate selective milieu favouring the evolution of vocal learning. Mouse lemurs have a patchy distribution in their forest habitats where particular population nuclei are spread over suitable forest sites (Harcourt and Thornback, 1990). They are suggested to have a pre-gregarious social structure and a polygynous mating system (Martin, 1972). Home ranges in heavy-bodied breeding males are larger than in breeding females and overlap within and between sexes. Lighter-bodied males are found at the peripheries of these population nuclei and may contribute to the genetic exchange among different population nuclei. Females are often found together in holes of the same sleeping tree and may be related to each other (Barre et al., 1988; Pages-Feuillade,

1988). Nothing is known, however, about the social dynamics of these population nuclei or about the genetic structure and the genetic relations within and between them. Our behavioral studies in captivity discerned that females show active mate choice (Kugel, 1993; Zimmermann, in press) and seem to prefer partners with whom they had particular social bonds during the nonbreeding season (Schuchardt and Zimmermann, in preparation.). Females seem to distinguish males on the basis of differences in their loud calls (Zimmermann, in press). Dominant males try to monopolize females during their pre-estrous and estrous period (Perret, 1992; Zimmermann and Lerch, 1993). Lighter-bodied males may profit, if they adapt their calls to locale-specific variants of heavy-bodied males, because they then become more accessible for potential mates and receive less aggression from heavy-bodied males.

These eco-ethological prerequisites may cause learned dialect-like vocal differences in mouse lemurs or non-human primates with similar life styles and require further experimental and comparative studies, both in the field and in captivity.

CONCLUSION AND PERSPECTIVES

Diversity in Loud Call Structures

In contrast to anthropoids (e.g., Gautier and Gautier-Hion, 1977, 1988), the structure of the loud call in nocturnal strepsirhines shows a high diversity among the different taxa. Because the loud call is the most distinctive and species-specific call type within the adult vocal repertoire, evokes call-specific responses in visually separated conspecifics, and is produced with high rates by males during the mating season (see Tab. 1 , for references), it is possible to compare its structural differentiation among 17 different species of four nocturnal strepsirhine families. Unfortunately, detailed information is lacking for three bushbaby species, one lorisid species, three cheirogaleid species, one indrid species as well as for all species of the Malagasy family Lepilemuridae and Daubentonidae.

Among the species examined, contrasts in phylogeny, morphology (body size, locomotion) and eco-ethology (habitat, predation, home-range size) occur which allow their effect on the acoustic structure of the loud call to be studied. Three major differentiations in loud call structure were revealed, characterized by differences in Fo, and the complexity of frequency and amplitude modulation, intonation and temporal pattern. Fo was not a simple allometric function of body weight, but also depends upon phylogeny and differences in predation. A hypothesis was put forth to explain the rich diversity in acoustic design of the loud call among different species. It postulates that the evolution of loud call structure was primarily influenced by predation risk and home-range size. Highly vulnerable species or species with small home-ranges should evolve loud call structures with shorter propagation distances than less vulnerable species or species with larger home-range size. It may also be that predation risk alone is sufficient to explain the diversity in loud call structure and that this, in turn, may have an effect on the distribution of individuals in time and space and thereby on home-range sizes. This hypothesis was based on analysis of 17 different nocturnal strepsirhine species. It may also help to explain the divergences in the loud call structure among tamarins and marmosets, where the acoustic pattern of the loud call in some species was found to be maladaptive (Snowdon, 1989). The generality of this hypothesis, however, has to be tested in a broader array of both strepsirhine and anthropoid primates.

To date, there is a great lack of information of how the different loud call structures in non-human primates are produced. Results on squirrel monkeys (Jürgens et al., 1978) imply that the production of low-pitched sounds involves a totally different set of laryngeal muscles than in high-pitched sounds. Comparative investigations about the involvement of laryngeal and supralaryngeal structures and their respective motor control areas in the brain during vocal production, may not only help to clarify the proximate mechanisms of loud call diversity in primates but may also provide a much deeper knowledge about the evolution of primates in general.

Convergent and Divergent Patterns in the Differentiation of the Adult Vocal Repertoire

Nocturnal strepsirhines display five major ontogenetic patterns in the differentiation of calls in the adult vocal repertoire:

1. There are reflexive calls with almost fixed acoustic patterns and usage from birth onwards. The appropriate stimulus evokes the appropriate vocal response. Calls of this category may reflect the actual emotional state of the sender ("comfort", "discomfort", "anger"). The stereotyped stimulus-response pattern of these calls imply that they are innate, hard-wired and belong to fixed action patterns. Similar calls can be found in primate-related animals such as tree shrews (Benson et al., 1992) and even in humans (see Jürgens, this volume). Perhaps these vocalizations exhibit a conservative pattern among the different taxa because they regulate basic social interactions between mother and infant, critical for the survival of the infant.
2. There are calls which appear suddenly in the adult repertoire (e.g., "whistle" calls in slow loris) in the appropriate social situation. Since there is only descriptive information, it is impossible to assess the relative contribution of genetic and experiential effects. Vocalizations that do not appear before adulthood have been described in a variety of primate and primate-related groups (e.g., Benson et al., 1992; Gautier, 1989; Gautier and Gautier-Hion, 1977; Hohmann, 1991). Their production might be dependent on the hormonal state and/or particular age-specific social interactions.
3. There are vocalizations with group-specific acoustic structure in infants which develop gradually toward the species-specific adult contact calls (e.g., infant isolation calls) without major modifications in usage. Whether the time course of their maturation and the differentiation of their structure is triggered solely by a fixed genetic plan or additionally dependent upon external influences is yet unknown. The group-specific similarity in infant isolation calls of galagids and lorisids on one side and cheirogaleids on the other imply a common phylogenetic heritage.
4. There are vocalizations specific for infants which do not correspond to any vocal pattern in adults. They are found only in bushbabies and Malagasy lemurs and could not be detected in the less vocal lorisids. Such vocalizations do not seem to have a particular function. They are produced by all infants during affiliative social interactions with social companions and may be intermingled in vocal streams with vocalizations resembling elements of the adult vocal repertoire.
5. There are vocal elements resembling elements of particular adult social calls which are infant-specific and produced in the same behavioral situations as in 4. They also seem to occur in bushbabies and Malagasy lemurs, but not in lorisids. In the most thorougly studied species, the mouse lemur, the development of these vocal patterns was investigated during social play interactions with the mother. It seems that at their appearance, they are interspersed in vocal sequences with infant-specific vocalizations. As soon as the animals show sexual play behavior, they are organized in an adult-similar order, but with a high degree of structural and intonation flexibility. A stereotypic, individual-specific structure does not emerge before adulthood. To date, similar differentiations are unknown in other primates, besides humans (see e.g., Elowson et al., 1992). They are widespread, however, in male oscines who learn songs from social companions (e.g., Marler, 1990).

This summary of the different ontogenetic profiles within and among the different nocturnal strepsirhine groups indicates that there is no uniform mechanism controlling the time course and differentiation of social calls. Segregated mechanisms for each developmental pattern occur. The investigation of a broader variety of taxa from lemurs to apes is necessary to reveal whether these patterns are peculiar for the respective species or groups and their socio-ecology, or distributed among primates in general. Quantitative data on vocal development of individually known infants reared in social groups with known social and genetic relationships are necessary but so are experiments with infants reared under selective socio-acoustical experience (e.g., reversible deafness by ear-plugs, acoustical deprivation, cross-fostering of monozygotic twins among care-givers communicating with different dialects). They should be conducted on a broader spectrum of taxa to determine the extent of modification in vocal structure during development and whether such changes are guided by auditory feedback and self-stimulation or socio-acoustical learning. Additionally, in the future, the rapid improvement of the resolution of modern brain imaging techniques (e.g., positron-emission tomography coupled with nuclear magnetic resonance techniques) and their application in behavioral research will probably allow more insight into brain mechanisms controlling the differentiation of vocal repertoires of primates in an almost non-invasive manner.

Flexibility in the Acoustic Pattern of Social Calls in Adults

Until recently, it was generally accepted that, in contrast to humans, the acoustic pattern of social calls in non-human primates show almost no modification during development (see Biben and Bernhards, this volume; Newman, this volume). Recent findings, however, suggest that the socio-acoustical environment may have an impact on the structure of particular social calls. Thus, thoroughly conducted acoustic analyses on vocal turn-taking sequences in squirrel monkeys and Japanese macaques indicate, that the acoustic structure of calls of responding individuals is dependent upon the structure of the preceding one (e.g., Maurus, 1988; Masataka, this volume). Particular acoustic characteristics in "coo" calls of rhesus monkeys typical for one matriline gradually disappear as animals join another matriline (Hauser, 1992). Within this review, I have provided evidence that the loud call in mouse lemur males is influenced by the socio-acoustical environment. The acoustic pattern of this call is more similar within the same than among different social groups. Most interestingly, dialect-like vocal differentiations in the loud call structure were revealed in a reproducible, context-homologous behavioral situation. Males of three different mouse lemur colonies display distinct deme-specific loud call structures. Three different hypotheses were provided which may explain this high flexibility in a phenotypically homogenous population. One of them (motivation hypothesis) could be rejected, two (genetic hypothesis, social adaptation hypothesis) require further experiments. We are currently about to include more captive individuals of different demes in such an analysis and to determine their degree of genetic concordance. Additionally, we are examining the distribution of such "dialects" in neighboring and geographically isolated mouse lemur populations in Madagascar, why they emerged and what their consequences are for social life (Hafen and Zimmermann, in press, in preparation).

To date, there is only limited information about context-homologous, locale-specifc differentiations of social calls. In only one species, the red- chested moustached tamarin (Maeda and Masataka, 1987), is there reliable evidence that a locale-specific long call occurs in phenotypically similar individuals within the same behavioral context. "Dialects", locale-specific differences in social calls among neighboring demes or geographically isolated populations (Marler, 1990) occur in diverse groups of the animal kingdom such as crickets, frogs, birds, viverrids, whales, and humans (Bailey,1991; Baker and Cunnigham, 1985; Payne, 1981; Wickler, 1986), whereby learned dialects are restricted to oscine birds and humans (Wickler, 1986).

More information about the distribution, origin and consequences of "dialects" in primates are urgently needed for a deeper insight as to where and why they appeared, why they are maintained, and the extent to which they are innate or learned in the different primate groups. Such a comparative approach will probably help us to gain a new understanding of the diversity and evolution of primate vocal communication.

ACKNOWLEDGMENTS

I thank Hans-Jörg Kuhn, Hubert Markl and Hinrich Rahmann for support and encouragement and the German Primate Center, the Animal Research Facility, Konstanz, and Helmut and Marianne Zimmermann, Stuttgart, for providing space for conducting research. I would like to thank Robin Dunbar, Heinrich Güttinger, Mauvis Gore, Marc Hauser, Gottfried Hohmann and Barbara Fruth for critical comments and discussions and M. Gore for checking the English version of this manuscript. Field studies at Madagascar were logistically supported by the C.F.P.F. at Morondava and the Ministère des Eaux et Forêts at Antananarivo, Morondava and Andasibe. I am especially grateful to Martine Perret and Christian Welker for giving us access to their mouse lemur colonies, to the Duke Primate Center for giving me access to lemur colonies, to Simon Bearder, Holger Preuschoft and Helena Fitch-Snyder, for comparative sound recordings on bushbabies, slender loris and pygmy slow loris and to Thomas Hafen, Carola Lerch, Simone Kugel and Petra Schuchardt for their help in data collecting and analysing. Financial support was provided by several grants of the German Research Council (DFG Zi 345/1,3,5,8).

REFERENCES

Andrew, R.J., 1963, The origin and evolution of the calls and facial expressions of the primates, *Behaviour* 20: 1-109.

Bailey, W.J., 1991, "The Acoustic Behavior of Insects", Chapman and Hall, London.

Baker, M.C. and Cunningham, M.A., 1985, The biology of bird-song dialects, *Behav. Brain Sci.* 8: 85-133.

Barre, V., Lebec, A., Petter, J.-J., and Albingnac, R., 1988, Etude du Microcèbe par radiotracking dans le forêt de l'Ankarafantsika, *in*: "L'Equilibre des Ecosystèmes Forêstiers à Madagascar", L. Rakotovao, V. Barre and J. Sayer, eds. , IUCN, Gland.

Barrett, E.B.M., 1984, "The ecology of some nocturnal arboreal mammals in the rain forest of peninsular Malaysia", PhD thesis, University of Cambridge, unpubl.

Bearder, S.K., 1974, "Aspects of the Ecology and Behaviour of the Thick-Tailed Bushbaby (*Galago crassicaudatus*)", PhD thesis, Univ. of Witwatersrand, Johannesburg, unpubl.

Bearder, S.K., 1987, Lorises, bushbabies and tarsiers: Diverse societies in solitary foragers, *in*: "Primate Societies", B.B. Smuts, D.L. Cheney, R.M. Seyfarth, R.W.Wrangham and T.T.Struhsaker, eds., Univ. Chicago Press, Chicago.

Benson, B., Binz, H., and Zimmermann, E., 1992, Vocalizations of infant and developing tree shrews (*Tupaia belangeri*), *J. Mamm.* 73:106-119.

Boehm, Ch., 1989, Methods for isolating chimpanzee vocal communication, *in*: "Understanding Chimpanzees", P.G. Heltne and L.A. Marquardt, eds., Harvard Univ. Press, Cambridge.

Brown, C.H., 1989, The active space of blue monkey and grey-cheeked mangabey vocalizations, *Anim. Behav.* 37, 1023-1034.

Brown, C.H. and Gomez, R., 1992, Functional design features in primate vocal signals: The acoustic habitat and sound distortion, *in*: "Topics in Primatology, Vol. 1", T. Nishida, W. C. McGrew, P. Marler, M. Pickfort and F. de Waal, eds., Univ. Tokyo Press, Tokyo.

Byrne, R.W., 1982, Distance vocalizations of Guinea baboons (*Papio papio*) in Senegal: An analysis of function, *Behaviour 75*: 283-312.

Cavalli-Sforza, L., Piazza, A., Menozzi, P., and Mountain, J.L., 1988, Reconstruction of human evolution: Bringing together genetic, archaeologic and linguistic data, *Proc. Nat. Acad. Sci.* 85: 6002-6006.

Charles-Dominique, P., 1977, "Ecology and Behaviour of Nocturnal Primates", New York Columbia Univ. Press, New York.

Cheney, D. L. and Seyfarth, R.M., 1990, "How Monkeys See the World", Univ. Chicago Press, Chicago.
Cheney, D.L. and Wrangham, R.W., 1987, Predation, *in*: "Primate Societies", B.B. Smuts, D.L. Cheney, R.M. Seyfarth, R.W. Wrangham and T.T.Struhsaker, eds., Univ. Chicago Press, Chicago.
Crystal, D., 1992, "The Cambridge Encyclopedia of Language", Cambridge Univ. Press, Cambridge.
Deputte, B.L., 1982, Duetting in male and female songs of the white-cheeked gibbon (*Hylobates concolor*), *in:* "Primate Communication", Ch. T. Snowdon, Ch.H. Brown and M.R. Petersen, eds., Cambridge Univ. Press, Cambridge.
Fay, R.R., 1988, "Hearing in Vertebrates. A Psychophysics Databook", Hill Fay Ass., Winnetka.
Elowson, A.M., Snowdon, C.T., and Sweet, C.J.,1992, Ontogeny of trill and J-call vocalizations in the pygmy marmoset, *Cebuella pygmaea*, *Anim. Behav.* 43: 703-715.
Gautier, J.P., 1988, Interspecific affinities among guenons as deduced from vocalizations, *in*: "A Primate Radiation: Evolutionary Biology of the African Guenons", A. Gautier-Hion, F. Bourliere, J.-P. Gautier and J. Kingdon, eds., Cambridge Univ. Press, Cambridge.
Gautier, J.P. and Gautier, A., 1977, Communication in Old World monkeys, *in:* "How Animals Communicate", T.A. Seboek, ed., Indiana Univ. Press, Bloomington.
Gautier, J.P. and Gautier, A., 1988, Vocal quavering: A basis for recognition in forest guenons, *in*: "Primate Vocal Communication", D. Todt, P. Goedeking and D. Symmes, eds., Springer, Heidelberg.
Goodman, S.M., O'Connor, S., and Langrand, O., 1993, A review of predation on lemurs: Implications for the evolution of social behavior in small nocturnal primates, *in*: "Lemur Social Systems and their Ecological Basis", P.M. Kappeler and J.U. Ganzhorn, eds., Plenum Press, New York.
Gould, S.J. and Lewontin, R.P.C., 1979, The spandrels of San Marco and the Panglossian paradigm: A critique of the adaptive program, *Proc. R. Soc. London* B 205: 581-598.
Gouzoules, S., Gouzoules, H., and Marler, P., 1984, Rhesus monkey (*Macaca mulatta*) screams: Representational signalling in the recruitment of agonistic aid, *Anim. Behav.* 32: 182-192.
Green, S., 1975, Variation of vocal pattern with social situation in the Japanese monkey (*Macaca fuscata*). A field study, *in:* "Primate Behavior, Vol. 4", L.A. Rosenblum, Academic Press, New York.
Green, S., 1975, Dialects in Japanese monkeys: Vocal learning and cultural transmission of locale-specific vocal behavior? *Z. Tierpsychol.* 38, 304-314.
Haimoff, E.H., Chivers, D.J., Gittins, S.P., and Whitten, A.J., 1982, A phylogeny of gibbons based on morphological and behavioural characters, *Folia Primatol.* 39: 213-237.
Harcourt, C. and Thornback, J., 1990, "Lemurs of Madagascar and the Comores", IUCN, Gland.
Harvey, P.H., Martin, R.D., and Clutton-Brock, T.H., 1987, Life histories in a comparative perspective, *in*: "Primate Societies", B.B. Smuts, D.L. Cheney, R.M. Seyfarth, R.W.Wrangham and T.T.Struhsaker, eds., Univ. Chicago Press, Chicago.
Hauser, M.D., 1991, Sources of acoustic variation in rhesus macaques (*Macaca mulatta*) vocalizations, *Ethology* 89: 29-46.
Hauser, M.D., 1992, Articulatory and social factors influence the acoustic structure of rhesus monkey vocalizations: A learned mode of production, *J. Acoust. Soc. Amer.* 91: 2175-2179.
Hohmann, G., 1991, Comparative analysis of age- and sex-specific patterns of vocal behavior in four species of Old World monkeys, *Folia Primatol.* 56: 133-156.
Hodun, A., Snowdon, C.T., and Soini, P., 1981, Subspecific variation in the long calls of the tamarin, *Saguinus fuscicollis*, *Z. Tierpsychol.* 57: 97-110.
Jürgens, U., Hast, M., and Pratt, R., 1978, Effects of laryngeal nerve transection on squirrel monkey calls, *J. Comp. Physiol. A* 123: 23-29.
Krebs, J.R. and Davies, N.B., 1993, "An Introduction to Behavioural Ecology", Blackwell Scientific Publ., Oxford.
Kroodsma, D.E., 1982, Learning and the ontogeny of sound signals in birds, *in*: "Acoustic Communication in Birds, Vol. 2, Song Learning and its Consequences", D.E. Kroodsma and E.H. Miller, eds., Academic Press, New York.
Konishi, M. ,1973, Locatable and nonlocatable acoustic signals for barn owls, *American Nat.* 107: 775-785.
Konishi, M., 1985, Birdsong: from behavior to neuron, *Ann. Rev. Neurosci.* 8: 125-170.
Kugel, S., 1993, "Verhaltensbiologische und bioakustische Untersuchungen zur strukturellen Variabilität des Soziallauts "Triller" beim Mausmaki (*Microcebus murinus*), Diplomarbeit, Universität Stuttgart-Hohenheim, unpubl.
Kuhn, M., 1989, "Verhaltensbiologische Untersuchungen zur Ontogenese von Mausmakis (*Microcebus murinus*) mit besonderer Berücksichtigung der akustischen Kommunikation", Diplomarbeit, Universität Stuttgart-Hohenheim, unpubl.
Lieblich, A.K., Symmes, D., Newman, J.D., and Shapiro, M., 1980, Development of the isolation peep in laboratory-bred squirrel monkeys, *Anim. Behav.* 28: 1-9.

Macedonia, J.M. and Stanger, K.F., submitted, Phylogeny of the Lemuridae revisited: Evidence from communication signal structure, *Folia Primatol.*

Maeda, T. and Masataka, N., 1987, Locale-specific vocal behaviour of the tamarin (*Saguinus l. labiatus*), *Ethology* 75: 25-30.

Martin, R.D., 1972, A preliminary field-study of the Lesser Mouse Lemur (*Microcebus murinus*), *Advances in Ethology* 9: 43-90.

Marler, P., 1955, Characteristics of some animal calls, *Nature* 176: 6-8.

Marler, P, 1975, Bird song and speech development: Could there be parallels? *Amer. Scient.* 58: 669-673.

Marler, P., 1990, Song learning: The interface between behavior and neuroethology. *Phil. Trans. R.. Soc. Lond.* B 329, 109-114.

Marler, P. and Tenaza, R., 1977, Signalling behavior of apes with special reference to vocalization, *in*: "How animals communicate ", T.A. Seboek, ed., Indiana Univ. Press, Bloomington.

Marshall, J.T. and Marshall, E.R., 1976, Gibbons and their territorial songs, *Science* 193: 235-237.

Marshall, J.T. and Sugardjito, J., 1986, Gibbon systematics, *in*: "Comparative Primate Biology, Vol.1: Systematics, Evolution and Anatomy", D. Swindler and J. Erwin, eds., Alan R. Liss, New York.

Mascagni, O. and Doyle, G.A., 1993, Infant distress vocalizations in the South African Lesser bushbaby (*Galago moholi*), *Int. J. Primatol.* 14: 41-60.

Maurus, M., Streit, K.M., Barclay, D., Wiesner, E., and Kühlmorgen, B., 1988, A new approach to finding components essential for intraspecific communication, *in*: "Primate Vocal Communication", D. Todt, P. Goedeking and D. Symmes, Springer, London.

Mitani, J.C., 1992, Singing behavior of male gibbons: Field observations and experiments, *in*: "Topics in Primatology, Vol.1, Human Origins", T. Nishida, W.C. McGrew, P. Marler, M. Pickford and F.B.M. de Waal, eds., Univ. Tokyo Press, Tokyo.

Mitani, J.C., Hasegawa, T., Gros-Louis, J., Marler, P., and Byrne, R., 1992, Dialects in wild chimpanzees? *Am. J. Primatol.* 27: 233-243.

Mitani, J.C.and Nishida, T., 1993, Contexts and social correlates of long-distance calling by male chimpanzees, *Anim. Behav.* 45: 735-746.

Nash, L.T., Bearder, S.K., and Olson, T.R., 1989, Synopsis of Galago species characteristics, *Int. J. Primatol.* 10: 57-80.

Niaussat, M.M. and Petter, J.J., 1980, Etude de la sensibilité auditive d'un lémurien malgache: *Microcebus murinus*, *Mammalia* 44: 553-558.

Niemitz, C., 1984, "The Biology of Tarsiers", Gustav Fischer Verlag, Stuttgart.

Niemitz, C., Nietsch, A., Warter, S., and Rumpler, Y., 1991, *Tarsius dianae*: A new primate species from Central Sulawesi (Indonesia), *Folia Primatol.* 56: 105-116.

Oates, J.F. and Trocco, T.F., 1983, Taxonomy and phylogeny of black- and -white colobus monkeys: Inferences from an analysis of loud call variation, *Folia Primatol.* 40: 83-113.

Oller, D.K., Wieman, L.A., Doyle, W.J., and Ross, C., 1976, Infant babbling and speech, *J. Child Lang.* 3: 1-11.

Oppenheimer, J.R., 1977, Communication in New World monkeys, *in*: "How Animals Communicate", T.A. Seboek, ed., Indiana Univ. Press, Bloomington.

Oster, G.E. and Wilson, E.O., 1979," Caste and Ecology in Social Insects", Princeton Univ. Press, Princeton.

Payne, R.B., 1981, Population structure and social behavior. Models for testing the ecological significance of song dialects, *in*: "Natural Selection and Social Behavior", R.D. Alexander and D.W. Tinkle, eds., Chiron Press, New York.

Pages-Feuillade, E., 1988, Modalités de l'occupation de l'espace et relation interindividuelles chez un prosimien nocturne malgache, *Microcebus murinus*, *Folia Primatol.* 50: 204-220.

Perret, M., 1992, Environmental and social determinants of sexual function in the male Lesser Mouse Lemur (*Microcebus murinus*), *Folia Primatol.* 59: 1-25..

Petter, J.-J. and Charles-Dominique, P., 1979, Vocal communication in prosimians, *in*: "Prosimian Behaviour", G.A. Doyle and R.D. Martin, eds., Academic Press, New York.

Petter-Rousseaux, A., 1975, Activité sexuelle de *Microcebus murinus* soumis à des regimes photoperiodiques experimentaux, *Ann. Biol. Animal. Biochim. Biophys.* 153: 503-508.

Ploog, D., 1992, The evolution of vocal communication, *in*:" Nonverbal Vocal Communication. Comparative and Developmental Approaches", H. Papousek, U. Jürgens and M. Papousek, eds., Cambridge Univ. Press, Cambridge.

Richard, A.F., 1987, Malagasy prosimians: Female dominance, *in*: "Primate Societies", B.B. Smuts, D.L. Cheney, R.M. Seyfarth, R.W. Wrangham and T.T. Struhsaker, eds., Univ. Chicago Press, Chicago.

Robinson, J.G. ,1981, The regulation of inter- and intra-group spacing during boundary encounters in the titi monkey, *Callicebus moloch*, *Primates* 22: 161-172.

Rohlf, R.R. and Sokal, F.J., 1981, "Biometry", W.H.Freeman, New York.
Ryan, M.J. and Brenowitz, E.A., 1985, The role of body size, phylogeny and ambient noise in the evolution of bird song, *Amer. Natur.* 126: 87-100.
Sekulic, R., 1982, The function of howling in red howler monkeys (*Alouatta seniculus*), *Behaviour* 81: 38-54.
Snowdon, C.T., 1989, The structure and usage of long calls in marmosets and tamarins, *in*: "Perspectives in Primate Biology, Vol. 3", P.K.Seth and S. Seth, eds., Today and Tomorrow's Printers and Publ., New Delhi.
Snowdon, C.T. and Elowson, M., 1992, Ontogeny of primate vocal communication, *in*: "Topics in Primatology, Vol.1: Human Origins", T. Nishida, W.C. McGrew, P. Marler, M. Pickford and F.B.M. de Waal, eds., Univ. Tokyo Press, Tokyo.
Snowdon, C.T., French, J.A., and Cleveland, J., 1986, Ontogeny of primate vocalizations: Models from bird song and human speech, *in*: "Current Perspectives in Primate Social Behavior", D.M. Taub and F.A. King, eds., Van Nostrand Reinhold, New York.
Snowdon, C.T., Hodun, A.L., Rosenberger, A.L., and Coimbra-Filho, A.,1986, Long call structures and its relationships to taxonomy in lion-tailed tamarins. *Amer. J. Primatol.* 11: 253-262.
Stanger, K.F., 1988, "Lautrepertoire und begleitendes Verhalten von nachtaktiven Halbaffen", Diplomarbeit, University of Tübingen, unpubl.
Stanger, K.F., 1993, "Structure and Function of Vocalizations of Nocturnal Prosimians", PhD thesis, University of Tübingen, unpubl.
Struhsaker, T.T., 1970, Phylogenetic implications of some vocalizations of Cercopithecus monkeys, *in*: "Old World Monkeys, Evolution, Systematics and Behavior", J.R. Napier and P.H. Napier, eds., Academic Press, New York.
Struhsaker, T.T., 1981, Vocalizations, phylogeny and paleogeography of red colobus monkeys (*Colobus badius*), *Afr. J. Ecol.* 19: 265-283.
Vogel, C., 1973, Acoustical communication among free-ranging common Indian langurs (*Presbytis entellus*) in two different habitats of North India, *Am. J. Phys. Anthrop.* 38: 469-480.
Waser, P.M. and Waser, M.S., 1977, Experimental studies of primate vocalization: Specializations for long-distance propagation, *Z. Tierpsychol.* 43: 239-263.
Waser , P.M., 1982, The evolution of male loud calls among mangabeys and baboons, *in*: "Primate Communication", C.T. Snowdon, C.H. Brown and M.R. Petersen, eds., Cambridge Univ. Press, Cambridge.
Wickler, W., 1986, "Dialekte im Tierreich - Ihre Ursachen und Konsequenzen ". Aschendorffsche Verlagsbuchhandlung, Münster.
Wiley, R.H. and Richards, D.G., 1978, Physical constraints on acoustic communication in the atmosphere: Implications for animal vocalizations, *Behav. Ecol. Sociobiol.* 3: 69-94.
Winter, P., 1969, Dialects in squirrel monkeys: Vocalizations of the Roman arch type, *Folia Primatol.* 10: 216-229.
Zimmermann, E., 1981, First record of ultrasound in two prosimian species, *Naturwissenschaften* 68: 531.
Zimmermann, E., 1985a, Vocalizations and associated behaviours in adult slow loris (*Nycticebus coucang*), *Folia primatol.* 44: 52-64.
Zimmermann, E.,1985b, The vocal repertoire of the adult Senegal bushbaby (*Galago senegalensis senegalensis*), *Behaviour* 94: 211-233.
Zimmermann, E., 1989a, Reproduction and behavioral development in slow loris (*Nycticebus coucang*), *Human Evol.* 4: 171-179.
Zimmermann, E.,1989b, Aspects of reproduction, behavioral and vocal development in Senegal bushbabies (*Galago senegalensis*), *Intern. J. Primatol.* 10: 1-16.
Zimmermann, E., 1990, Differentiation of vocalizations in bushbabies (Galaginae) and the significance for assessing phylogenetic relationships, *Z.Zool. Syst. Evol.* 28: 217-239.
Zimmermann, E., 1991, Ontogeny of acoustic communication in prosimian primates, *in*: "Primatology today", A. Ehara, T. Kimura, O. Takenaka and M. Iwamoto, eds., Elsevier, Amsterdam.
Zimmermann, E., in press, Acoustic communication in nocturnal prosimians, *in*: "Creatures of the Dark. The biology of nocturnal prosimians", K. Izard, L. Altermann and G.A. Doyle, eds., Plenum Press, New, York.
Zimmermann, E., Bearder, S.K., Doyle, G.A., and Andersson, A., 1988. Variations in vocal patterns of Senegal and South African lesser bushbabies and their implications for taxonomic relationships, *Folia Primatol.* 51: 87-105.
Zimmermann, E. and Lerch, C., 1993, The complex acoustic design of an advertisement call in male mouse lemurs (*Microcebus murinus*) and sources of its variation, *Ethology* 93: 211-224.

VOCAL ONTOGENY IN MACAQUES AND MARMOSETS: CONVERGENT AND DIVERGENT LINES OF DEVELOPMENT

John D. Newman

Laboratory of Comparative Ethology
National Institute of Child Health and Human Development
National Institutes of Health
Poolesville, MD 20837, USA

INTRODUCTION

Of the many published studies of behavioral development in non-human primates (NHP), very few have had a comparative emphasis in which the developmental course of a behavior found in two taxa is compared. Where comparisons have been made, they consist of rather broad leaps between taxa, such as between NHP and songbirds or humans (e.g, Snowdon et al., 1986). The emphasis in these comparisons is on the emergence of apparently similar behaviorally relevant phenomena such as imitative learning. Within the published literature on NHP, developmental studies of individual species are becoming increasingly prominent, particularly for species which do well in captivity and can, therefore, be observed in great detail from birth. Communicative behavior, particularly vocalizations, have come under close scrutiny in several species, and, with the availability of instruments which can graphically represent vocalization time-histories (sound spectrographs and related forms of time history displays), some of the developmental details of a species' vocal repertoire are being documented. One group of related species, the squirrel monkeys (*Saimiri*), has been studied over more than twenty years, and considerable information about vocal development is now available (see contribution by Biben, this volume). Representatives of two other primate groups, the genus *Macaca* and the callitrichids (marmosets and tamarins) also have been the subjects of detailed analyses of vocal development. Interestingly, in both of these primate taxa, efforts have been made to demonstrate an important role for learning in the acquisition of vocal communication skills (macaques: Gouzoules and Gouzoules, 1989; Masataka and Fujita, 1989; callitrichids: Elowson et al., 1992; Snowdon et al.,1986). While controversial, these pioneering efforts have kept alive the hope among primatologists that a non-human primate model for studying language development in human children may yet be found.

As interesting as behavioral studies of vocal communication in non-human primates are, there is even more inherent interest in determining the neural substrates of vocal communication in non-human primates, given the great similarities of brain structure in human and non-human primates (Brodmann, 1909), and the possibility of understanding how the human brain might have evolved to mediate communication by speech (Lieberman, 1984).

Brain systems mediating vocal production have been most thoroughly characterized in the squirrel monkey (Jürgens and Müller-Preuss, 1977; Jürgens and Ploog, 1970; Jürgens and Pratt, 1979; Kirzinger, 1985; Kirzinger and Jürgens, 1985; MacLean and Newman, 1988; Newman and MacLean, 1982). A few studies have been directed at brain areas of macaques relevant to vocal production (Aitken, 1981; Larson et al., 1978; Larson and Kistler, 1986; Robinson, 1967; Sutton et al., 1985; Vogt and Barbas, 1988). However, there is presently no information at all regarding brain substrates for vocal production in callitrichids, although vocalizations accompanied flight and aggressive components resulting from electrical stimulation of the ventromedial hypothalamus (Lipp and Hunsperger, 1978). This is a significant gap of information, given the fact that many callitrichids are highly vocal, have been the subject of numerous studies in captivity, and for which a possible role for learning in vocal development may exist in some species (see below). There are no studies of brain development as it relates to vocal development in any species of NHP, although a few studies have begun to explore early neural substrates mediating development of cognitive skills and emotional expression (Bachevalier, 1991).

The present paper reviews previous studies of vocal development in macaques and callitrichids, then outlines a research strategy that may be fruitful in identifying critical periods of vocal development, an important step in the process of determining where in the brain relevant substrates and neural mechanisms that mediate the maturation of vocal skills might be sought. I will argue that there are emerging principles of convergence in the development of macaque and marmoset vocalizations that suggest a conservative general plan transcending phylogenetic differences. This plan is apparent in the gradual loss of variable sounds originating in the neonatal period. At the same time, however, differences between the vocal behavior of macaque and callitrichid infants suggest that the vocal skills of developing individuals are the expression of a complex set of underlying factors, including maturation of brain systems and the substantial differences in social organization and ecological requirements that may exist between primate species. Given the evidence that early brain insults can influence neonatal vocal performance in human children and subsequent language development, there is a critical need for NHP models of the neural substrates underlying vocal development.

PREVIOUS STUDIES OF MACAQUE VOCAL DEVELOPMENT

Very little information exists on the vocal behavior of any macaque species during the newborn period (but see Grimm, 1967). Typically, studies of the vocal behavior of developing macaques focus on individuals who are socially interacting with other troop members, and on vocalizations which are discernible as part of the adult repertoire. Furthermore, there is little information on the emergence of categories or subtypes of vocalization with age, or the age-related disappearance of vocalizations occurring in the neonatal period but not in older individuals. Hence, on the basis of available published information, it would be impossible to identify critical periods with respect to early vocal development in any species of macaque.

What follows is a review of the published accounts of the vocalizations of immature macaques. An attempt will be made to indicate the evidence that similar vocalizations are used by developing individuals of different species (Table 1).

Stumptail Macaque (*Macaca arctoides*)

Chevalier-Skolnikoff (1974) attempted to describe the ontogeny of infantile communication in captive individuals of this species, to discover if infants communicate differentially with various group members, how infant communication changes with maturation, and how the adult communication system develops. Twenty-eight different call types were described, based on inspection of sonagrams, 11 of which were found mainly or only in infants. The author noted that several characteristics of infant vocalization change fairly consistently during ontogeny. Infantile vocalizations (particularly those of very young infants) tend to be of lower intensity and higher pitch. Emphasis is usually on the higher of two bands (i.e., the first harmonic of the fundamental), and there is an unsteady or quavering pitch. Only 3 vocalizations (chirp, gecker and infantile scream, plus some intermediates) were heard during the first week of life in the 3 infants observed. Only 4 (chirp, gecker, infantile scream and trilled whistle) had been heard by the end of the second week. Most infantile vocalizations appeared to be replaced by acoustically homologous adult vocalizations (e. g., the trilled whistle by the whistle and the girn by the staccato grunt), but a few vocalizations (chirp, twit) apparently simply dropped out of the repertoire as an individual approached adulthood. Even by the end of the sixth month, only 12 vocalizations (and some intermediates) could be distinguished, in comparison with the adult repertoire of 17 vocal types.

Lillehei and Snowdon (1978) focused on individual differences in captive stumptail macaque coo vocalizations. A range of acoustic parameters were measured. The study found that vocalizations of all but 2 pairs of subjects (out of 15 possible pairs) were differentiated "by a unique set of parameters." This finding strongly supports the notion of individual differences in vocalization structure. No differences in acoustic characteristics could be found in pairwise comparisons between youngest and oldest or youngest and intermediate age (although these individuals seemed to be at similar developmental stages of social maturation).

Pigtail Macaques (*Macaca nemestrina*)

Grimm (1967) provided a catalogue of sounds produced by captive pigtail macaques (*Macaca nemestrina*). When contact is lost with the mother, newborns emit a lengthy train of intense sounds while clutching and lurching spasmodically against the cage floor. The sounds stop when the mother returns. This "long cry" is classified with the harsh sounds and may last 3 seconds. A two-week old infant made an unmistakable clear or harmonic call, termed the "neo coo". The "squeak" is produced by the newborn when roughly handled by the mother. Among so-called "clear sounds", the "coo" is the most frequent call of pigtail macaques in the laboratory. The principal frequency band is 300-400 Hz. The coo usually occurs singly. The "cry coo" is produced by isolated juveniles. It is intense and inflective, and usually repeated in trains of 3-5, especially if the mother is responding vocally . The "leap coo" begins as a clear call, and is rapidly altered by a fast-rising transient, carrying the sound to a higher pitch. This sound, repeated during excitement, was not recorded from adults. The "harsh coo" contains a scatter of noise of variable intensity and duration in addition to a clear signal. The "screech" has harsh densities of noise continuously distributed over long, high-pitched signals. The "long screech" has a vibrato quality, with the fundamental below 1 kHz. The most intense sound of juvenile or adult is the "scream". It is similar to the screech, but all parameters are intensified.

Simons and Bielert (1973) studied vocal communication in captive, feral-born mothers and their 3- to 4-month old infants. The mothers' locomotor activity was measured as activity counts, and vocalizations between mothers and infants recorded under 3 conditions: mothers in 2-way contact with own infants; 2-way contact with other infants of matched sex and age; and no contact. Mothers' activity counts were higher with auditory access to an infant. However, neither activity counts nor vocalizations of mothers differed when in communication with their own infants from the same measures when in communication with matched infants. Neither mothers nor infants vocalized differently, either quantitatively or qualitatively, in the three conditions. Infants vocalized frequently in all conditions. With one exception, the only vocalization was the harsh coo. This vocalization was usually given in trains of 2 to 14 calls. No qualitative or quantitative difference could be discerned between number or pattern of infant vocalizations in the three conditions. Like the infants, mothers gave only a limited number of different vocalizations. The most frequent was the "great coo", usually given singly (rarely, in brief trains).

Gouzoules and Gouzoules (1989) studied the ontogeny of recruitment screams in captive pigtail groups. Individuals 3 years of age and older (considered 'adults' in this study) produced screams that were acoustically and contextually divisible into 5 categories. Using the statistical profiles established for adult screams, the authors then tested whether the screams of developing individuals from the same social groups could be assigned to the same acoustic and contextual categories. The results of this analysis suggested that there is a developmental process, possibly involving learning, during which the screams of maturing individuals acquire adult acoustic and contextual distinctiveness.

Lion-Tailed Macaque (*Macaca silenus*)

Hohmann and Herzog (1985) studied captive groups of lion-tailed macaques, and described a repertoire of 17 basic call types. These results were verified in a field study. The vocal repertoire differs from terrestrial macaque species in having fewer variations. Also, new call types not found in other macaque species were described. Call types found in immature animals were infant whoo, scream, shriek, twit, weeping, milk grumble, sneeze, and yap. These sounds were produced largely in contexts involving distress or fear of potential danger.

Rhesus Macaque (*Macaca mulatta*)

Rowell and Hinde (1962) produced the first detailed acoustic analysis of the rhesus macaque vocal repertoire in captivity. While most of their attention was devoted to the vocalizations of adults, they did describe some infant vocalizations, including chirp-like clicks when a young baby was left by its mother and a long, high, plaintive sound when the infants were separated from their group.

Newman and Symmes (1974) described abnormal features of the vocalizations of isolation-reared rhesus macaques. Two males and 2 females were deprived of physical contact with other monkeys from birth, and received only minimal handling from humans. They were raised in incubators in the same room as other isolates to 1 month of age, then transferred to individual cages permitting visual and auditory contact with feral adult macaques and mother-reared macaques of all ages. The control group consisted of 3 rhesus macaques, 2 males and a female, each housed with its mother from birth to age 6 or 7 months in a large animal room. The 7 monkeys in the longitudinal study were between 8 and 9 months old when data collection began, and were 24 months old when the study was terminated. The primary data consisted of recordings of vocalizations during formal observation periods made at ages 9, 12, 16 and 24 months, in which the monkey was placed in a nylon mesh cage fitted with a clear plexiglas front inside an acoustic isolation chamber. Comparison of vocal structure was made by visual inspection of sound spectrograms. Vocalizations classified as pant-threats, geckers, squeaks, and screeches (as defined by Rowell and Hinde, 1962) were structurally similar in both groups. These were associated with arousing situations, such as the observer entering the room. However, the most frequent calls were coos. Coos from the controls were 0.4-0.8 sec in duration, with a smooth, gradually inflected harmonic structure. More than 90 % of the coos recorded from isolates had several unusual structural features, which readily separated them from the control group. These included abrupt pitch changes, harmonic emphasis shifts, and discontinuities in the call's progression. Other calls had a conspicuous lack of inflection, and sounded 'hollow'. Another kind of call given by isolates was unusually long (to 2 sec) and possessed variable abnormal structural features (lack of inflection, abrupt pitch changes, temporal disintegration). Although all of the isolates produced calls with more than one of these structural abnormalities, each individual had a distinctive set of abnormalities which characterized most of its clear calls. The abnormal features characterizing isolate vocalizations were found at all ages analyzed, and persisted through the subjects' second year. These results suggest that early social isolation can result in the development of vocal abnormalities which persist until at least 2 years of age. They also suggest that an acoustic environment rich in species-specific sounds is not, by itself, sufficient to produce completely normal vocal development. Several attributes of the isolates' coos are suggestive of other, nonvocal aspects of the isolation-rearing syndrome: the calls are stereotyped (both in physical structure and in their repeated occurrence); the form of vocal abnormality is an individual characteristic; the number of structural abnormalities is also an individual peculiarity. These attributes indicate that the structural abnormalities of coos and acoustically similar vocalizations are a heretofore undescribed set of behavioral manifestations of maternal deprivation.

Kalin et al. (1992) compared differences in call structure between coos and girns in infants approximately 5-8 months old. Significant differences were found for nine acoustic measures. Girns were shorter, had lower fundamental frequencies, and less peak-minus-start frequency change. Girns had significantly more energy bands and more inflections with the greatest intensity of energy in the second or third band (first or second harmonic). Girns were rarely emitted by infants when they were alone, but occurred more frequently when they were united with an unfamiliar adult male or reunited with the mother. Infants continued to girn longer with their mothers, paralleling the pattern for contact clinging. Only 2 animals ever cooed when reunited with their mother. Infants emitted the greatest number of coos when with the male (although not significantly different from when alone). Infants never barked when with their mothers, 3/9 barked when with the male, while 5/9 barked when alone. For shrieking, 6/9 shrieked when with the male, 1 when alone, and none when with the mother. Shrieks were emitted most often when the male either pushed the infants away or bit them. Shrieking infants also frequently displayed fear grimaces.

Levine et al. (1987) argued that two types of "whoo" calls (equivalent to the 'coo' of other authors) are emitted when an infant is adjacent to its mother (but unable to make physical contact), whereas a third, totally different, type of whoo call is emitted during total isolation. The authors state that the former are shorter in duration, lower in intensity, and have a shorter modulation, while the isolation calls have a long duration, high intensity, and high frequency modulation. The two types of whoo call elicited when the mother is present are never emitted in total isolation, and vice versa.

Erwin (1975) reviewed the vocal repertoire of captive rhesus macaques, and presented a classification scheme for harsh noises. Erwin noted that the earliest sounds heard in this species after birth are shrill and high-pitched, sounding somewhat like bird calls. They occur upon loss of nipple contact, loss of ventro-ventral proximity, or complete loss of physical contact. These sounds seem to be especially closely related to squeaks, and the spasmodic body jerk which accompanies geckering is often present, particularly upon total loss of contact with the mother. Within the last half of the first year of rhesus life, the frequency of occurrence of shrill harsh sounds begins to decrease, while the incidence of clear calls and bark complex sounds continues to increase into the second year. Clear calls (coos and related sounds) reach a relatively low rate by the onset of puberty.

Japanese Macaque (*Macaca fuscata*)

Green (1975) developed a classification scheme for the calls of this species around acoustic characteristics and the contexts in which the calls were most frequently given. Class I, girneys, are articulations alone or superimposed on voiced sounds. Class II, coos, are simple, nonplosive, and uniformly tonal. Class III, whistles and warbles, are nonplosive, modified tonal and predominantly tonal, with a duration greater than 0.18 sec. Class IV, squawks and squeaks, have a duration less than 0.18 sec, have a rich harmonic structure and are predominantly tonal. Class V, chirps and barks, have a duration less than 0.18 sec or are multiunit, are plosive or noisy, predominantly tonal, and some have clicks or dropouts. Class VI, squeals and screeches, have a duration greater than 0.10 sec, or may be multiunit, and vary from noisy to predominantly tonal with a dominant frequency around 2400 Hz. Class VII, shrieks and screams, have a duration greater than 0.10 sec, are predominantly atonal, and exhibit dominant energy above 100 and below 2400 Hz. Class VIII, whines, are predominantly tonal, have a duration greater than 0.18 sec, overtones more intense than the fundamental, and major energy around 2400 Hz. Class IX, geckers, are predominantly atonal, or plosive, and made up of units less than or equal to 0.10 sec. Class X, growled sounds and roars, are pulsed and/or uniformly atonal, with dominant energy at base line. Infants were observed to use calls of classes II-X.

Green (1981) reported that an infant male *M. fuscata* whose mother was out of sight vocalized in a sequence showing transitions from coos to screeches and screams. They ceased when she approached him later. Infant Japanese macaques compound two patterns to form a call that begins tonally, rises sharply in pitch, and becomes an atonal screech all within a single vocalization. It is used when an infant has been temporarily abandoned by its mother and either has been alone for some time or is disturbed by the approach of an unrelated animal.

Masataka and Fujita (1989) cross-fostered one Japanese macaque and two rhesus macaque infants with a mother of the other species. The cross-fostered animals could hear the calls given by other members of both species. The signal chosen for study was the 'food call' of Green (1975), a specific and distinctive vocalization given during foraging and group progression in free-ranging animals. In captivity, they are heard in the context of food distribution. Vocalizations were recorded when the infants were 1 year old. Acoustically, these sounds are basically tonal in character, represented on sonagrams as an intense band with harmonic overtones. The frequency of the lowest tonal band in each call was measured, as well as call duration. The frequency of the lowest tonal band completely separated the two species. The mean frequency for Japanese macaques was 1234.5 Hz, whereas for rhesus it was 756.2 Hz . The lowest value for a Japanese macaque was 1025 Hz, above the highest rhesus value of 987 Hz.This study showed that the foraging calls of Japanese monkeys reared by their biological mothers were acoustically dissimilar to those of a conspecific cross-fostered to a rhesus mother, but similar to those of rhesus infants cross-fostered to Japanese macaque mothers. Similarly, the foraging calls of rhesus monkeys reared by their biological mothers were acoustically dissimilar to those of their cross-fostered conspecifics, but similar to those of a Japanese macaque infant cross-fostered to a rhesus mother.

Owren et al. (1992) attempted to replicate the findings of Masataka and Fujita (1989), cited above. However, Owren and colleagues failed to find a significant difference in most acoustic measures between conspecific-reared rhesus and Japanese macaque food calls, hence were unable to establish a good cross-species model to test for a role for individual experience in establishment of species-specific food call attributes. These authors also found no difference in the structural variability of the food calls of cross-fostered rhesus or Japanese macaques compared to those of conspecific-reared individuals.

Long-Tailed Macaque (*Macaca fascicularis*)

Palombit (1992) studied vocal communication in wild long-tailed macaques, and developed a vocal classification scheme useful in the field. Screams are a group of loud, usually plosive vocal discharges whose total energy is distributed over a wide range of frequencies between 1 and 16 kHz but is typically concentrated in the 2- to 7-kHz range. The acoustic structure of screams varied. The khreeet screech, a loud, partially tonal scream, had a sharply ascending frequency 'sweep' extending over as many as 10 kHz, usually, but not always, followed by an equivalent descent (similar to the 'arched scream' of Gouzoules et al., 1974, for rhesus macaques). Noisy elements often occur as part of both the ascending and descending sweeps. Typically, the call ends in a 'harsh' pulse, yielding a bisyllabic 'khree-*eeet*!". This screech is significantly longer than the other combined screams. The khreeet call was given by distressed and agitated juveniles and possibly by older infants. Squeals are calls of varying acoustic structure generally organized around predominant tonal components that are disrupted by wide-band noise. These are essentially very low amplitude, high-pitched calls composed of multiple units of often dissimilar form and that characteristically experience marked changes in pitch. Only young macaques give these relatively rare calls. High Long Modulated (HLM) coos are long, frequency-modulated clear calls of high pitch. The whimper is a specific, consistently heard pattern of HLM coo call units with a distinctive pronounced upward modulation in frequency, with its highest peak in the last two-thirds to three-fourths of the call. HLM units with these characteristics are combined with alternating rapid audible breaths into a highly repetitive and loud series that is the whimper. Whimpers were performed by distressed infant or adolescent macaques, such as young juveniles left behind by their mothers and unable to immediately follow because of obstacles. Young monkeys engaged in tantrum behavior also performed these calls. Whimpers were usually accompanied by screams or khreeet screeches. A long series of whimpers, increasing dramatically in rate and amplitude, often culminated in screams or khreeet screams. Mothers usually responded to such calls with retrieval when the calling animal was an infant but did so less often when the vocalizer was older.

Vocal Development in Macaques: Summary and Conclusions

From the overview of macaque vocal development presented above, some of the same vocalizations (as defined acoustically) occur in all species of the genus *Macaca* for which information is available (see Table 1). Contextually, too, there appear to be commonalities. In particular, vocalizations of young infants are produced during distress (such as when being roughly handled or aggressed upon by a conspecific), or during circumstances when the mother either leaves or rejects the infant. Since there have been so few studies in which the effects of playing macaque infant vocalizations to conspecifics have been assessed, it is not possible at the present time to make any general statements about the signal function of most infant vocalizations. Evidence of a largely anecdotal nature suggest that coos and screams function to solicit aid from the mother or matrilineal group. Some sounds, such as the gecker and squeak, have a less clear function, and may largely be a reflection of the immature state of the central neural systems underlying vocal expression. Still other vocalizations, such as the girn, may emerge relatively late in ontogeny, and appear to function in close-range affiliative interactions at all ages.

In general, there is a need for future studies of vocal development in macaques to track the acoustic and contextual details of specific call types or groups of closely related calls from the age at which they first appear in the repertoire to adulthood. Given the acoustic variability that is inherent in primate vocalizations, these studies should focus on specific, identified individuals within a social group, so that the respective roles of individual experience and genetic background in the maturation of vocal skills can be assessed.

Perhaps the greatest need is to identify contexts in which most infants are likely to vocalize throughout significant portions of their development, so that the maturation of vocal performance can be compared with other developmental milestones. Further on in this chapter, I will present a behavioral paradigm in which this issue can be successfully addressed.

Table 1. Summary of Infant Macaque Vocalizations.

Species[1]	coo	scream	gecker	squeak
Stumptail	x[2,3]	x[2]	x[2]	x[2]
Pigtail	x[4]	x[4,5]	x[4,5]	x[4,5]
Lion-tailed	x[6]	x[6]	x[6]	x[6]
Rhesus	x[7,9-11]	x[7,8,10-11]	x[7,10-11]	x[7,10-11]
Japanese	x[12]	x[12]	x[12]	x[12]
Long-tailed	x[13]	x[13]	x[13]	x[13]

[1] stumptail, *M. arctoides*; pigtail, *M. nemestrina*; lion-tailed, *M. silenus*; rhesus, *M. mulatta*; Japanese, *M. fuscata*; long-tailed, *M. fascicularis*.
[2] Chevalier-Skolnikoff, 1974; [3] Lillehei and Snowdon, 1978; [4] Grimm, 1967; [5] Simons and Bielert, 1973; [6] Hohmann and Herzog, 1985; [7] Erwin, 1975; [8] Gouzoules et al., 1984; [9] Kalin et al., 1992; [10] Newman and Symmes, 1974; [11] Rowell and Hinde, 1962; [12] Green, 1975; [13] Palombit, 1992.

VOCAL DEVELOPMENT IN MARMOSETS AND TAMARINS

The Callitrichidae (marmosets and tamarins) are small New World primates that live in small social groups. Many species are monogamous, and both parents, along with older brothers and sisters, share responsibility for infant care. Furthermore, the species in this family characteristically give birth to twins. An overview of callitrichid biology can be found in Hershkowitz (1977) and in Mittermeier et al. (1988). Another unusual attribute of this group is the suppression of reproductive activity in mature sons and daughters who remain with their parents (Abbott and Hearn, 1978). The presence of twinning, reproduction restricted to one pair within a social group, and infant care by an entire family unit sets the callitrichids apart from other primates, and it would not be surprising if special forms of vocal communication evolved to subserve this array of reproductive and social organization attributes. Like other forest-dwelling primates, callitrichid species generally have one or more loud, penetrating vocalization in their repertoire that is used to maintain spacing between unrelated social units, as well as to promote and maintain contact within each social unit. These 'loud calls' are typically whistle-like in acoustic structure and species-specific (Snowdon, 1989). Close-range communication within a family unit is subserved by other vocalizations, typically referred to as 'trills' or 'chirps.' A particularly interesting aspect of vocal development in callitrichids is the fact that infants are quite vocal from an early age, and the context in which much of this vocalizing occurs involves transfer of an infant from one caregiver to another, wherein a family member carrying one or both infants will force the infant off by rolling on its back and rubbing against a substrate (e.g., Epple, 1968; Pola and Snowdon, 1975). Periods of up to several minutes in which an infant is not retrieved have been documented in the pygmy marmoset (Snowdon, 1987), during which time the infant will vocalize almost constantly. Snowdon (1987) has suggested that these periods of intense vocalizing by infants actually are parts of an adaptive process by which an infant acquires, through practice, the vocal capabilities of an adult. Perhaps because of this early vocality, the neonatal vocalizations of captive callitrichids have been more thoroughly described than for the comparable developmental stage in macaques.

Overview of Callitrichid Vocal Development

Cotton-Top Tamarin (*Saguinus oedipus*)

Snowdon, French and Cleveland (1986) summarized cotton-top tamarin long call development. The only form of long call given by tamarins up to two years of age is the Combination Long Call, a highly variable form. Normal and Quiet Long Calls were rarely, if ever, heard in unmated adult animals. Only in mated adults were these two forms of long call recorded. The ontogeny of another tonal call, the Chirp, was also summarized by these authors: in the adult there are 8 variants of short, frequency-modulated calls referred to as Chirps. Type A chirps are used in mobbing; Type B chirps are used in general investigation; Type C chirps are used as animals approach food; Type D chirps are given by animals that have just taken food; Type E chirps are general alarm calls; Type F chirps are given prior to long calls when animals hear members of strange groups; Type G chirps are mild investigation calls; Type H chirps are given in situations of mild alarm. In addition, a ninth chirp type is given by young animals as they approach mothers to nurse. The very first chirps heard in development are the nursing chirps and Type G chirps. The first differentiation of chirp types appears to be two forms of investigative chirps (B and G), and somewhat later comes the emergence of food chirps (C, D). Older infants more integrated into the family group give alarm chirps (A, E, H). Type F does not appear until well after puberty, and is usually restricted to mated adults. The authors concluded that the appearance of different chirp types at different points in development is likely to be a manifestation of learning or at least of an epigenetic mechanism mediating call structure.

Snowdon (1987) analyzed chirp development in the cotton-top tamarin through the first year of life. Twice-weekly vocal samples were collected from several infants, some reared in normal family groups and others hand-reared in an incubator during the first month, and then gradually introduced to an adoptive family during the second month. One-year old cotton-top infants failed to show the context-specific chirps found in adults. Measurement of three different chirp acoustic parameters indicated that each parameter had a separate course of maturation. End frequency reached adult levels of variation between the fourth and sixth month, while start frequency and call duration failed to consistently show adult levels of variation over the first year of life.

Roush and Snowdon (1994) further analyzed chirp development in the context of the presence of food. About half of the chirps produced by juveniles and subadults are structurally different from adult chirps in the same context. No association between specific foods and chirp structure was discerned. The authors suggest that the persistence of subadult vocal attributes may be advantageous to the vocalizers as they mature by making them appear to be less of a threat to the adult breeding pair.

McConnell and Snowdon (1986) showed that some of the chirp variants used in territorial disputes are uttered only by reproductively mature adults. In their study of developing tamarins, there were no consistent differences between chirps uttered in different situations, even in 12-month old infants. The chirps were also highly variable in structure compared with those of adults. End frequency reached adult levels of variation by the fourth to sixth month. Start frequency showed a gradual reduction in coefficient of variation over the course of the first year but had not reached adult levels by the end of the year.

Snowdon and colleagues have developed techniques for measuring estrogens, testosterone, and LH/CGH in the urine of cotton-tops (Snowdon, 1987). While subordinate males and females lived as family members, there was no evidence of orderly cycles indicative of ovulation. In contrast, within as few as eight days after being removed from the family and paired with males, females began to ovulate, and often conceived on the first or second ovulation. Males living in family groups show low levels of testosterone and dihydrotestosterone, and these levels became significantly higher within the first day or two after removal from the family and paired with a new mate. Thus, the social environment tamarins live in has a profound effect on reproductive physiology and reproductive function. Social environment plays an important role in vocal development, as well. Each of the calls used in intertroop encounters and intratroop cohesion had very low frequencies of occurrence while the tamarins lived in family units. The calls occurred at significantly higher rates after the animals were removed from the family and mated. This study was unable to determine if there was a causal relationship between hormonal changes and the appearance of vocalizations or whether both changed independently as a function of the change in social environment.

Snowdon (1989) further summarized the structure and usage of long calls in cotton-top tamarins. Combination Long Calls are the only long call subtype to occur prior to 1 year (reaching a peak at 1.5 years). Quiet and Normal Long calls emerge at 2 years and reach a peak in adulthood, at which time Combination Long Calls reach their low point (although remaining higher than the other two types of long calls). An immature version of the long call is one of the earliest calls to appear. These precursor calls are used in a wide variety of settings where the infant appears to be aroused or disturbed. At first, the calls consist of very brief whistle segments (around 100 ms compared with 900-1500 ms later on). However, these brief segments have the same frequency and are preceded by the same types of chirps that accompany Combination Long Calls. The brief calls get progressively longer, reaching normal duration when the infant is about three months old. The calls are highly variable in terms of the preceding syllables. In an experiment in which subadults were separated from their groups for 45 minutes, up to 30 different variants were recorded.

Goedeking and Immelmann (1986) reported on vocalizations given during social play in cotton-top tamarins (*Saguinus oedipus*). When there was increased frequency of occurrence of playful wrestling, the associated vocalizations were characterized by a high degree of instability in the contour of the fundamental frequency. Approximately 20 sec after these unstable vocalizations, however, frequency of occurrence of playful wrestling was significantly reduced.

Saddle-Backed Tamarin (*Saguinus fuscicollis*)

Moody and Menzel (1976) described the vocal repertoire of this species in captivity. Calls given by infants included: twitter-hooks, slow (isolated or hungry infant); twitter-hooks, modified (older infant not being carried); twitter-hooks, fast (infant distress often resulting from carrier's attempts to remove it from its back; squawk (infant seeking solid food). Twitter-hooks were used almost exclusively by infants, beginning around 4 days after birth. As the infants were very frequently carried by an individual other than the mother, hunger was the main motivation for the twitter-hook and squawk vocalizations. The call given just prior to nursing (rattle) may have a pacifying effect on the mother. A burst of rapid twitter-hooks denotes an emergency situation arising as a consequence of sudden loss of physical contact or having been nipped.

Pygmy Marmoset (*Cebuella pygmaea*)

Pola and Snowdon (1975) described the vocal repertoire of captive individuals living in family groups. The vocalizations of infants corresponded to a developmental sequence based on degree of parental dependence: stage a, complete dependence (constant parental care, lasting approximately 2 weeks); stage b, parent-initiated independence, lasting through the fourth week; stage c, infant and parent-initiated independence; stage d, complete independence. Complete independence appeared to coincide with the birth of a new set of twins at 5 months. Infant calls were emitted in a fixed 'tsik-phee' sequence, with randomly placed squeals. As infants became older, this combination of calls was heard more frequently and was correlated with an increased shifting of positions of the infant on the parent's body. During stage b, these vocalizations increased, and when the infants were separated from their parents, extended into lengthy bouts (from a few seconds to 10 minutes). While the calls seemed to indicate distress on the part of the infants, parents never were observed to respond. However, in threatening situations parents were able to retrieve their infants immediately. Upon retrieval, the vocalizations ceased. Toward the end of stage b the infant J-call emerged. This call only appeared when the infants were separated from their parents. With increased mobility during stage c the infant J-call was emitted without any evident distress, and became the most frequent vocalization heard. At this time the tsik, phee and squeal vocalizations were no longer noted. At 4-5 months, infants still gave the J-call, in most cases in conjunction with grooming adults or juveniles, during play, and in other interactions with animals. Playbacks of infant tsik, phee, squeal and J-calls to the entire colony elicited trills from the adults for the first two trials, but habituation was rapid. Older infants responded to both spontaneous and played back adult trills with J-calls, although adult-like trills never were observed throughout the first year of life.

Snowdon (1987) summarized trill development in pygmy marmosets, focusing on the closed mouth trill. This call first was recorded at three weeks of age. However, the earliest trills were much higher in pitch and much shorter in duration, and the frequency envelope was much more irregular than for adult calls. The parameters of pitch, duration, and regularity of frequency envelope approached adult levels with independent time courses. Pitch was the first to reach adult levels, by the sixth week of life. Call duration fell within the adult range between the 10th and 15th week , and the frequency envelope approached that of adults gradually over the entire 40 weeks of study. Typically, adults use closed mouth trills as affiliative contact or location calls, and the trills were never found in conjunction with other types of vocalizations. In contrast, almost all of the infant closed mouth trills were found in conjunction with other types of calls, and the trills appeared to be uttered indiscriminately, without regard to the proper context. Not until the 38th week did any individual show both adult structure and adult usage of the trills. Trills regressed to a more infantile form at around 30 weeks, with the birth of a new set of infants. The independent changes in call parameters and the regression towards more infantile calls with the birth of new infants suggests some flexibility in call development. During the period before the birth of new infants, the infants frequently monopolized all vocal activity in the colony. These 'babbling' bouts occurred when the parents placed infants on a branch or rope by themselves while the adults went off to forage or groom. During the entire period of separation from their caretakers, the infants vocalized continuously. Close analyses of these vocal bouts revealed a wide array of calls, many of which could be identified as imperfect versions of adult calls, although many idiosyncratic calls could be found. The calls were mixed together in highly inappropriate juxtaposition relative to adult usage. The author proposes that these 'babbling' bouts inform the caretakers of the location of the infants and also give the young extensive vocal practice. This study also presented preliminary data on the ontogeny of the J-call (another form of trill). This form appears much later in infancy (between week 14 and week 22), and the J-calls do not appear to be as variable as the early closed mouth trills. However, J-calls at this age are given in inappropriate contexts just as frequently as are closed mouth trills. "The extensive babbling seen in young pygmy marmosets may represent a time of vocal practice whereby adult structure and contexts of different call types can be acquired" (p. 186).

Elowson, Snowdon and Sweet (1992) presented a more detailed description of the ontogeny of trill and J-call vocalizations in pygmy marmosets. Five litters (nine individuals) were taped from infancy to adulthood. There was considerable variation among litters in the correlation of acoustic parameters with age, even in litters from the same parents. Duration of trills and rate of frequency modulation correlated significantly with age in most litters. Only two variables, cycles per second and lowest crest (frequency peak), showed increasing stereotypy as measured by progressive reduction of the coefficient of variation. J-calls were given less often than trills and few J-call variables showed significant correlations with age. Increased stereotypy of call structure did not occur. Although some changes in acoustic structure took place over time, the most consistent aspect of vocal ontogeny was the acquisition of adult usage. J-calls were defined as two or more 'J'-shaped notes in series with not more than 200 msec between adjacent notes. The highest note frequency, similar to the highest trough frequency in trills, was included as a measure of the consistency of the baseline of the J-note series. Note and inter-note duration were measured for the first four and last four notes in a J-call. J-call derived measures included bandwidth and notes per second. Juvenile trills, also termed the 'phee' (Epple, 1968; Pola and Snowdon, 1975), are not a separate category of trills, but a developmental stage, and were the focus of this paper. Individual J-notes tended to become shorter with age, whereas the internote intervals changed more irregularly over time. The last J-note tended to be longer in duration than the first J-note for all litters. One J-call variable, bandwidth, showed a significant positive coefficient of variation correlation with age. The authors concluded that the vocal structure of both trills and J-calls is not fixed at birth. Trill duration increased significantly in four litters and, for most litters, frequency measures of trills were negatively correlated with age. The inconsistent pattern of variation among litters of the same parents and comparisons of duration among trills and J-calls suggest that physical maturation alone is insufficient to explain the development of trills. The contact calls of the pygmy marmoset did not show increased stereotypy with age.

Common Marmoset (*Callithrix jacchus jacchus*)

Winter (1978) compared the frequency and time course of infant vocalizations on the day after birth with the vocal repertoire of adults. The purpose of this paper was to show the great variability in the vocalizations of 1-day old babies. The call groups that were present at 1 day of age and their relative incidence was as follows: phee calls, mean of 79.5%; twitters, 12%; tsik calls, 3.6%; ock calls, 0.2%; squeals, 4.7%. The phee calls had a duration of 40-800 ms and the fundamental frequency was between 6 and 12 kHz. Within these limits the shape of the frequency was very variable. Many intermediate patterns between these call types, as well as in the twitter and tsik calls, were found. Twitter calls had a duration of 20-200 ms. They start between 8 and 10 kHz, rise slowly and then increase rapidly to between 10 and 13 kHz. They often descend to about 1-2 kHz. The calls follow each other at intervals of 100-200 ms. Tsik calls have a similar shape to adult calls, but the frequency at the beginning of the calls ranged up to 15 kHz. Squeal calls are similar to those of adults, but are usually combined with phee, twitter and tsik components. The infant calls were not produced in a standard sequence and it was not possible to relate a call group to a special context. They seemed to be general indications of distress. Undisturbed babies uttered only a very few faint phee calls. When comfort was not forthcoming, the phee calls became more and more similar to twitter and tsik calls. The baby vocalizations were, on average, higher in frequency and more variable in shape than the corresponding calls of older individuals.

Winter and Rothe (1979) described the vocalizations of marmosets that were separated from conspecifics shortly after birth and raised by humans. The vocalizations described in this study were recorded from individuals when 9-20 months old. The authors were able to create 11 call-type categories, based on acoustic characteristics. These categories, however, were not discretely separable from each other because there was considerable variability within the groups as well as different types of continuous and summative transitions between them. Therefore, the authors were unable to correlate particular tonal patterns exclusively with specific contexts or functions. The call groups were: phii-calls, short to long (0.02-2 s), in frequency range from 5.5-10.5 kHz, straight or with sinusoid frequency course, with tonal character; zwitscher-calls, short (0.02-0.08 s) calls with tonal character, homotypical call units with 2-21 calls in sequence, of 0.08-0.14 s total length, beginning in the frequency range of 5.5-12.5 kHz, rising slowly (at a rate of 2.5-100 kHz/s) to 6.5-13.5 kHz and then rapidly (100-500 kHz/s) to 7.0-15.5 kHz/s; kecker-calls, short (0.06-0.08 s) calls with noisy character, in a frequency range from 250 Hz to over 16 kHz, consisting of homotypical units, 3-84 calls in series, over 0.08-0.12 s; tsik-calls, short (0.02-0.10 s) calls with tonal character, in a frequency range from 1-13 kHz, consisting of homotypical calls (up to 16) in series over 0.09-0.17 s, beginning in the frequency range of 5-13 kHz, then falling to 0.5-2.5 kHz at a rate of 10-200 kHz/s; ultrasounds, short (0.02-0.15 s) calls in the frequency range 14-20 kHz, with tonal character, and a variable frequency course over a range of 0.2-4.0 kHz; egg-calls, short (0.02-0.12 s) calls with tonal character, with at least 3 harmonics over 0.5-1.5 kHz; ock-calls, short (0.03-0.07 s) calls with noisy character, a frequency range starting no lower than 5 kHz, and reaching over 16 kHz; pfiff-calls, short (0.03-0.04 s) calls with tonal character, a frequency range from 9.5-15 kHz, and rising at a rate of 100-300 kHz/s; quärr-calls, 0.3-0.9 s in duration with frequency extending to over 16 kHz; schrei-calls, around 0.5 s long, with noisy character, and ranging from 0.5-12.0 kHz. The 11 call groups are distinctive only in their most extreme forms, with transitions occurring between all calls.

Epple (1968) described some aspects of vocal ontogeny in common marmosets. Babies riding contently on the back of a family member are almost always silent, sounds always being indications of dissatisfaction. Babies will vocalize upon loss of physical contact, when very hungry, cold and wet or in pain. As soon as an adult starts to rub a baby off, the young gives very faint, monosyllabic high pitched calls, transcribed as 'phee.' These calls become louder, longer and more frequent if the adult continues or intensifies the rubbing movements. With increasing distress, series of short, rather sharp and loud 'tsik, tsik,tsik' calls, twitters and long squeals are also given. Babies as young as two days were heard to give twitters. A twitter is a sequence of short (0.05-0.09 s), high-pitched notes, uttered at intervals of 0.08-0.11 s. The number of single notes given in a twitter varies from 2 to 5 in babies under 10 days of age, from 2 to about 12 in older infants, and from 2 to about 20 in adults. The vocal signals of babies are highly variable, with many intermediate patterns. The short 'phee' calls, in particular, are extremely variable; some look like twitter notes or tsik calls.

Silvery Marmoset (*Callithrix argentata melanura*)

Omedes (1985) studied the ontogeny of the gecker (an infant vocalization) in relationship to other calls during the first 10 weeks of life. The gecker consists of a series of elements which are present in adults, along with intermediate forms, and uttered in quick succession. These elements are ngå, phee, tsak, and trill. The ngå is a squeal sound, with tonal bands of energy that are possibly harmonically related. Phee is a whistle-like sound of constant frequency and very variable duration. Tsak is a short and loud sound that consists of a rapidly decreasing fundamental frequency. Trill is an element with frequency-modulated fundamental and of variable duration. The duration and composition of a gecker call varies according to the age of the individual and its situation. Geckers can last up to 4-5 minutes. The elements of infant calls are higher in pitch and have intermediate forms, unlike their adult counterparts. Gecker calls are mainly uttered by infants while being transferred from one member of the family to another, or when they are rejected. Around the age of 4-5 weeks, infants begin to be ignored when emitting gecker calls. From this age, gecker is basically composed of a series of soft phees and high pitched ngås. During week 6, gecker calls become mainly "ngå" and the resulting call is to show submissiveness to older animals. Around the age of 4-5 weeks, infants are carried less by other members of the family, and they start to ingest solid food. Also, play movements first appear. Four to six week old twins wrestle with each other while emitting tsaks and mixtures of ngå and trill calls.

Marmoset and Tamarin Vocal Development: Summary and Conclusions

The emergence of specific call types during vocal ontogeny is similar in marmosets and tamarins. In marmosets, early vocalizations include the tsik, squeal, phee (or J-call), twitter and trill. In tamarins, early vocalizations include the chirp, squawk, twitter-hook and long call. Although more detailed analyses are required to make careful comparisons, it would seem that there are shared acoustic features between squeal and squawk, twitter and twitter-hook, tsik and some chirps, and phee and long call. The only exception to this is the trill of marmosets, for which there does not appear to be an acoustic counterpart in the tamarins. This difference may reflect a fundamental difference in social signalling between marmosets and tamarins. However, since trills are very soft vocalizations, it remains possible that they have been missed in studies of tamarin communication signals. In addition to clarifying the elements of the vocal repertoires of most species, there remains a need to quantify the acoustic and contextual variability in most infant vocalizations with ontogeny.

A BEHAVIORAL PARADIGM FOR STUDYING VOCAL DEVELOPMENT

Nearly every study on the vocal development of NHP has remarked on the difficulty of collecting suitable data for analysis. The reasons for this are that infants generally are quiet during the neonatal period, vocalize softly, and are protected by the mother or other family members even when older, making close access for recording difficult. There is one context, however, in which infant primates are highly vocal from birth, vocalize loudly, and are readily identified as the source of the vocal data. This is during periods of separation from their caregivers. Such occurrences are rare during early development in most NHP, but in callitrichids these separations occur regularly, albeit usually briefly, when an infant is 'wiped off' onto the substrate by whatever family member is carrying it. This appears to be a normal part of the process by which family members share in infant care, and by which an infant begins to learn some degree of independence. It also presents an opportunity to record infant vocalizations, since infants vocalize continuously until retrieved by another family member. In other species of NHP, also, infants are described as being very vocal during weaning. A major difficulty with these natural separations is that they occur sporadically, and occupy only a small part of the period of maturation to adulthood. However, we have developed a paradigm that takes advantage of the high degree of vocality in separated infants, by separating an infant from its family group for brief periods at regular intervals throughout development. The duration of each separation is usually 5 to 30 minutes, depending on the infant's age. The frequency of separations typically is weekly for the first month, and monthly thereafter. These brief and infrequent separations minimize stress on the family, yet provide a longitudinal sample for the analysis of vocal development.

The advantages of studying vocal development in the context of social separation are several: normal, healthy infants are robust vocalizers throughout development in this context; the dynamics of interacting with other conspecifics is eliminated, thereby simplifying the association between vocal behavior and context; the types of vocalizations produced in this context are relatively limited, and one type, the species-typical isolation call, is invariably produced regardless of the age of the infant. Despite the artificiality of a forced separation in captivity, the context of social separation can also occur in nature, and, given the strong vocal response of the infant, this response is likely to have been under strong selection pressure during evolution to effect a rapid reunion between the separated infant and its caregivers. Since tracking the acoustic character of specific vocalizations during ontogeny is of interest, it is important that the same call type, the species-typical isolation call, is produced in the context of social separation throughout ontogeny (and throughout adulthood in squirrel monkeys and common marmosets). The species-typical isolation call shares similar acoustic characteristics across a wide range of primate species (Newman, 1985, 1992), although structural details may differ between species (Newman and Symmes, 1982), subspecies (Boinski and Newman, 1988), and even local populations (Newman and Symmes, 1982). The potential value of studying a recognizably distinct vocalization in the same context throughout development is that changes in the acoustic details of this vocalization may prove to be useful as indices of maturation, and correlate with other measures of development. Since the isolation call is a relatively stereotyped vocalization, deviations in normal structure may be useful in assessing abnormalities in neurological development and underlying neurological damage. Clearly this has proven to be the case in studies of adult squirrel monkeys (MacLean and Newman, 1988; Newman and MacLean, 1982). Furthermore, studies of the crying sounds of human neonates have shown a clear association between acoustic abnormalities of the cries and neurological status (Lester and Boukydis, 1985; Newman, 1990).

STUDIES AT POOLESVILLE ON COMMON MARMOSETS AND RHESUS MACAQUES

Although our studies of vocal development in rhesus macaques and common marmosets began with different aims, they have begun to converge both methodologically and conceptually with the goal of examining developmental processes in two species representing widely different taxa, social organizations and ecologies.

Common Marmoset

The common marmoset (*Callithrix jacchus*) was selected as a species of interest for several reasons. It does well in captivity, serving as the primary primate species in Europe in studies on neurotoxicity, neuropharmacology and reproductive physiology. Their size (adults weigh around 400 g) permits housing family units in relatively small quarters without seriously impeding social behavior, and they are highly vocal. Epple (1968) gave an overview of the vocal repertoire of captive common marmosets that continues to be useful. A brain atlas, showing details of internal brain organization, is available (Stephan et al., 1980). Numerous studies have been published on the social behavior of captive groups (e.g., Box, 1978). Our studies of the species-typical isolation call ('phee') in adults of this species indicate that acoustic details correlate with gender (Norcross and Newman, 1993), with time of separation from a social partner (Newman and Goedeking, 1992), and with the identity of the vocalizer (Newman and Goedeking, 1992; see also Jones et al., 1993). Our studies on vocal development in this species have also focused on the context of social separation. Since 5-month old infants will participate in caring for their newborn siblings

(Newman et al., 1993), infants by this age must have well-developed nervous sytems. Thus, we hypothesize that their vocal signalling abilities also are well-developed by this age.

While the phee call is present from shortly after birth (see Fig. 1), there are numerous other types of vocalizations given by isolated neonates, as well (Fig. 2, Table 2). These non-phee vocalizations continue to be produced during development, but most eventually disappear in this context (Fig. 3). Although we have not, as yet, made detailed measurements of individual vocalizations, sonagraphic analyses by Athena Baldwin and Julia Zehr indicate that the phee calls produced during the neonatal period are more variable than those produced by older individuals. The variability may either be in the form of increased FM 'jitter' along the course of a call, juxtaposition with non-phee vocalizations (especially trills), or both. At what age during development these infantile characteristics disappear is still under study. However, trill-phees and other juxtaposed calls are essentially absent during social separation tests from 4 months of age onward (Fig. 3).

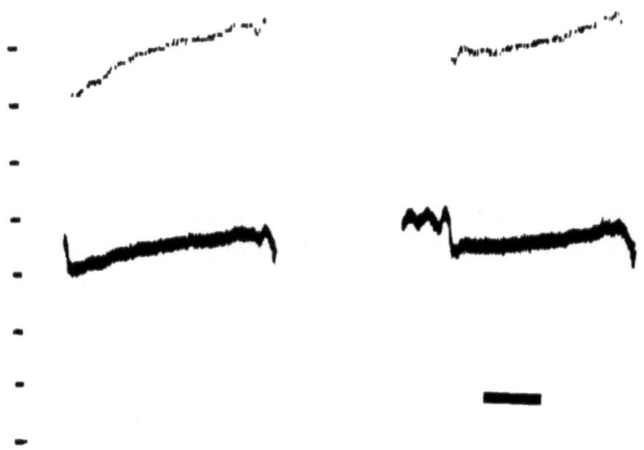

Figure 1. Sonagram example of isolation phee calls (the second being a 'trill-phee') from a 2-day old common marmoset. This infant was separated from its family group, and recorded in a weighing basket out of visual contact with the family. These examples were taken from a 1-minute recording, in which nonphee calls were also produced. Frequency scale: 2 kHz. Time scale: 200 msec.

Rhesus Macaque

Our recent studies with rhesus macaque infants began initially as a collaborative project with Jocelyne Bachevalier on the role of neonatal forebrain ablations on vocal expression. This work suggested that neonatal bilateral ablations of the amygdala (Newman and Bachevalier, 1988) or inferotemporal cortex (Newman et al., 1990) led to changes in vocal expression during brief periods of social separation when tested later in their first year of life. These studies also suggested that the vocal response to separation was mediated by brain structures that were involved in nonvocal aspects of emotional behavior (Bachevalier, 1991). In addition, we have embarked on a detailed longitudinal analysis of the behavioral and physiological response to social separation in infant rhesus macaques, with the primary goal of determining the relationship between vocal and physiological (heart rate change) measures in this context, and the maturation of these responses during development.

Table 2. Overview of Marmoset Vocalizations Given during Social Separation.

Phee: Calls with tonal structure, no abrupt FM except at start and end of call, fundamental frequency between 5.5 and 10.5 kHz, and syllable duration of 0.02-2.0 sec. Sinusoidal FM may be present, but cycles do not exceed 0.24 kHz in frequency range. May have more than 1 syllable (Winter, 1978; Norcross and Newman, 1993; this study).

Twitter: ("zwitscher-calls" of Winter and Rothe, 1979): tonal calls, with 2-21 similar syllables in sequence, each 0.02-0.08 s in duration and ascending rapidly in frequency from a minimum of 5.5 kHz to a maximum of 15.5 kHz.

Trill: Two or more uninterrupted cyclic FM units connected to each other, each of which is > 0.24 kHz in frequency range (lowest to highest frequency in the cycle).

Squeal: Tonal call with low fundamental (below 0.5 kHz) and many harmonics. May be nasalized, in which case fundamental and lower harmonics are missing. Tonal structure may be partially obscured by broad-band 'noise.'

Tsik: a brief call, typically consisting of 'flag,' 'mast,' and 'cackle' in that order. The flag is a brief tonal element that starts above 10 kHz, rising to around 12 kHz. The mast is a brief tonal element starting above 10 kHz, then descending rapidly to below 5 kHz. The cackle may be tonal (with a fundamental below 0.5 kHz) with many harmonics, or a broad-band 'noisy' sound ranging from below 0.5 kHz to above 6 kHz. The flag-mast-cackle units together are less than 0.05 sec in total duration.

Cough: like the Tsik, but the flag and mast are missing, leaving only a brief, plosive 'cackle'.

Heh: a brief tonal call, with a fundamental higher than the cough and squeal, and a smooth frequency course (rises steadily to a peak, then falls to about the start frequency). Often occurs with Tsiks.

Trill-Phee: Any combination (without interruption) of 'trill' and 'phee' elements.

Since we collect vocal samples at several ages (2 weeks, 1, 2, 3, 4 and 5 months), the opportunity to study vocal development in this species has manifested itself. Studies of the vocal response to separation by infant rhesus macaques have had a long history. However, the number of studies documenting the acoustic character of this vocal response is relatively few. Most studies refer to the typical vocal response as the 'coo' call (also called 'whoo' or 'clear call' by various authors). Our analyses indicate that other vocalizations are commonly given during social separation, as well, particularly by younger infants. Sonagraphic analyses of infant vocalizations by Susan Jones indicate that there are a range of acoustically disparate types of calls given during brief separations (Fig. 4, Table 3). As was the case with common marmosets with respect to non-phee calls, non-coo vocalizations are only infrequently given by older infants, and, by 5 months, coos make up most of the vocal response in this context (Fig. 5).

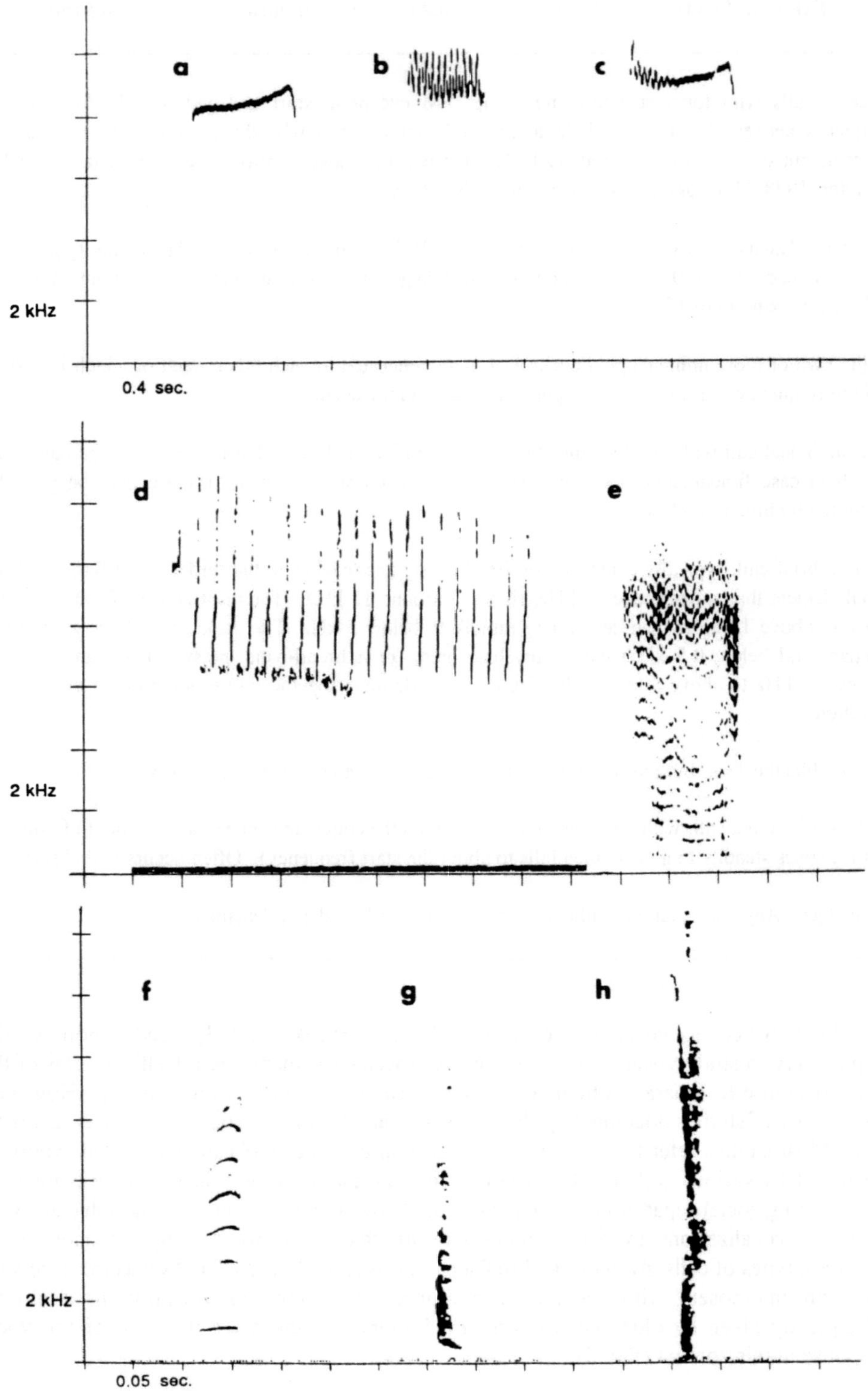

Figure 2. Overview of call types produced by infant common marmosets during social separation. Calls were recorded from several infants around 2 months of age, and were chosen to illustrate typical acoustic features of each call type. Key: a, phee; b, trill; c, trill-phee; d, twitter; e, squeal; f, heh; g, cough; h, tsik.

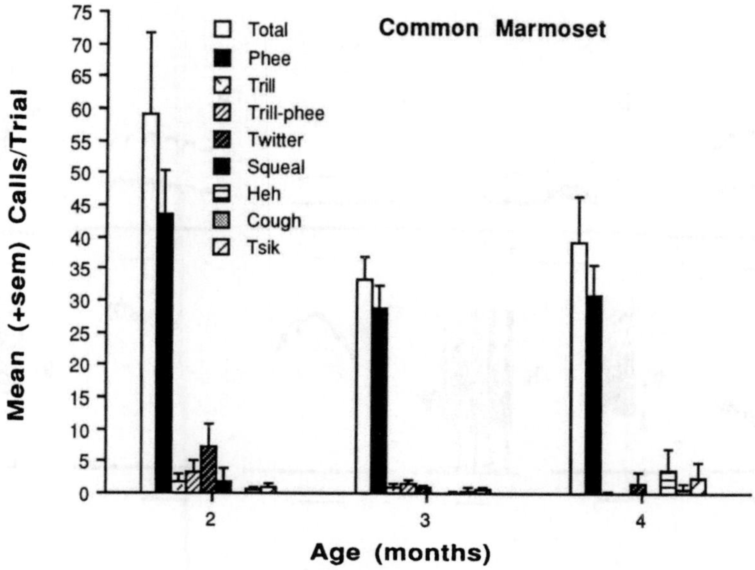

Figure 3. Distribution of call types by age for four infant common marmosets. Phees remain a relatively constant proportion of total vocalization across the 3 ages (73% at 3 months, 86% at 4 months, 79% at 5 months). Calls that are neither phees nor twitters likewise represent a relatively constant proportion of the total over these 3 ages (14.5% at 3 months, 13.7% at 4 months, 17% at 5 months). Twitters are most common during month 2 (12% of the total), and hehs most common during month 4 (4% of the total).

DISCUSSION

Convergent Lines of Vocal Ontogeny in Marmosets and Macaques

Although representing disparate branches of the primate phylogenetic tree, common marmoset and rhesus macaque infants exhibit some interesting similarities in the ontogeny of their vocalizations. When separated from their caregivers, neonates of both species produce a continuous stream of vocalizations that are highly variable. One form of variability is the wide excursion in frequency modulation of calls (phee and coo), that become much more smoothly and slowly changing in frequency later in infancy. Another form of variability is the intermixing of acoustically disparate call types (tsik, twitter, trill, squeal, cough, heh and phee in marmosets; coo, leap, scream, chirp, bark, and squeak in macaques). Most of these call types drop out of this particular context with development, leaving the phee (marmosets) and coo (macaques) as the main signals of vocal expression when socially separated. Another feature of developmental convergence is that phees and coos given in the first month of life have more FM variability during the course of a vocalization than is found in the comparable calls of older individuals.

Divergent Areas of Vocal Ontogeny in Marmosets and Macaques

Based on our current stage of analysis, there appear to be several differences in the process by which vocalizations achieve a mature or adult-like character in the two taxa. These differences largely correlate with differences in social organization and ecology. Marmosets, on the one hand, live in social groups in which there are minimal gender

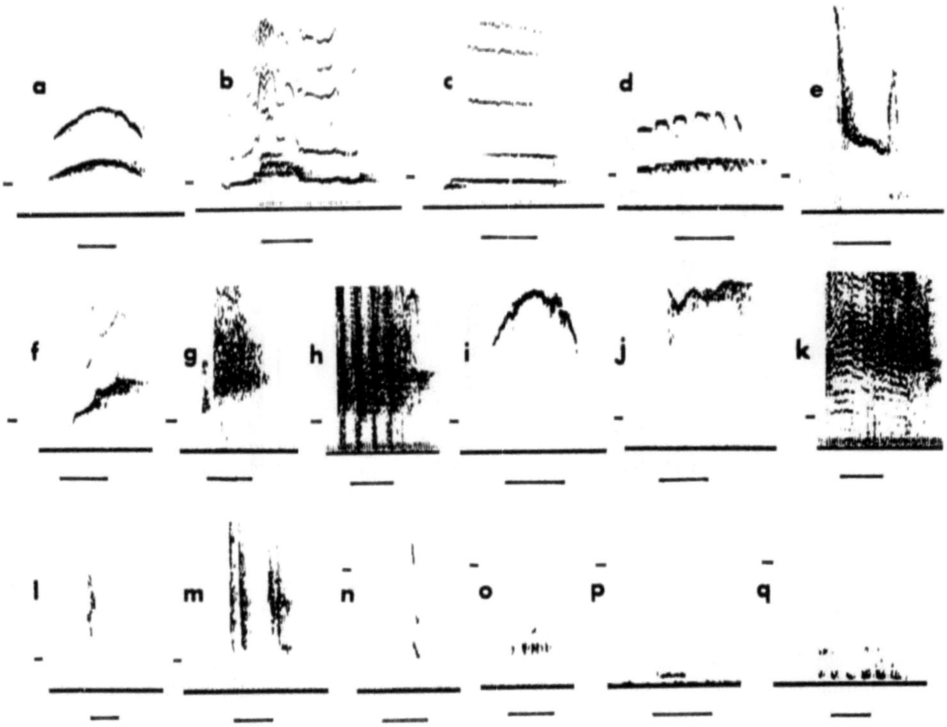

Figure 4. Overview of call types produced by infant rhesus macaques during social separation. Calls were recorded from infants of various ages. Key: a, coo; b, modulated coo; c, flat coo; d, broken coo; e, glissade; f, leap; g, noisy scream; h, pulsed scream; i, arched scream; j, tonal scream; k, undulating scream; l, squeak; m, gecker; n, chirp; o, trill; p, grunt; q, pant. Frequency scale: a-k, 0-8 kHz (mark indicates 2 kHz); l-q, 0-4 kHz (mark indicates 1 kHz). Time scale (horizontal line beneath each print): a-d, i, j, l, n-q, all 100 msec; e-g, 200 msec; h, k, m, 500 msec.

differences in social roles. Both sexes, even as subadults, participate in caring for subsequent litters. Both reproductively active adults of the group participate in territorial defense, and in warding off intruders (Harrison and Tardiff, 1989). Correlated with this is the apparent absence of pronounced sexual dimorphism in vocal signals. While it is true that the phee call of adults has some gender-specific acoustic details (Norcross and Newman, 1993), both sexes use the phee in the same contexts. And there is no evidence so far for gender differences in vocal ontogeny. Rhesus macaques, on the other hand, show strong gender differences in behavior, including infant care, and live in social units in which the adults are largely sexually segregated. Male and female juveniles exhibit different social tendencies prior to achieving adulthood, in that young females tend to stay with their mother and sisters, and participate in matriline defense and infant care, while males emigrate out of their natal group and play minimal roles in matriline defense and infant care. Gender differences are also seen in the response to social separation, at least in older infants. Females vocalize more and scream more during the second half of the first year of life (Newman et al., 1990). Screams become an important signal used to recruit aid from matriline members in agonistic encounters (Gouzoules et al., 1974), which males but not females emigrate from (e.g., Colvin, 1986). It is, therefore, of potential significance that

Table 3. Descriptions of Rhesus Macaque Infant Vocalizations Given During Social Separation[1].

Glissade. Tonal, peak frequency at start of call, sweeping downward. May end in steady tone or upsweep.

Coo. Tonal or tonal/atonal call, with start and end frequencies lower than peak frequency. End frequency closer to start frequency than to peak frequency (to distinguish from Leap). Start frequency below 3 kHz (to distinguish from Arched scream).

Girn. Similar to coo but lower fundamental frequency and variable FM throughout call. May be broken up into several discrete sound elements.

Leap. Tonal or tonal/atonal; starts like a coo, but rises rapidly to a peak frequency. May return to a lower frequency, but end frequency is closer to peak frequency than to start frequency.

Scream. Variable; tonal or atonal. We distinguish four subtypes (after Gouzoules et al., 1984):

> Tonal screams: start above 3 kHz, progress with variable frequency modulation, containing 2 or more slope reversals, start/peak frequency range less than 4 kHz.
>
> Arched screams: tonal or mixed tonal/noisy, with start and end frequencies above 3 kHz; distinguished from tonal screams by duration (less than 0.2 sec), and by greater start/peak frequency range.
>
> Noisy screams: continuous broad-band energy for all or most of call (may have brief tonal component, or dominant frequency band embedded in noise).
>
> Pulsed screams: like noisy screams, but consisting of 2 or more repeated units, each less than 0.2 sec.
>
> Undulating screams: Predominantly tonal call, like tonal scream, but very low fundamental (often missing), nasalized, with many closely spaced harmonics.

Squeak. Mainly tonal, high-pitched (> 3 kHz), short (< 0.1 sec). May begin with click (plosive) element.

Gecker. A string of 2 or more squeaks.

Chirp. Mainly tonal, fundamental between 1 and 3 kHz, short (< 0.1 sec).

Trill. A string of 2 or more chirps.

Grunt. Low fundamental (below 1 kHz).

Pant. A string of 2 or more grunts.

---

[1]Tonal sounds (fundamental and/or harmonic(s) present); atonal sounds (wide-band; fundamental and/or harmonic(s) not clearly present over all or most of call).

older infant females also show this greater tendency to produce screams in a nonagonistic context.

Future Directions in Studies of NHP Vocal Development

The shared developmental progression of vocal maturation in macaques and marmosets suggests that it would be profitable to test whether a conservative plan of vocal development is common to a wide range of primate species. There may have been strong selection pressures to evolve this developmental process, in that the youngest and most vulnerable infants would produce the least adult-like vocalizations. These, in turn, may evoke a stronger protective response from caregivers (i. e., actively seeking out the source of the vocalizations). A possibly related situation appears to exist in humans, in that mothers judge cries that are more variable and higher in pitch to be coming from infants that are at greater risk to develop abnormally (Zeskind and Marshall, 1988). On the other hand, the

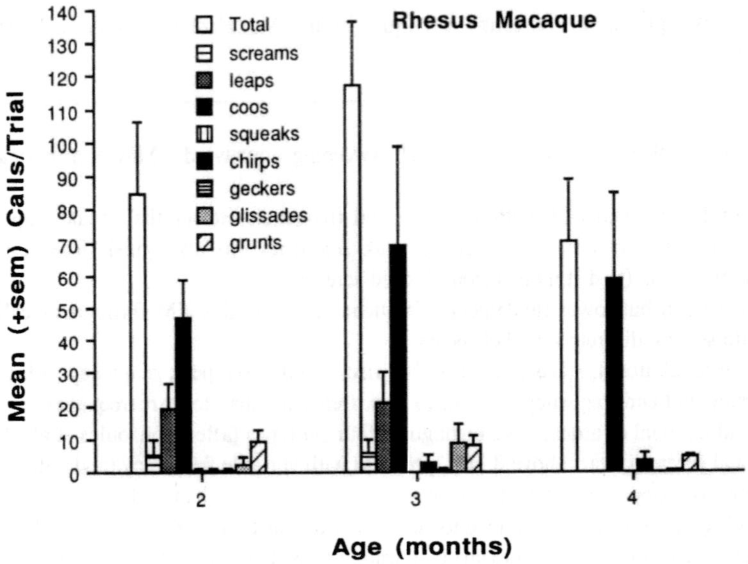

Figure 5. Distribution of call types by age for six infant rhesus macaques. Coos represent the most common vocalization given during 5 minutes separations at all 3 ages, but account for the greatest proportion of total calling at 4 months (83% of the total, while only 59% at 3 and 55% at 2 months). Leaps show the opposite trend, accounting for around 21% of the total during months 2 and 3, but dropping out entirely by month 4. Similarly, calls that are neither coos nor leaps drop off in incidence at month 4 (accounting for 21.3% of the total at 2 months, 22.1% at 3 months but only 10.6% at 4 months).

more variable vocal output from neonates may simply be a reflection of an immature nervous system without any functional correlate. Following a period of maturation (and, perhaps, individual experience), the neural control of vocal expression may become more orderly, resulting in a more predictable output. If this were true, there should be correlations between vocal maturation and other measures of brain development. At the moment, such evidence is lacking. However, this is a promising area of research for the future.

While the context of social separation has several attributes that make it useful in studies of vocal development, it also has clear limitations. Most notable is the fact that vocalizations used by older individuals in close-range affiliation and agonistic contexts are used only rarely by separated infants. Thus, to study the development of vocalizations used in close-range social encounters, one must observe developing individuals in that context. One possible way to increase the data from such studies might be to focus on the reunion process between infants and their caregivers. By permitting separated infants to see and hear nearby conspecifics, the range of vocalizations produced may increase, and by systematically documenting vocal interactions during reunions, it should be possible to track the maturation of call types other than the species-typical separation call.

CONCLUSION

This chapter has attempted to argue that the study of infant vocal signalling in the specific, well-defined context of social separation permits a relatively clear view of vocal ontogeny that may not be available to observers of infants in their social group. Studies of vocal behavior in this context in two species of non-human primate suggest the possibility that there may be a common plan of vocal development across a wide range of primate species, and that where differences can be found they reflect the ecology and social organization of the particular species in question.

ACKNOWLEDGEMENTS

I would like to thank Drs. Marc Hauser, Charles Snowdon and Elke Zimmermann for their comments on an earlier version of the manuscript.

REFERENCES

Abbott, D.H. and Hearn, J.P., 1978, Physical, hormonal and behavioural aspects of sexual development in the marmoset monkey, *Callithrix jacchus*, *J. Reprod. Fert.* 53:155-166.

Aitken, P.G., 1981, Cortical control of conditioned and spontaneous vocal behavior in rhesus monkeys, *Brain Lang.* 13:171-184.

Bachevalier, J., 1991, An animal model for childhood autism: memory loss and socioemotional disturbances following neonatal damage to the limbic system in monkeys, *in*: "Advances in Neuropsychiatry and Psychopharmacology, Vol. 1 (Schizophrenia Research)", C.A. Tamminga and S.C. Schulz, eds., Raven Press, New York.

Boinski, S. and Newman, J.D., 1988, Preliminary observations on squirrel monkey (*Saimiri oerstedi*) vocalizations in Costa Rica, *Amer. J. Primatol.* 14:329-343.

Box, H.O., 1978, Social behaviour in the common marmoset monkey (*Callithrix jacchus*), *Biol. Human Affairs* 43:51-64.

Brodmann, K., 1909, "Vergleichende Lokalisationslehre der Grosshirnrinde in ihren Prinzipien dargestellt auf Grund des Zellenbaues", J.A. Barth, Leipzig.

Chevalier-Skolnikoff, S., 1974, The ontogeny of communication in the stumptail macaque (*Macaca arctoides*), *Contrib. Primatol.* 2:1-174.

Colvin, J.D., 1986, Proximate causes of male emigration at puberty in Rhesus monkeys, *in*:"The Cayo Santiago Macaques", R.G. Rawlins and M.J. Kessler, eds., SUNY Press, Albany.

Elowson, A.M., Snowdon, C.T., and Sweet, C.J., 1992, Ontogeny of trill and J-call vocalizations in the pygmy marmoset, *Cebuella pygmaea*, *Anim. Behav.* 43: 703-715.

Epple, G., 1968, Comparative studies on vocalization in marmoset monkeys (Hapalidae), *Folia Primatol.* 8:1-40.

Erwin, J., 1975, Rhesus monkey vocal sounds, *in*: "The Rhesus Monkey, Vol. 1", G. H. Bourne, ed., Academic Press, New York.

Goedeking, P. and Immelmann, K., 1986, Vocal cues in cotton-top tamarin play vocalizations, *Ethology* 73:219-224.

Gouzoules, H. and Gouzoules, S., 1989, Design features and developmental modification of pigtail macaque, *Macaca nemestrina*, agonistic screams, *Anim. Behav.* 37:383-401.

Gouzoules, S., Gouzoules, H., and Marler, P., 1984, Rhesus monkey (*Macaca mulatta*) screams: representational signalling in the recruitment of agonistic aid, *Anim. Behav.* 32:182-193.

Green, S., 1975, Variation of vocal pattern with social situation in the Japanese monkey (*Macaca fuscata*), *in*: "Primate Behavior, vol. 4", L. A. Rosenblum, ed., Academic Press, New York.

Green, S.M., 1981, Sex differences and age gradations in vocalizations of Japanese and lion-tailed monkey, *Amer. Zool.* 21:165-184.

Grimm, R.J., 1967, Catalogue of sounds of the pigtailed macaque (*Macaca nemestrina*), *J. Zool. Lond.* 152:361-373.

Harrison, M.J. and Tardif, S.D., 1989, Species differences in response to conspecific intruders in *Callithrix jacchus* and *Saguinus oedipus*, *Intern. J. Primatol.* 10:343-362.

Hauser, M.D., 1989, Ontogenetic changes in the comprehension and production of vervet monkey (Cercopithecus aethiops) vocalizations, *J. Comp. Psychol.* 103:149-158.

Hershkowitz, P., 1977, "Living New World Monkeys (Platyrrhini), Volume 1", University of Chicago Press, Chicago.

Hohmann, G.M.and Herzog, M.O., 1985, Vocal communication in lion-tailed macaques (*Macaca silenus*), *Folia Primatol.* 45:148-178.

Jones, B.S., Harris, D.H.R., and Catchpole, C.K., 1993, The stability of the vocal signature in phee calls of the common marmoset, *Callithrix jacchus*, *Amer. J. Primatol.* 31: 67-75.

Jürgens, U. and Müller-Preuss, P., 1977, Convergent projections of different limbic vocalization areas in the squirrel monkey, *Exp. Brain Res.* 29:75-83.

Jürgens, U. and Ploog, D., 1970, Cerebral representation of vocalization in the squirrel monkey, *Exp. Brain Res.* 10: 532-554.

Jürgens, U. and Pratt, R., 1979, The cingular vocalization pathway in the squirrel monkey, *Exp. Brain Res.* 34: 499-510.

Kalin, N.H., Shelton, S.E., and Snowdon, C.T., 1992, Affiliative vocalizations in infant rhesus macaques (*Macaca mulatta*), *J. Comp. Psychol.* 106: 254-261.

Kirzinger, A., 1985, Cerebellar lesion effects on vocalization of the squirrel monkey, *Behav. Brain Res.* 16: 177-181.

Kirzinger, A. and Jürgens, U., 1985, The effects of brainstem lesions on vocalizations in the squirrel monkey, *Brain Res.* 358: 150-162.

Larson, C.R. and Kistler, M.K., 1986, The relationship of periaqueductal gray neurons to vocalization and laryngeal EMG in the behaving monkey, *Exp. Brain Res.* 63:596-606.

Larson, C.R., Sutton, D., and Lindeman, R.C., 1978, Cerebellar regulation of phonation in rhesus monkey (*Macaca mulatta*), *Exp. Brain Res.* 33:1-18.

Lester, B.M. and Boukydis, C.F.Z., 1985, "Infant Crying (Theoretical and Research Perspectives)", Plenum Press, New York.

Levine, S., Wiener, S.G., Coe, C.L., Bayart, F.E S., and Hayashi, K.T., 1987, Primate vocalization: a psychobiological approach, *Child Develop.* 58:1408-1419.

Lieberman, P., 1984, "The Biology and Evolution of Language", Harvard University Press, Cambridge.

Lieblich, A.K., Symmes, D., Newman, J.D., and Shapiro, M., 1980, Development of the isolation peep in laboratory-bred squirrel monkeys, *Anim. Behav.* 28:1-9.

Lillehei, R. A. and Snowdon, C. T., 1978, Individual and situational differences in the vocalizations of young stumptail macaques (*Macaca arctoides*), *Behaviour* 65:270-281.

Lipp, H.P. and Hunsperger, R.W., 1978, Threat, attack and flight elicited by electrical stimulation of the ventromedial hypothalamus of the marmoset monkey, *Callithrix jacchus*, *Brain Behav. Evol.* 15:249-282.

MacLean, P.D. and Newman, J.D., 1988, Role of midline frontolimbic cortex in production of the isolation call of squirrel monkeys, *Brain Res.* 450:111-123.

Masataka, N. and Fujita, K., 1989, Vocal learning of Japanese and rhesus monkeys, *Behaviour* 109:191-199.

McConnell, P. and Snowdon, C.T., 1986, Vocal interactions between unfamiliar groups of captive cotton-top tamarins, *Behaviour* 97:273-296.

Mittermeier, R.A., Rylands, A.B., Coimbra-Filho, A.F., and Da Fonseca, G.A.B., 1988, "Ecology and Behavior of Neotropical Primates, volume 2", World Wildlife Fund, Washington, D.C.

Moody, M.I. and Menzel, W.W.JR., 1976, Vocalizations and their behavioral contexts in the tamarin *Saguinus fuscicollis*, *Folia Primatol.* 25:73-94.

Newman, J.D., 1985, The infant cry of primates: an evolutionary perspective, *in*: "Infant Crying: Theoretical and Research Perspectives", B.M. Lester and C.F.Z. Boukydis, eds., Plenum Press, New York.

Newman, J.D., 1990, Crying in infants, *in*: "Principles and Practice of Pediatrics", F.A. Oski, C.D. DeAngelis, R.D. Feigin, and J.B. Warshaw, eds., Lippincott, Philadelphia.

Newman, J.D., 1992, The primate isolation call and the evolution and physiological control of human speech, *in*: "Language Origin: A Multidisciplinary Approach, NATO ASI Series D, vol. 61", J. Wind, B. Chiarelli, B. Bichakjian and A. Nocentini, eds., Kluwer, Dordrecht.

Newman, J.D. and Bachevalier, J., 1988, Acoustic differences in separation calls of rhesus monkeys following neonatal ablation of temporal lobe limbic areas, *Soc. Neurosci. Abstr.* 14:692.

Newman, J.D. and Goedeking, P., 1992, Noncategorical vocal communication in primates: The example of common marmoset phee calls, *in*: "Nonverbal Vocal Communication", H. Papousek, U. Jürgens and M. Papousek, eds., Cambridge University Press, Cambridge.

Newman, J.D. and MacLean, P.D., 1982, Effects of tegmental lesions on the isolation call of squirrel monkeys, *Brain Res.* 232:317-329.

Newman, J.D. and Symmes, D., 1974, Vocal pathology in socially deprived monkeys, *Dev. Psychobiol.* 7:351-358.

Newman, J.D. and Symmes, D., 1982, Inheritance and experience in the acquisition of primate acoustic behavior, *in*: "Primate Communication", C.T. Snowdon, C.H. Brown and M.R. Petersen, eds., Cambridge University Press, New York.

Newman, J.D., Bachevalier, J., Michjeda, M., and Suomi, S.J., 1990, A possible neural substrate for gender differences in vocal behavior by rhesus macaque infants during brief periods of social separation, *Soc. Neurosci. Abstr.* 16:599.

Newman, J.D., Baldwin, A., and Moody, K., 1993, Latency to retrieve isolated infant marmosets, *Amer. J. Primatol.* 30: 337-338.

Norcross, J.L. and Newman, J.D., 1993, Context and gender-specific differences in the acoustic structure of common marmoset (*Callithrix jacchus*) phee calls, *Amer. J. Primatol.* 30: 37-54.

Owren, M.J., Dieter, J.A., Seyfarth, R.M., and Cheney, D.L., 1992, 'Food' calls produced by adult female rhesus (*Macaca mulatta*) and Japanese (*M. fuscata*) macaques, their normally-raised offspring, and offspring cross-fostered between species, *Behaviour* 120: 218-231.

Omedes, A., 1985, Infantile calls of silvery marmosets (*Callithrix argentata melanura*) during the first ten weeks, *Miscell. Zool.* 9: 413-418.

Palombit, R.A., 1992, A preliminary study of vocal communication in wild long-tailed macaques (*Macaca fascicularis*). I. Vocal repertoire and call emission, *Intern. J. Primatol.* 13: 143-182.

Pola, Y.V. and Snowdon, C.T., 1975, The vocalizations of pygmy marmosets (*Cebuella pygmaea*), *Anim. Behav.* 23:826-842.

Robinson, B.W., 1967, Vocalization evoked from forebrain in *Macaca mulatta*, *Physiol. Behav.* 2:345-354.

Roush, R.S. and Snowdon, C.T., 1994, Ontogeny of food-associated calls in cotton-top tamarins, *Anim. Behav.* 47:263-273.

Rowell, T.E. and Hinde, R.A., 1962, Vocal communication by the rhesus monkey (*Macaca mulatta*), *Proc. Zool. Soc., Lond.* 138:279-294.

Simons, R.C. and Bielert, C.F., 1973, An experimental study of vocal communication between mother and infant monkeys (*Macaca nemestrina*), *Amer. J. Phys. Anthrop.* 38:455-462.

Snowdon C.T., 1987, A comparative approach to vocal communication, *Nebraska Symposium on Motivation* p. 145-197.

Snowdon, C.T., 1989, The structure and usage of long calls in marmosets and tamarins, *in*: "Perspectives in Primate Biology, vol. 3", P.K. Seth and S. Seth, eds., Today & Tomorrow's Printers, New Delhi.

Snowdon, C.T., French, J.A., and Cleveland, J., 1986, Ontogeny of primate vocalizations: models from bird song and human speech, *in*: "Current Perspectives in Primate Social Dynamics", D.M. Taub and F.A. King, eds., Van Nostrand Reinhold, New York.

Stephan, H., Baron, G., and Schwerdtfeger, W.K., 1980, "The Brain of the Common Marmoset", Springer-Verlag, Berlin.

Sutton, D., Trachy, R.E., and Lindeman, R.C., 1985, Discriminative phonation in macaques: effects of anterior mesial cortex damage, *Exp. Brain Res.* 59:410-413.

Vogt, B.A. and Barbas, H., 1988, Structure and connections of the cingulate vocalization region in the rhesus monkey, *in*: "The Physiological Control of Mammalian Vocalization", J.D. Newman, ed., Plenum Press, New York.

Winter, M., 1978, Some aspects of the ontogeny of vocalizations of hand-reared common marmosets, *in*: "Biology and Behaviour of Marmosets", H. Rothe, H.J. Wolters and J.P. Hearn, eds., Rothe, Göttingen.

Winter, M. and Rothe, H., 1979, Darstellung des Lautrepertoires handaufgezogener Weissbüscheläffchen (*Callithrix j. jacchus*) unter besonderer Berücksichtigung der fliessenden Übergänge zwischen den einzelnen Lautgruppen, *Primates* 20:259-276.

Zeskind, P.S. and Marshall, T.R., 1988, The relation between variations in pitch and maternal perceptions of infant crying, *Child Develop.* 59:193-196.

VOCAL ONTOGENY OF THE SQUIRREL MONKEY, *Saimiri boliviensis peruviensis*

Maxeen Biben and Deborah Bernhards

Laboratory of Comparative Ethology
National Institute of Child Health and Human Development
National Institutes of Health
Bethesda, MD 20892, USA

INTRODUCTION

The vocal behavior of the squirrel monkey (*Saimiri* sp.) is perhaps the best-studied of the nonhuman primates. The abundant vocal output of this small, easy-to-maintain species falls conveniently within the range of human hearing, making squirrel monkeys unusually amenable to study. Even the infants are very vocal from the first day of life (Biben, 1992; Lieblich et al., 1980). Despite the accessibility of this species, or perhaps because of it, there has historically been less than 100% agreement on both the makeup and usage of the repertoire. To a great extent, these differences represent biases towards a graded (e.g., Maurus et al.,1986; Schott, 1975) versus a categorically discrete output (e.g., Newman, 1985; Winter et al., 1966), with most researchers subscribing to the latter. Early work on individual call types was some of the first in the primate literature to establish (in the isolation peep) individuality, species specificity, and heritability (Newman and Symmes, 1982; Symmes et al.,1979).

Squirrel monkeys maintain their social and vocal habits well in the laboratory, which allowed close-range affiliative calls to become an early and major focus of study, in contrast to the long-range alarm and territorial calls studied in many other primates. We thus are privy to a wealth of detail in understanding how adults (particularly adult females) communicate and what information is available in their calls. Close-range affiliative calls are of great interest because of the importance of this type of communication to our own species. We know, for instance, that the most commonly used affiliative call, the chuck (Fig. 1), possesses redundant individuality and, as it is most commonly used, denotes the extent of affiliation between adult females (Biben et al.,1986; Smith et al., 1982). We know also that the structure of chucks changes depending on their position in an exchange, suggesting primitive syntactic usage (Biben et al., 1986; Masataka and Biben, 1987). Playback studies demonstrate that, somewhere in the complex melodic structure of this call resides information on individual and group or kin identity. Recent field work is contributing information on chuck usage in the wild (Boinski, 1991; Boinski and Mitchell, 1992). One

Current Topics in Primate Vocal Communication
Edited by E. Zimmermann *et al.*, Plenum Press, New York, 1995

major gap in our knowledge is an understanding of how these communicative abilities, or even some of the calls themselves, are acquired in ontogeny.

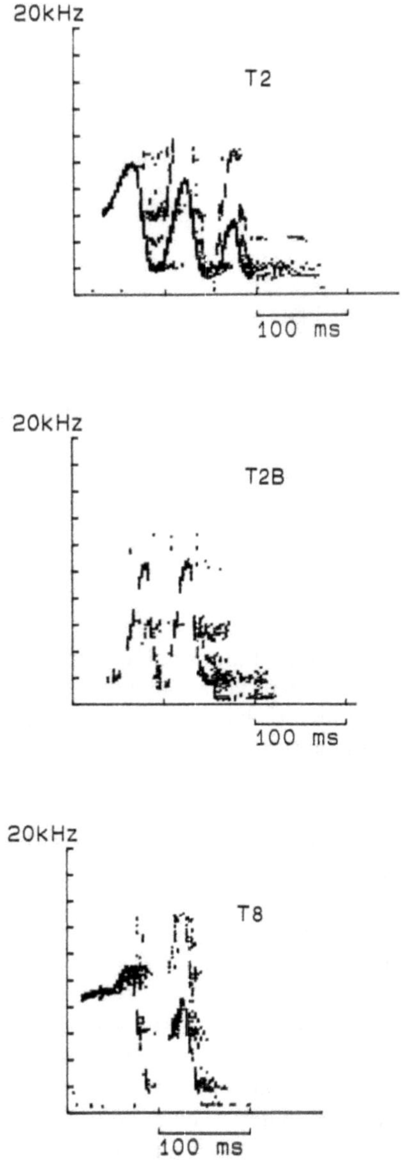

Figure 1. Representative chuck calls from 3 adult females, T2, T2B and T8. Chucks are usually made up of 2 or 3 repeated elements, less commonly a single element. The steeply descending frequency-modulated (FM) feature in each element is the "mast".

The abrupt discontinuity between human and nonhuman primates with respect to acquisition of the vocal repertoire has long been both a curiosity and a frustration to primatologists (Snowdon, 1988). The concerted efforts of modern primatologists equipped with excellent sound analysis technology have so far yielded almost no evidence of learning in vocal production (Newman and Symmes, 1982; Symmes and Biben, 1988), despite the dazzling abilities of human children to learn immense vocabularies in a proliferation of human languages and dialects. A limited role for learning is coming to light in abilities of usage and comprehension (Gouzoules and Gouzoules, 1989; Hauser, 1989; Seyfarth and Cheney, 1986), but here, too, non-human primates lag far behind humans.

A cornerstone of comparative ethology is the belief that similar behaviors can evolve in diverse species as responses to similar needs, demands or problems. However, much of the investigation into the origins of the closely-held human capacity to learn language has centered on phylogeny, in the belief or hope that our nearer relatives are more likely to share our abilities. Even outside of the primates, "intelligence", however defined, and language abilities are presumed to be linked, if not causally related. Perhaps one problem limiting our research approaches is that there is still little understanding of the forces that drove the emergence of language in early humans (coordination of hunting efforts among men? "small talk" among the women left behind?, or perhaps language was a convenient byproduct of increasing brain size?). Hence we have few clues as to which animal models may be most enlightening. At the present time, lack of flexibility in vocal development appears to be a largely universal trait in nonhuman primates, with no apparent phylogenetic trends in the manner in which vocal repertoires are acquired outside of the Hominoids. Still, there are four characteristics that place a species in an opportune position as a candidate for learning. Possibly necessary, but clearly not sufficient, these include a large vocal repertoire, frequent social communication (particularly affiliative interchanges), discontinuities between infant and adult vocal repertoires, and/or individual differences or dialects among adults.

The four favorable conditions for vocal learning have been documented for the squirrel monkey (*Saimiri sciureus*). Previous studies of infant vocal development in *Saimiri* concluded that vocal production developed "normally" in acoustic isolation, even in a deafened animal (Winter et al.,1973), leaving little apparent room for learning. Winter et al. also found the infant repertoire (at 9 months) to be virtually identical to that of adults. In fact, infant vocal production in the first week was reported to be equivalent to that of adults, with the exception of "twitter/trill" calls which were not observed until after the third month. Lieblich et al. (1980) demonstrated substantial conservativeness in the isolation peep from day 1 through the second year of life. Minor age-related changes in the call did not affect individual differences in the isolation peep that are apparent from birth, nor did they obscure structural distinctions between the isolation peeps of *Saimiri sciureus* and *S. boliviensis* species (Symmes et al., 1979) Newman and Symmes (1982) later demonstrated an inherited basis for at least some of these taxonomic differences. The body of squirrel monkey data is therefore some of the strongest evidence for essential immutability in the vocal repertoire of nonhuman primates.

Nonetheless, our accumulated observations of *Saimiri* infants, made over a period of 15 years, were not entirely in agreement with these earlier conclusions. Infants vocalized using mainly peeps and rarely used other calls, while peeps were relatively uncommon in adults. The most common adult vocal activity, the exchange of chuck calls (Smith et al., 1982; Symmes and Biben, 1992), was virtually nonexistent in observations we made of social and play behavior in groups of youngsters, aged approximately 12 months (Biben, 1986). Recently, we have shown that chuck calls of young female squirrel monkeys resemble chucks of adults of their natal troop more than those of conspecific strangers (Biben and Bernhards, 1994). The resemblance to a nonrelated troop member may be greater than that to the mother, suggesting that learning may contribute to the resemblance. On the perceptual side, playback experiments indicate that squirrel monkeys recognize the chucks of familiar animals (Biben and Symmes, 1991), an ability that is most likely learned.

Three approaches have dominated the study of vocal ontogeny in primates. One of the most powerful, the cross-fostering experiment, has not been feasible for squirrel monkeys, who rarely accept an infant assigned to them by a researcher. A variant, hand-rearing, produced a youngster with mildly aberrant behavior but apparently intact vocal repertoire (Hopf, 1970). Deprivation experiments (deafening, isolation) have likewise failed to produce a marked effect on vocal output (Herzog and Hopf, 1983; Winter et al., 1973). The failure of these experimental approaches to produce any noticeable changes in the vocal repertoire are hardly unique to *Saimiri*. But perhaps the key here is the failure to produce changes detectable to the human observer. This could be due to the scale studied (presence or absence of vocalizations) or a search for characteristics not relevant to the animals themselves.

A prerequisite of learning is the ability to change. If change is predictably responsive to experiential influences, then learning may be said to occur. Age-related changes in vocal production have recently been reported for several primate species (Elowson et al., 1992; Gouzoules and Gouzoules, 1989; Hauser, 1989;), with most modifications appearing to be the result of maturation or, alternatively, still open to interpretation. The present study documents changes in vocal production and usage in *Saimiri boliviensis peruviensis*, a subspecies for whom a wealth of vocal and behavioral data are available, during the first 18 months of life. Because of the importance of chuck exchanges to social communication among members of a squirrel monkey troop and because the apparent delay in the appearance of chuck exchanges suggests an opportunity for learning to occur, we paid particular attention to the developmental pattern of chuck emergence. The following analysis highlights developmental changes in the acoustical structure of calls with sharply descending FM ("masted calls"), of which chucks are the most prominent in the adult repertoire. The ontogeny of these calls is contrasted with that of the most frequent infant call category, peeps.

METHOD OF STUDY

Eight infant *Saimiri boliviensis peruviensis* (4 males, 4 females), born into our colony in 1988-1992, started in the study as newborns (Table 1). Only 3 of each sex were still alive at 3 months of age, with 3 males and 2 females surviving to 18 months of age. For two of these latter (E96 and E98), rates of production of masted calls (see below) were so low in months 13-18 that data for this time period were not used for those individuals. 3575 calls from the eight infants are included in the study.

Infants included in the study were observed with their mothers from the first day of life, or as early in the first week as possible, using the experimental approach described in Biben et al. (1989), which also gives details of sound recording and analyzing techniques. An observer followed a focal infant and recorded, remotely via closed-circuit monitor, 15-45 minutes of non-sleep activity weekly for the first 12 weeks of life. Thereafter, 30 minutes of data were collected monthly within 1 week of an infant's month anniversary. These longitudinal data were the basis for assessment of change in infant vocal production. As pointed out by Elowson et al. (1992) in one of the few longitudinal studies of primate vocal development, stability in measured acoustical variables over time is indicative of calls whose structure is fully formed at birth, while age-related changes indicate learning or maturation. In addition, for comparative purposes, seven 30-min observations were made (using the method described above) of call production by the adult females during a one-week period immediately preceding the birth season, including 4 females in 1989 and 5 in 1991.

Data analysis was a multi-step process. First, videotapes were scanned for time of occurrence and duration of infant positional, social, and vocal behaviors, and any social or vocal behaviors directed to the infant by the mother, other adult females, or other infants (Table 2).

Table 1. Number and types of calls from each infant, and ages at which calls were obtained.

Infant	Ages included in study (mos.)	N of masted calls	N of peeps
23D	1-18	109	848
76E	1-18	296	872
5D	1-18	168	887
E96	1-13	104	not used
E98	1-12	168	"
23C	1-3	56	"
15C	1-2	26	"
163C	1-2	41	"

Table 2. Behaviors of the infants and adult females.

INFANT POSITION AND BEHAVIORS	ADULT-INFANT BEHAVIORS
On mother	Carrying infant
Clinging ventral	
Nursing	Inspection of infant, including sniffing, touching, vocalizing
Clinging dorsal	
Off mother	
Nearby	Aggression to infant
Separated by at least a body length	
Infant-infant behaviors	Huddling with infant
Play	
Display	
Greeting	
Huddling	

Next, vocalizations of the focal infant were digitized, converted to color sound spectrograms, and stored as data files, by infant identity and age. Calls were then assigned, by visual inspection, to one of four categories (Table 3).

The squirrel monkey vocal repertoire has been studied extensively, but there is no broad consensus on where to split and where to lump the widely-ranging output (e.g., Maurus et al., 1986; Newman, 1985; Schott, 1975; Winter et al., 1966). To remain as open as possible to structural changes within a call or related calls, we elected to use only a few, broadly defined yet mutually exclusive, groupings and to follow structural changes within two of these groupings ("peeps" and "masted calls"). Noisy calls and purrs accounted for most of the remainder of the infant repertoire, but were less frequent and more variable throughout.

Included in the category "peeps" were tonal calls with no extensive wide-band elements and without any segment that could be defined as a "mast" (see below). While other studies have distinguished functional, structural and ontogenetic differences among different types of peeps in infants and juveniles (Biben and Symmes, 1986; Lieblich et al., 1980), we found it most conservative to ignore such distinctions. In so doing, we were aware that we would be lumping an individually-stable type of peep, the "isolation peep" (Symmes et al., 1979), with other peeps whose variability is unknown.

Table 3. Definitions of call categories.

1. Peeps: tonal calls with most energy above 5 kHz. Often considerable ascending FM, but limited or no descending FM (except for alarm peeps, which were not included in the study.) Duration varies between 30 - 1000 ms.

2. Masted calls: these calls are dominated by rapid FM, descending to 4 kHz or less, at a rate no less than 100 kHz/sec. All adult chucks would fall into this category, as well as the infantile "tuck" and "location trill" (Winter, 1968). Masted calls that were given in a clearly agonistic/defensive context (i.e., keckers and yaps (Winter et al., 1966)) were assigned to the following category.

3. Noisy calls: loud calls with dominant wide-band structure, temporally associated with conflict between two infants or an infant and adult (e.g., weaning conflict or being grabbed by an adult).
4. Purrs: soft, often pulsed, calls temporally associated with nursing bouts.

Because peeps were so numerous at some ages, we used for analysis only the first 100 peeps that occurred in any infant's sample for a particular month, taking care that these were not all from the same week or session. Peeps were analyzed only for the 3 infants for whom we also had good numbers of masted calls throughout the 18 month period; i.e., infants 23D, 76E, and 5D (Table 1). Masted calls were more infrequent; we used for analysis all available well-recorded examples. This approach, studying intensively the production of one or a few individuals, is similar to that used by many developmental psycholinguists (e.g., Papousek and Papousek, 1981; Stark, 1978; Stark et al., 1975)

The category of masted calls (Newman et al., 1983) included all vocalizations having the defining characteristic of a mast: an element with rapidly (300-500 kHz/sec) descending FM. The mast of a chuck call is preceded by a tonal element of variable length (the flag), and followed by a brief harmonic or noisy element (the cackle). Among adults, chucks are the most common masted call, but the agonistic "yap" and "kecker" and the infantile "tuck" and "location trill" also contain (or are limited to) a mast element (Newman et al., 1983). Because the relationships (structural, functional, and most particularly ontogenetic) among these calls were unclear, we found it most conservative to lump them, as we did for peeps.The two remaining categories, noisy calls and purrs, were clearly structurally distinct from either a peep or a masted call.

The structure of calls was determined on the basis of several acoustic parameters (Table 4) deemed important with regard to characterising the spectral shapes of both peeps and masted calls; these parameters have been used in previous studies of these calls (Biben et al. 1986; Smith et al., 1982).

Univariate longitudinal trends in acoustic parameters were evaluated by correlating infant age with the sample mean in age blocks of one month (for the first 12 months) or two months (for months 13-18), using Pearson product-moment correlation coefficients. Although the sample size was small, data were not pooled; i.e., a mean was calculated for each individual for each age block, with data points representing the mean of these individual means.

Changes in the stereotypy of calls, as estimated by coefficients of variation (C.V.), were examined both within and between individuals. Between-individual changes in the C.V. of the age-block sample means of all infants in the study were observed to ascertain whether the calls of infants converged on a species-typical norm (as might happen with an alarm call or a call encoding group identity). For each of the 3 infants for whom the data were most complete over this time period, we calculated within-individual C.V.'s for each month in the 18 month period, to determine whether calls became more individually distinct (as would be the case in calls having individual signatures). Both types of measurement are necessary because an increase in the between-individual C.V. may come about either because individuals' calls become more variable or because individuals' calls become more stereotyped, but individuals diverge from each other as they establish individual signatures.

Table 4. Measured acoustic parameters of masted calls and peeps.

Peak frequency: highest frequency attained by the call, measured in Hz

Peak frequency location: where in the call the peak frequency is found, given as percentage of temporal distance along the call. For example, if a call starts low and ends high, its peak is at 100%.

Minimum frequency: lowest frequency attained by the call, measured in Hz

Minimum frequency location: where in the call the minimum frequency is found, given as percentage of temporal distance along the call.

Frequency range: calculated in Hz as peak frequency minus minimum frequency

Duration: measured in msec

Frequency jitter: instability of pitch, calculated as total number of positive peaks and negative peaks per 100 msec

Frequency leaps: number of slopes steeper than 60 kHz/sec (either up or down) that continue for at least 2 kHz

RESULTS

Vocal Production: Call Rates

The most immediately noticeable departure of infant vocal behavior from that of adults is the sheer magnitude (i.e., rate) of infant vocalizing. Compared to adults, all but the youngest infants are considerably more vocal (Fig. 2).

Baseline data from adults yielded a rate of 59.6 (±4.3 SE) calls per hour, but it was not uncommon for infants to call at 10 times that rate. The highest infant rate recorded was 513 calls from a single 8-week-old infant in one 17.5 min observation session, almost 1 call every 2 seconds! Infant vocalizing peaked during the second month of life, coinciding with two activities: 1) infants began spending more time away from their mothers, their wanderings accompanied by almost constant peeping, and 2) infants began to play together, emitting streams of play peeps. While calling decreased somewhat after the second month, it was still highly variable and exceeded the adult rate throughout the duration of the study.

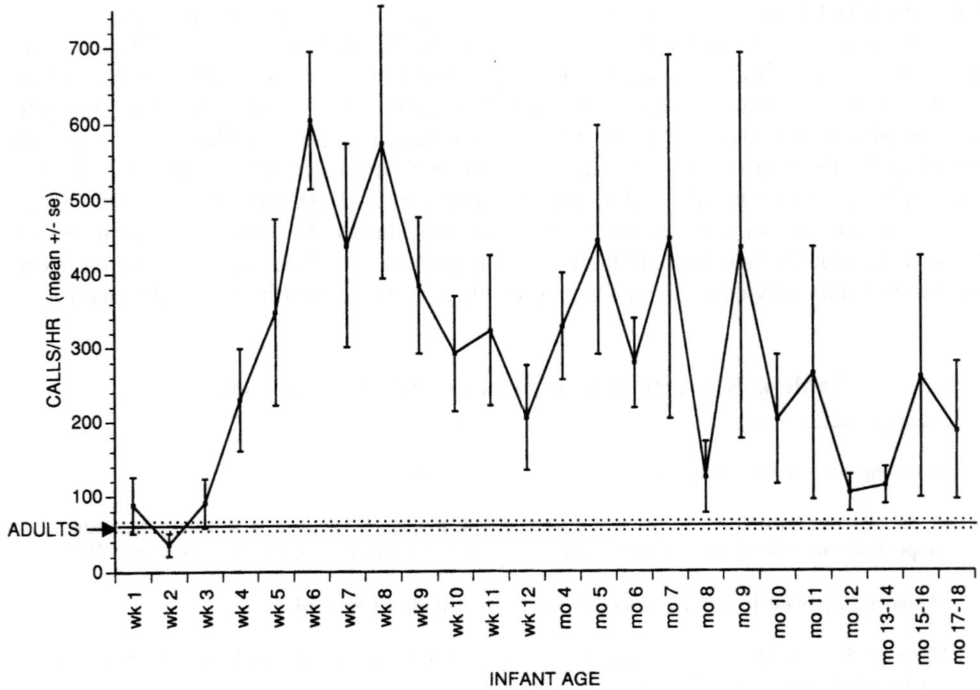

Figure 2. Infant noisiness: Number of calls per hour emitted by infant squirrel monkeys housed with their mothers in a social group. For comparison, solid line labelled "adults" is the mean (and dotted lines are standard error) number of calls emitted by adult females housed in a social group with no infants present. Adult data were pooled within a session; the reported value is a mean of 7 session means.

Vocal Production: Call Types

Peeps are clearly the predominant infant vocalization at each age sampled (Fig. 3). For comparison, data from 9 adult females in two different social groups, taken just before the birthing season, shows chucks to be the most common vocalization. While we made no distinction in these data among various types of peeps, most fit the description of "short peeps" (Newman, 1985). Infants used short peeps in conjunction with both social and nonsocial behaviors, including during greeting and display with other infants, prior to nursing, and when separated from the mother. We recorded relatively few "isolation peeps", probably because our infants rarely experienced the type of stressful separation that prompts the emission of isolation peeps. "Play peeps" were also uncommon in the sample, perhaps because the youngsters were not particularly playful at the usual data collection times.

Vocal Production: Acoustic Structure

Masted Calls: Peak Frequency. The peak frequency of adult chucks averaged slightly over 8 kHz (Fig. 4A). Infants' masted calls were initially considerably higher than

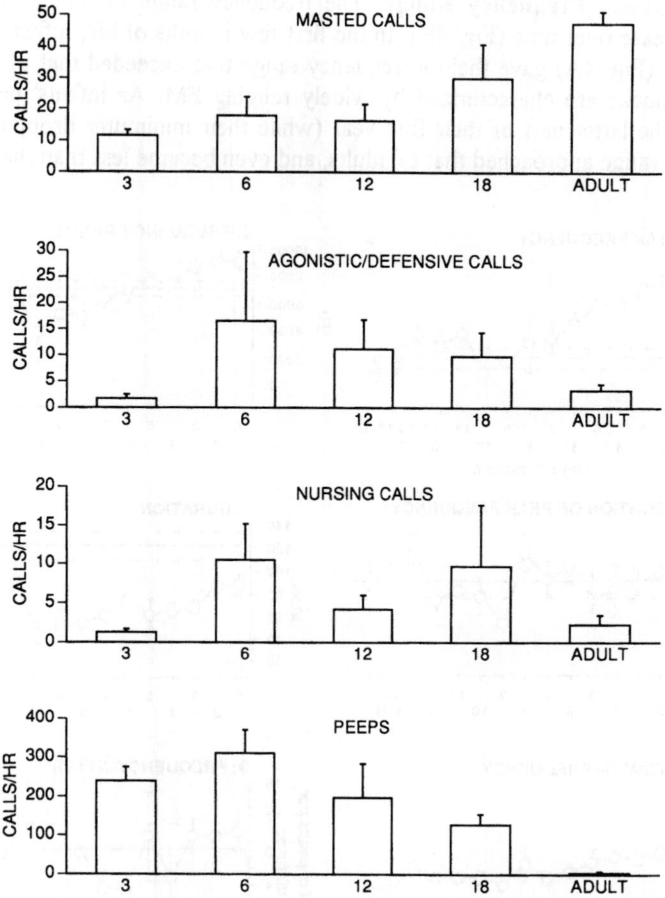

Figure 3. Mean (+SE) of different call types emitted by infants at progressive developmental ages (3, 6, 12 and 18 months) with mean (+SE) of the adult female sample for comparison. Note differing intervals on Y axes.

this, but showed a steady drop in mean peak frequency until approximating the adult value at about 8-10 months (Fig. 3, $R_{1\text{-}10}$ = -0.98, p<0.0001), after which point infants' mean peak frequency fluctuated about the adult range.

Because the timing of various features in a call is crucial in defining that call, we also looked at where in the call the peak frequency occurred. In masted calls, the temporal location of the peak frequency was variable, but remained in the first half of the call (Fig. 4B), averaging about one-third of the way through, as it did for adult chucks.

Masted Calls: Minimum Frequency. A minimum frequency of 2kHz or less is characteristic of adult chucks. In this sample of infants, the minimum frequency attained in the call declined slowly but fairly steadily (and significantly), approaching but not yet reaching adult-like values in the latest months (Fig. 4C).

The location of the minimum frequency attained by infant masted calls remained very late into the call, without showing any significant trends, throughout the time period studied (Fig. 4D). It was, however, somewhat later than was observed for the adult chucks.

Masted Calls: Frequency Range. The frequency range of infant calls showed a significant decrease over time (Fig. 4E). In the first few months of life, infants' much higher peak frequency (Fig. 4A) gave them a frequency range that exceeded that of adult chucks, even though chucks are characterized by widely ranging FM. As infants' peak frequency plummeted in the latter part of their first year (while their minimums dropped much less), their frequency range approached that of adults, and even became less than that of adults.

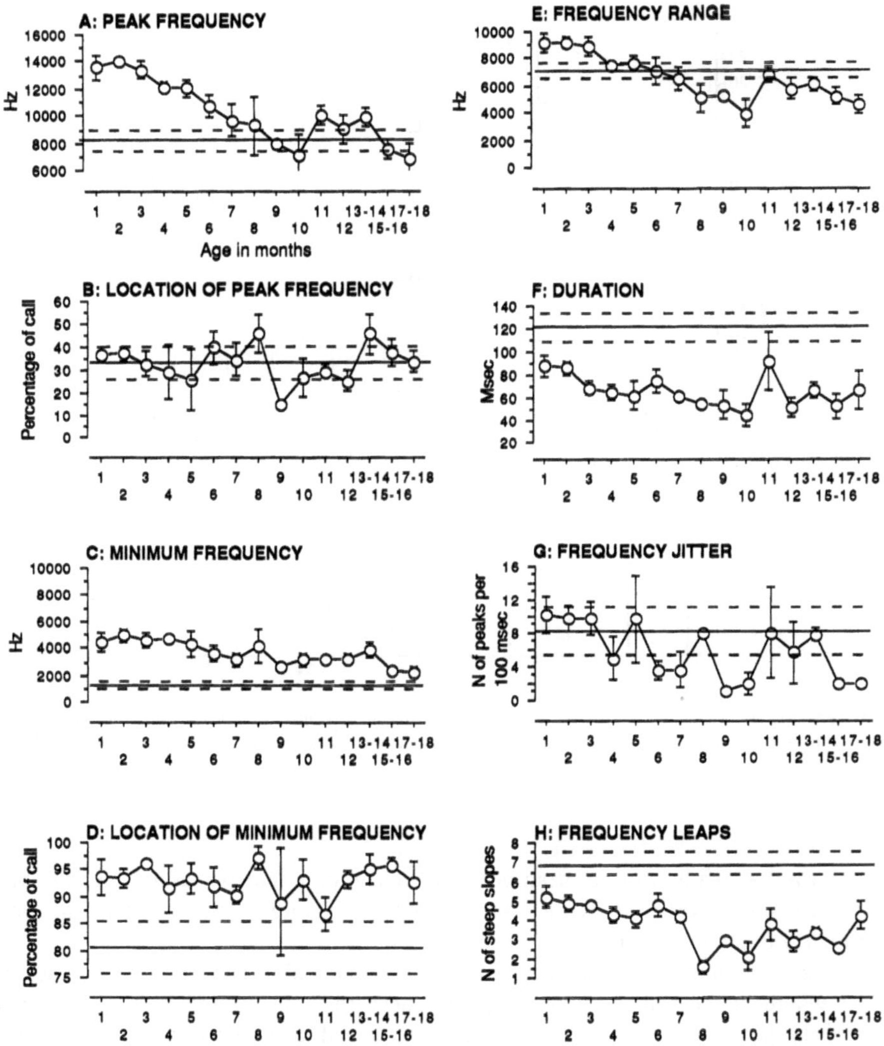

Figure 4. **A**: Change in peak frequency attained by masted calls (Hz, mean ± SE) as a function of infant age ($R_{1\text{-}18} = -0.88$, $p<0.0001$). For comparison, solid line is the mean (and dotted lines are standard error) of the same measure or adult females' masted calls. **B**: Location of the peak frequency of masted calls, given as percentage (mean ± SE) of temporal distance along the call. $R_{1\text{-}18} = -0.01$, NS. **C**: Change in minimum frequency attained by masted calls (Hz, mean ± SE) as a function of infant age ($R_{1\text{-}18} = -0.83$, $p<0.001$). **D**: Location of the minimum frequency of masted calls, given as percentage (mean ± SE) of temporal distance along the call. $R_{1\text{-}18} = -0.05$, NS. **E**: Frequency range of masted calls (Hz, mean± SE). $R_{1\text{-}18} = -0.83$, $p=0.0001$. **F**: Change in duration of masted calls (msec, mean ± SE) as a function of infant age. $R_{1\text{-}18} = -0.45$, NS. $R_{1\text{-}10} = -0.90$, $p<0.0001$. **G**: Instability of pitch (frequency jitter) in masted calls. $R_{1\text{-}18} = -0.58$, $p=0.021$. **H**: Frequency leaps in masted calls. Mean number (± SE) of steep slopes in the call. $R_{1\text{-}18} = -0.60$, $p=0.016$.

Masted Calls: Duration. Duration of masted calls (Fig. 4F) was shorter (never averaging more than 100 ms) throughout the period of observation than the duration observed for chuck calls of adult squirrel monkeys. The data showed no significant trend over the 18 month period, although durations dropped steadily during most of the first year, from a high of almost 90 msec in the first month to almost half that by the 10th month (Spearman rho = -0.93, p=0.005). Thereafter, durations became quite variable. The period of greater variability coincides with the appearance in the repertoire of masted calls occurring in groups of 2-5 brief (generally, 20-80 msec), ˄ or ∧ shaped, low peak frequency repeating elements following an initial, usually crook shaped, element of longer duration and higher peak frequency. In most cases, these elements, which were indistinguishable from some masted calls that occurred singly, were clearly separated by short periods of vocal silence. In some cases however, the elements were connected at their lower frequencies (˄˄˄ or ⋀⋀⋀). Connected groupings and unconnected groupings were obviously related but because it was not obvious that one or the other should prevail, unconnected calls were measured as separate brief calls and connected calls were measured as a single long call, engendering much of the variability in the latter months of Fig. 4F.

Masted Calls: Measures of Frequency Modulation. Two independent measures of FM were obtained from each spectrogram. Frequency jitter, a measure of pitch instability, generally stayed within the range of adult scores (Fig. 4G), although there was an overall age-related decline. The last measure of frequency modulation, frequency leaps, measured steeper, more sustained slopes. Infants never achieved adult levels of this measure and, in fact, showed a significant decline over time (Fig. 4H). Frequency leaps and frequency range were highly correlated (R= 0.82).

Peeps: Peak and Minimum Frequencies. Frequencies of peeps ranged over no more than 4000 Hz during the 18 month period, and declined significantly overall (Fig. 5A). Both the minimum and maximum frequencies attained in peeps showed a small but significant decrease over time (Fig. 5B), although the difference between highest and lowest values was not as great as for masted calls. There were too few adult peeps produced in the sampling period to make a comparison.

The gross structure of peeps remained constant over time: calls showed a general rising trend, with the minimum frequency occurring within the first 10-15% of the call, and the maximum frequency occurring within the final 10-15% (Fig. 5C). The location of the minimum advanced a small but significant amount over time, while the location of the maximum showed no trend.

Peeps: Duration. The mean duration of peeps did not change in any consistent way over time (Fig. 5D), and was comparable to the durations of masted calls over the same period. Unlike masted calls, peeps had clear terminal boundaries without trailing elements.

Peeps: Measures of Frequency Modulation. Measures of FM in peeps showed calls becoming more restricted in their major frequency excursions over time (Fig. 5E) while remaining instable on a finer scale (Fig. 5F). Peeps were about twice as instable as masted calls and, unlike the values for masted calls, did not decline over time. The number of frequency leaps in peeps and masted calls were similar, and both declined significantly over time.

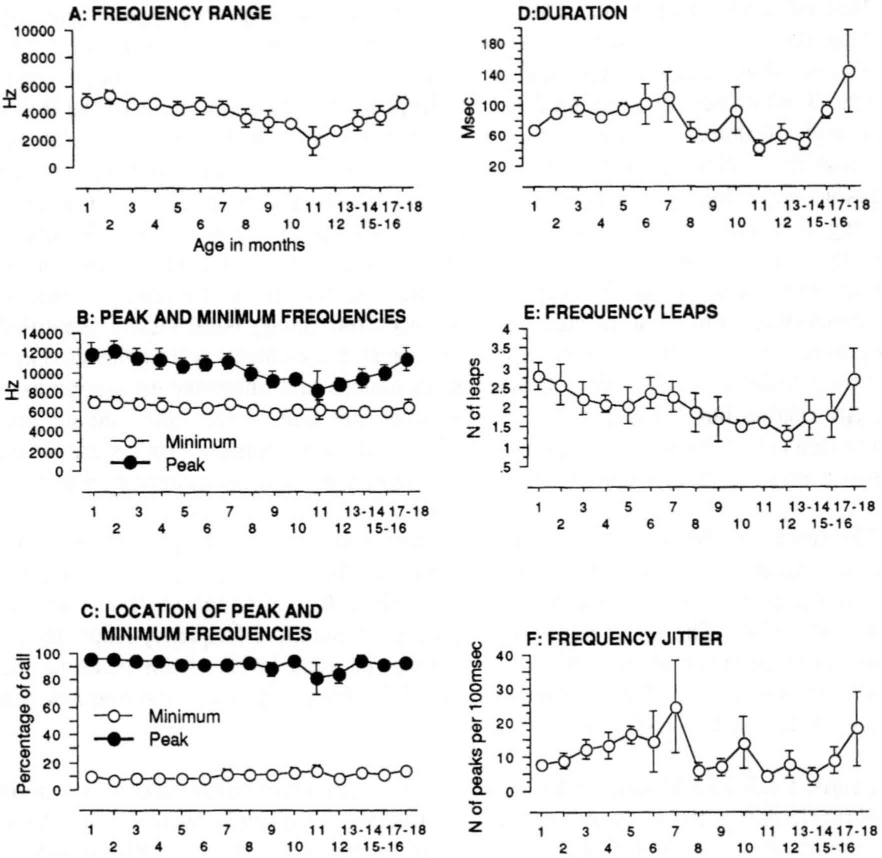

Figure 5. **A**: Frequency range of peeps (Hz, mean± SE). R_{1-18} =-0.61, p=0.0001. **B**: Change in minimum and peak frequencies attained by peep calls (Hz, mean ± SE) as a function of infant age. R_{min} = -0.76, p<0.001; R_{peak} = -0.68, p<0.01. **C**: Location, given as percentage (mean ± SE) of temporal distance along the call, of the minimum and peak frequencies. For example, if the call starts low and ends high, the minimum is at 0% and the maximum is at 100%. $R_{min\ loc}$ = 0.64, p<0.01; $R_{peak\ loc}$ = -0.45, NS. **D**: Change in duration of peep calls (msec, mean ± SE) as a function of infant age. R_{1-18} =0.01, NS. **E**: Frequency leaps in peeps. Mean number (± SE) of steep slopes in the call. R_{1-18} = -0.52, p=0.046. **F**: Frequency jitter of peeps. Total number of positive peaks and negative peaks per 100 msec. R_{1-18} =-0.12, NS.

Vocal Production: Developmental Changes in Individuality

Individuality was evaluated by comparing mean values of each acoustic parameter at 6-month intervals, for each of the 3 infants for whom the data were most complete. Individual differences were much more prevalent in peeps than in masted calls.

Masted calls: Individual differences were, at best, weakly expressed in most parameters, with some parameters showing no evidence (Table 5). No age-related changes in individuality were apparent.

Peeps: Individuality was most apparent in the parameters peak frequency, minimum frequency, and duration, where significant differences were present in each of the four sampling periods (Table 6). The parameters showing the least individuality were the locations of the peak and minimum frequencies. These parameters are probably more important in defining the call itself. Frequency range, jitter, and leaps, all parameters expressing frequency modulation, also showed considerable individuality. Individual differences were about equally apparent at all ages.

Vocal Production: Developmental Changes in Stereotypy

For each acoustic parameter, coefficients of variation were derived for the values in each age-block. Age-related tendencies for the call parameters of individual infants to become more stereotyped (versus becoming more variable) were analyzed for the 3 infants for whom the data were most complete. Trends across individuals used all available data: 7 infants in the case of masted calls, 3 infants for peeps.

Table 5. Individual differences between masted call parameter means, for selected age-blocks. Entries in boldface indicate means that differed significantly (2-tailed t test, with a <0.05) from both other means during the month indicated.

	Infant			Infant		
Month	76E	5D	23D	76E	5D	23D
Peak Frequency (Hz)				Duration (msec)		
1	15147	8733	14063	74.0	82.2	72.8
6	8363	10463	9737	70.4	79.5	**38.7**
12	**5070**	10254	9234	**25.6**	77.9	54.4
18	**8966**	5820	5886	95.6	64.8	38.9
Peak Frequency Location (% of call)				Frequency Range (Hz)		
1	44.8	39.3	40.0	**8152**	**5809**	**10430**
6	32.8	47.1	**18.1**	5586	5301	5940
12	13.2	38.9	21.2	**2960**	7044	6104
18	35.6	**25.1**	40.0	6019	3925	3972
Minimum Frequency (Hz)				Frequency Jitter (see text)		
1	**6995**	2924	3633	14.48	9.13	20.10
6	2777	**5162**	3797	1.18	5.68	2.19
12	2110	3210	3130	0.00	0.55	1.39
18	2947	1895	1914	2.53	1.31	2.14
Min. Frequency Location (% of call)				Frequency Leaps (N, see text)		
1	97.9	88.7	97.8	4.57	3.60	4.00
6	90.8	93.9	95.6	4.27	4.71	3.11
12	97.8	91.4	92.6	**1.40**	4.50	3.25
18	87.2	90.7	100.0	5.70	3.64	3.33

Masted calls: There were no significant across-infant age-related changes in stereotypy of any measured parameter of masted calls (Table 7). One infant, 76E, showed significant *increases* in variability, in 4 of 8 parameters, while the other infants showed few changes.

Peeps: Two parameters, duration and frequency jitter, showed significant tendencies to become *more* variable over time, across infants, while data for individuals showed virtually no evidence of age-related trends (Table 8).

Vocal Usage: Masted Calls

37% of infants' masted calls were made responsively, i.e., followed within 2 seconds the vocalization of another animal who was either within reach of the infant or who was looking at the infant. This percentage did not change significantly over time. In the first month of life, all responsive vocalizing was to the calls of mothers or allomothers, who used caregiver calls (Biben et al., 1989) or chucks when addressing infants. For older infants, the percentage of responsive vocalizing that involved two infants, rather than an adult and an infant, averaged 36% until about 12 months of age, when the peer-directed percentage began to climb steadily to over 80% at 18 months of age.

Table 6. Individual differences between peep parameter means, for selected age-blocks. Entries in boldface indicate means that differed significantly (2-tailed t test, with a <0.05) from both other means during the month indicated.

	Infant			Infant		
Month	76E	5D	23D	76E	5D	23D
Peak Frequency (Hz)				Duration (msec)		
1	**13857**	**11490**	**10438**	67.7	62.4	**47.8**
6	**12341**	10152	10181	**82.3**	**154.6**	**67.9**
12	9070	9403	**7898**	**40.3**	**87.5**	**55.4**
18	**12407**	10060	9336	89.0	**194.1**	93.7
Peak Frequency Location (% of call)				Frequency Range (Hz)		
1	95.8	95.0	95.2	5537	5196	**4031**
6	95.0	89.9	91.7	**5670**	**4182**	**3710**
12	95.5	73.5	85.0	3074	2732	2245
18	94.8	90.9	94.1	**5182**	4277	3736
Minimum Frequency (Hz)				Frequency Jitter (see text)		
1	**8320**	6294	6407	9.39	5.82	8.03
6	6671	**5970**	6471	6.72	**31.99**	4.97
12	**5996**	**6671**	**5653**	**0.72**	13.77	8.42
18	**7225**	5783	5600	7.48	**29.16**	6.38
Min. Frequency Location (% of call)				Frequency Leaps (N, see text)		
1	10.7	8.5	10.0	3.29	2.94	**2.13**
6	10.2	**4.0**	10.1	**3.08**	1.82	2.12
12	9.4	**5.2**	8.9	1.77	1.17	0.86
18	14.3	12.4	**7.4**	**3.46**	**1.90**	**1.45**

Infant vocal exchanges were heterogeneous throughout the study period. Overall, only 25% involved both infants using masted calls. In 54% of exchanges, the focal infant's masted call was a response to another infant's peep; other infant vocalizations were mainly noisy calls. There were no significant age-related changes. On the other hand, infant vocal exchanges with adults changed dramatically with age. Initially, all infants' masted call responses to adults involved caregiver calls made by mothers or allomothers. By the second month, infants were responding to adult chucks, and by month 6, most responses were to chucks. The correlation between infant age and response, with masted calls, to adults' chucks (rather than other calls) was 0.83 ($p<0.01$). However, adults' use of caregiver calls drops sharply during the first year, and chucks become more commonly used when addressing infants (Biben, 1992). The ontogeny of masted call exchanges is summarized in Table 9.

The social context of each infant utterance was noted, but not specifically categorized, by the observers. Categories of context were established post hoc after examination of these contexts. Categories included exploration, caregiving (nursing or retrieval), interaction, and alone. In 50% of infant masted call utterances, infants were sitting alone, not looking at another animal. 3% of masted calls were made by infants who were exploring, and 2% in caregiving situations. "Interaction" was the most common category of

Table 7. Correlation of age and coefficient of variation, for acoustic parameters of masted calls.

PARAMETER	R	PARAMETER	R
Peak Frequency		Duration	
76E	.788***	76E	.594*
5D	.163	5D	.287
23D	.419	23D	-.008
all infants	.345	all infants	.407
Peak Frequency Location		Frequency Range	
76E	.085	76E	.695**
5D	.510	5D	.268
23D	-.355	23D	.123
all infants	-.212	all infants	.275
Minimum Frequency		Frequency Jitter	
76E	.241	76E	-.228
5D	.280	5D	.097
23D	-.191	23D	-.159
all infants	-.260	all infants	-.037
Min. Frequency Location		Frequency Leaps	
76E	.496	76E	.533*
5D	.206	5D	.574*
23D	-.468	23D	-.027
all infants	-.104	all infants	.207

* $p<0.05$; ** $p<0.01$; *** $p<0.001$.

context for masted calls (45% of occurrences). Interactions ranged from mildly agonistic grabbing of infants by adults, to affiliative acts like huddling. Most interactions lacked any evident physical component, however, beyond the vocal exchange itself.

Vocal Usage: Peeps

Infant peeps were very responsive calls; in the first month of life, more than 70% of infants' peeps were made in response to mainly caregiver or chuck calls from mothers and allomothers. Responding to the calls of other infants began early in the second month, but averaged only 27% (±7% S.D.) of all responsive peeps, with no age-related changes after the first month. Most peep exchanges between infants were part of prolonged vocal interactions which accompanied grabbing, displaying, or playing. Most exchanges involving adults had no obvious nonvocal behavioral context, although a small percentage preceded nursing or accompanied retrieval by the mother.

Table 8. Correlation of age and coefficient of variation, for acoustic parameters of peeps.

PARAMETER	R	PARAMETER	R
Peak Frequency		Duration	
76E	.123	76E	-.147
5D	-.100	5D	.039
23D	.289	23D	.128
all infants	.134	all infants	.516*
Peak Frequency Location		Frequency Range	
76E	-.093	76E	.209
5D	.471	5D	.254
23D	.326	23D	.281
all infants	.498	all infants	.348
Minimum Frequency		Frequency Jitter	
76E	-.281	76E	.000
5D	-.564*	5D	.147
23D	.469	23D	.138
all infants	-.191	all infants	.591*
Min. Frequency Location		Frequency Leaps	
76E	-.170	76E	-.085
5D	.164	5D	.083
23D	-.092	23D	.308
all infants	-.023	all infants	.492

* $p<0.05$

Table 9. Summary of masted call exchanges.

	Infant age (in months)			
	1	6	12	18
A) % of infant masted calls that were responsive	52.3	36.2	27.3	41.7
B) % of responses that involved 2 infants	00.0	33.3	55.7	84.3
C) % of infant-infant exchanges involving 2 masted calls	-	11.1	14.0	37.0
D) incidence of masted call exchange between infants (A x B x C)	00.0	1.3	2.1	13.0
E) % of responses that involved infant and adult	100.0	66.7	44.3	15.7
F) % of adult-infant exchanges involving 2 masted calls (adult's is always a chuck)	00.0	87.5	64.3	100.0
G) incidence of masted call exchange between infants and adults (A x E x F)	00.0	21.1	7.8	6.5
H) total incidence of infants' masted call exchanges (D + G)	00.0	22.4	9.9	19.5

CONCLUSION AND FUTURE DIRECTIONS

What do these results suggest about how we should view the present status of primate vocal ontogeny, and how we should direct our future investigations? Research in the vocal development of primates has languished for many years for several reasons. First and, it would seem, most easily put aside, is the assumption that we are working with essentially closed systems, proceeding in an inevitable and immutable manner from birth to maturity.

It is evident, both from data in the present study and from previous reports in the literature, that infant squirrel monkeys are vocally distinct in many ways from adults, and it thus follows that a transition, whether gradual or abrupt, must occur. Previous studies (Lieblich et al., 1980; Ploog et al., 1967) documented unique calls (Baby-Piepen) in the infant repertoire but there has been some disagreement about whether all adult calls were also present. The evidence presented here argues against the assertion of Winter et al. (1973) that all calls of the adult repertoire are heard from infants by three months of age (with most present at day one). Winter et al. recorded from newborn squirrel monkey infants what they considered to be the full range of vocalizations in the adult repertoire, the only exception being the "twitter/trill", a multiple element call made up of 0.1 sec units with bidirectional FM sweeps between 3 and 7.5 kHz. Individual units ("twits") were present. It is clear from these findings, as well as from the structural data presented here, that FM, range, and other acoustic features of adult vocalizations can be produced by infants. Despite this apparent capability, and the occasional observation of infantile calls that met some of the criteria of adult chucks, the central tendencies, variability, and usage patterns of infants' masted calls diverged greatly from those of adults.

An initial problem we faced when defining the terms of the present study was identifying how chuck ontogeny would be studied. In adults, chucks are common, readily identified, and easily distinguished from other calls. There was no evidence in vocal samples from very young infants of calls that met all the criteria of mature chucks (Newman et al., 1983). However, all component acoustic segments characteristic of chucks were separately present in infants' utterances, indicating that youngsters were capable of producing these sounds in at least some vocal contexts (although not necessarily capable of stringing them together into one call). Infants' vocal output is not babbling; like adults, they produce sounds that can be classified into particular call types and in some cases ("tucks", "location trills": Winter, 1968) these share acoustic characteristics with chucks. Unlike chucks, these very early masted calls were invariably single element (1 mast) calls, had abbreviated and fairly invariant (or absent) tonal segments (flags) preceding the mast, and lacked a low-pitched harmonic or noisy segment (cackle) following the mast.

The rate of utterance of these chuck-like calls was never great but was always variable, making it most expedient to lump them into the single category "masted calls" and follow their ontogeny as a class. Our approach was thus relatively conservative and unbiased, similar to that used by Maurus and others (Maurus et al., 1985, 1986). Results of such structurally-based studies, which use acoustic measures from the total vocal output, can be surprisingly similar to those of studies focussing on discrete call types (Biben and Symmes, 1986; Goedeking and Immelmann, 1986).

Our developmental tracking of masted calls found disparities in call structure between the calls of youngsters and those of adult females up to at least 18 months (the terminus of the study). The significance of these differences is summarized in Table 10. In contrast, the shape of peep calls, their mean duration, and the frequency range changed only slightly, if at all, in the first 18 months of life.

Table 10. Summary of infant-adult differences in pitch contour of masted calls.

ACOUSTIC FEATURE	DIFFERENCE BETWEEN INFANTS AND ADULTS	SIGNIFICANCE OF DIFFERENCE
Peak frequency	3 mos: 60% higher in infants 18 mos: no difference	3 mo. calls sound distinctly higher
Location of peak	no significant difference	pitch contour is distinct in repertoire at all ages
Min. frequency	3 mos: twice as high in infants 18 mos: approaching adults'	infants lack the rough-sounding, noisy, low-frequency cackle element at end of call
Location of min.	infant minimum is at end of call, adults' is somewhat earlier	difference is due to lack of cackle in infants
Frequency range	3 mos: slightly higher in infants 18 mos: slightly lower in infants	3 mos: due to higher peak 18 mos: mainly due to drop in peak
Duration	infants about 40% shorter	infants have shorter flag and lack repeating elements
Frequency jitter	no significant difference	calls are strong and crisp at all ages
Frequency leaps	30-60% fewer in infants	mainly due to lack of flag

Because chuck calls show both structural and functional evidence of individual signature in several measurable parameters, it is of interest to assess possible individuality in infantile masted calls, and, lacking that, any evidence of increasing stereotypy over time which might indicate movement toward individuality with age. However, the data were disappointing in this respect. There was little evidence of individual differences, even in such parameters as peak frequency and duration, which are highly individual in adult chucks (Biben et al., 1986). With regard to stereotypy, significant trends in the data were few, erratic, and generally in the "wrong" direction, that is, they indicated increases in variability. It is probably safe to say that the production of individually identifiable chuck calls is not complete at 18 months.

Peep calls, on the other hand, showed considerable evidence of individuality from an early age and in many acoustic parameters. Individual signature appears early and remains stable in at least one type of peep call, the isolation peep (Symmes et al., 1979; Symmes and Biben, 1985), but most peeps in our samples would not be classified as isolation peeps. The shorter peeps used most often by youngsters are uttered either responsively with the calls of adults, or by infants venturing away from their mothers. Both of these circumstances would be well-served by individuality allowing mothers to recognize their infants, and it is apparently present in these short peeps. As with masted calls, there were no significant age-related changes in the stereotypy of peeps calls, perhaps reflecting the fact that individuality was present at all ages. Lack of change in stereotypy may also be taken to indicate that the form of peep calls (and of masted calls) did not change appreciably over time.

Usage of peeps also diverged greatly from that of adults. This may, in part, be attributed to the different physical and social circumstances of infants and adults. Youngsters were as much as ten times noisier than adults, and their output heavily favored peeps. Adults emit short peeps while exploring, short steeply sloped peeps when alarmed, and longer isolation peeps when separated (Newman, 1985). These contexts were associated with youngsters' peeps as well, but because youngsters explored more, were more readily alarmed, and more likely to experience separation, they used peep calls much more frequently than did adults. One usage of peeps by infants was clearly different from adults. Infants used peeps responsively in vocal exchanges with mothers and allomothers (Biben, 1992). Very young infants responded to caregiver calls, mainly from allomothers, by uttering one or two brief, soft peeps, while 3- to 12-month olds prompted and/or responded to caregiver calls, mainly from mothers, with strings of 2-5 brief but loud peeps.

Usage of masted calls was more difficult to categorize. The earliest masted calls were brief, with very steep slopes, resembling the "tucks" first described by Winter (1968) and considered by Newman (1985) to be an infantile form of the chuck. Winter ascribed a "searching" function to tucks and found them in association with peeps. Here, we found exploration to be an infrequent context of masted calls. Half of masted calls were emitted by infants sitting alone. Most of the remaining half of calls were made by animals interacting physically or, most often, just vocally. This scenario of usage is not unlike that described for the adult usage of chucks, found by Smith et al. (1982) to be generally affiliative but specifically independent of any particular affiliative or nonaffiliative response context.

The exchange of chuck calls, which occurs at rates over 40%, is the most common and important affiliative communication between adult female squirrel monkeys (Biben et al., 1986; Biben and Symmes, 1991; Smith et al., 1982). Infants' responsive use of masted calls, accounting for about 37% of utterances throughout the first 18 months of life, indicates a fairly strong responsive role for masted calls right from the start. In adults, the most frequent response to a chuck is another chuck, accounting for better than 40% of chucks uttered (Biben et al., 1986; Biben and Symmes, 1991). However, infant-infant or adult-infant exchanges involving two masted calls were not well-represented in the sample. Even at 18 months, where masted calls had become quite chuck-like in structure, such exchanges accounted for only about 20% of masted calls uttered, even though the youngsters had been together all their lives and had good affiliative bonds. Thus, while the affiliative context was present, the rate of appropriate vocal exchange remained below expected levels, and it must be noted that no other vocalization took their place.

It is clear from these data that both production and usage of masted calls and peeps undergo many types of change as squirrel monkey infants develop. Some of these may build individuality into a call; others may encode membership in a small group of affiliated animals. It is these latter that are most likely to be learned. Even at 18 months, chuck calls could not be considered mature, allowing ample time for infants to acquire group-specific characteristics such as those demonstrated in Biben and Bernhards (1994). While an open system such as this affords many opportunities for learning, specific demonstrations of learning are as yet out of reach. The extent of variability at an early age, demonstrated here by an unusually detailed analysis, is, in all probability, not unique to Saimiri. Shall we view this variability as mere noise in the system, or as the raw material for a determinative shaping and honing into the adult repertoire? Although the chasm between human and nonhuman primates remains large, we must be open to explanations for which learning may be the most reasonable accounting. All the same, we must choose our subjects wisely. Most primate calls under study were chosen for their obvious survival value or identified referrents. We now recommend greater scrutiny of close-range affiliative communications which, like chucks, show considerable flexibility and individuality on many levels.

A second deterrent to research is primarily methodological. Our best tool, the deprivation experiment in its many forms, fails here as it does in most questions involving social behavior. Another potentially powerful approach, cross-fostering, has had only limited, and equivocal, success (Masataka and Fujita, 1989; Owren et al., 1992). It is time to try novel, if indirect, approaches, aimed at alterating acoustic input to the infant. These could include raising infants with their natural mothers (muted or not) but in social groups with dialect differences from the mother, or in an environment of artificially altered sound input (extreme reverberation or masking).

A 10 year series of playback studies performed in our laboratory has culminated in an example of noninvasive alteration of acoustic input. Like most social mammals, squirrel monkeys respond differentially to the calls of different individuals. Playback studies demonstrate that differential responses can be elicited to the isolated calls of individuals (maternal recognition of infant isolation peep: Symmes and Biben, 1985), to group members versus outsiders (chuck calls: Symmes and Biben, 1992), and even to intraindividual differences related to chuck usage (Biben, 1993). Comparable usage differences were present in the calls of both familiar animals and strangers but only the former elicited a differential response. Group-based differences are most expediently attributed to genetic commonalities; individual differences may also have a genetic basis, or they may have a variety of epigenetic origins. Usage differences, if viewed as linguistic conventions, are more difficult to explain as other than learned phenomena. Squirrel monkeys normally grow up in groups containing many relatives but, in the laboratory, groups of unrelated females do well. In an analysis of the mature chuck calls of daughters raised with their mothers and unrelated females, daughters' chucks in some cases showed a greater resemblance to those of an unrelated female than to those of the mother (Biben and Bernhards, 1994). In Fig. 1, T2B is the daughter of T2, but her chuck resembles more closely that of female T8 based on discriminant analysis using 3 major acoustic variables from 10 chucks of T8 and 15 from T2 (posterior probability scores of 0.54 for T2B and T8, 0.002 for T2B and T2). DNA fingerprinting analysis showed that, although T2B and T8 had been members of the same captive group, they were not closely related (% of shared genetic material = 0.42 for T2B and T8, 0.42 ± 0.15 SD for T2B and 8 known nonrelatives, and 0.74 ± 0.10 SD for T2B and her parents). Such a finding is a tantalizing suggestion that learning may in some way contribute to the resemblance. If so, learning would have had to take place within the first year of life, as offspring were removed from the natal group by 12 months of age - well before the chuck matures. The mechanism for information transfer may be the early vocal allomothering that is common in this species (Biben, 1992), allowing infants lots of exposure to the calls of group members other than their mothers.

Where do we go from here? Our suggestions are four:

1. Start from an assumption of openness.
2. Choose an animal subject not for reasons of phylogeny or intelligence; rather, based on the quality of its communication.
3. Choose a vocalization based on its flexibility of usage rather than ease of identifying a referrent.
4. Look for novel and indirect ways of detecting where modification may have entered the system.

ACKNOWLEDGMENTS

We thank D. Pavon and R. Tumlirova for assistance in data collection, and W. Gergits for DNA analysis.

REFERENCES

Biben, M., 1986, Individual and sex-related strategies of play wrestling in squirrel monkeys, *Ethology* 71:229-242.

Biben, M., 1992, Allomaternal vocal behavior in squirrel monkeys, *Develop. Psychobiol.* 25:79-92.

Biben, M., 1993, Recognition of order effects in squirrel monkey antiphonal call sequences, *Amer. J. Primatol.* 29:109-124.

Biben, M. and Bernhards, D., 1994, Naive recognition of chuck calls in squirrel monkeys (*Saimiri sciureus macrodon*), *Lang. Commun.* 14:167-181.

Biben, M. and Symmes, D., 1986, Play vocalizations of squirrel monkeys, *Folia Primatol.* 46: 173-182.

Biben, M. and Symmes, D., 1991, Affiliative vocalizing in squirrel monkeys: Familiarity as a cue to playback response, *Behaviour* 117:1-19.

Biben, M., Symmes, D., and Bernhards, D., 1989, Contour variables in vocal communication between squirrel monkey mothers and infants, *Develop. Psychobiol.* 22:617-631.

Biben, M., Symmes, D., and Masataka, N., 1986, Temporal and structural analysis of affiliative vocal exchanges in squirrel monkeys (*Saimiri sciureus*), *Behaviour* 98:259-273.

Boinski, S., 1991, The coordination of spatial position: a field study of the vocal behavior of adult female squirrel monkeys, *Anim. Behav.* 41: 89-102.

Boinski, S. and Mitchell, C., 1992, Ecological and social factors affecting the vocal behavior of adult female squirrel monkeys, *Ethology* 92: 316-330.

Elowson, A. M., Snowdon, C. T., and Sweet, C. J., 1992, Ontogeny of trill and J-call vocalizations in the pygmy marmoset, *Cebuella pygmaea*, *Anim. Behav.* 43: 703-715.

Goedeking, P. and Immelmann, K., 1986, Vocal cues in cotton-top tamarin play vocalizations, *Ethology* 74: 219-224.

Gouzoules, H. and Gouzoules, S., 1989, Design features and developmental modification of pigtail macaque, *Macaca nemestrina*, agonistic screams, *Anim. Behav.*37: 383-401.

Hauser, M. D., 1989, Ontogenetic changes in the comprehension and production of vervet monkey (*Cercopithecus aethiops*) vocalizations, *J. Comp. Psychol.* 103: 149-158.

Herzog, M. and Hopf, S., 1983, Effects of species-specific vocalizations on the behaviour of surrogate-reared squirrel monkeys, *Behaviour* 86: 197-214.

Hopf, S., 1970, Report on a hand-reared squirrel monkey (*Saimiri sciureus*), *Zeitschr. Tierpsychol.* 27: 610-621.

Lieblich, A. K., Symmes, D., Newman, J. D., and Shapiro, M., 1980, Development of the isolation peep in laboratory-bred squirrel monkeys, *Anim. Behav.* 28: 1-9.

Masataka, N. and Biben, M., 1987, Temporal rules regulating vocal exchanges of squirrel monkeys, *Behaviour* 101: 311-319.

Masataka, N. and Fujita, K., 1989, Vocal learning of Japanese and rhesus monkeys, *Behaviour* 109: 191-199.

Maurus, M., Kühlmorgen, B., Wiesner, E., Barclay, D., and Streit, K.-M., 1985, "Dialogues" between squirrel monkeys, *Lang. Commun.* 5: 185-191.

Maurus, M., Streit, K.-M., Barclay, D., Wiesner, E., and Kuehlmorgen, B., 1986, Interrelations between structure and function in the vocal repertoire of *Saimiri*; asking the monkeys themselves where to split and where to lump, *Europ. Arch. Psychiat. Neurol. Sci.* 236: 35-39.

Newman, J. D., 1985, Squirrel monkey communication, *in*: "Handbook of Squirrel Monkey Research", L. A.Rosenblum and C. L. Coe, eds., Plenum, New York.

Newman, J. D. and Symmes, D., 1982, Inheritance and experience in the acquisition of primate acoustic behavior, *in*: " Primate Communication", C. T. Snowdon, C. H. Brown, and M. R. Petersen, eds., Cambridge Univ. Press, Cambridge.

Newman, J. D., Smith, H. J.,and Talmage-Riggs, G., 1983, Structural variability in primate vocalizations and its functional significance: an analysis of squirrel monkey chuck calls, *Folia Primatol.* 40: 114-124.

Owren, M., Dieter, J., Seyfarth, R., and Cheney, D., 1992, 'Food' calls produced by adult female rhesus (*Macaca mulatta*) and Japanese (*M. fuscata*) macaques, their normally-raised offspring, and offspring cross-fostered between species, *Behaviour* 120: 218-231.

Papousek, M. and Papousek, H., 1981, Musical elements in the infant's vocalization: their significance for communication, cognition, and creativity, *in*: "Advances in Infancy Research", Vol. 1, L. P. Lipsett and C. Rovee-Collier, eds., Ablex, Norwood, N. J.

Ploog, D., Hopf, S., and Winter, P., 1967, Ontogenese des Verhaltens von Totenkopfaffen (*Saimiri sciureus*), *Psychol. Forsch.* 31: 1-41.

Schott, D., 1975, Quantitative analysis of the vocal repertoire of squirrel monkeys (*Saimiri sciureus*), *Zeitschr. Tierpsychol.* 38: 225-250.

Seyfarth, R. M. and Cheney, D., 1986. Vocal development in vervet monkeys, *Anim. Behav.* 34: 1640-1658.

Smith, H. J., Newman, J. D., Hoffman, H., and Fetterly, K., 1982, Statistical discrimination among vocalizations of individual squirrel monkeys (*Saimiri sciureus*), *Folia Primatol.* 37: 267-279.

Smith, H. J., Newman, J. D., and Symmes, D., 1982, Vocal concomitants of affiliative behavior in squirrel monkeys, *in*: " Primate Communication", C. T. Snowdon, C. H. Brown, and M. R. Petersen, eds., Cambridge Univ. Press, Cambridge.

Snowdon, C. T., 1988, Communications as social interaction: its importance in ontogeny and adult behavior, *in*: "Primate Vocal Communication", D. Todt, P. Goedeking, and D. Symmes, eds., Springer-Verlag, Berlin.

Stark, R. E., 1978, Features of infant sounds: the emergence of cooing, *J. Child Lang.* 5:379-390.

Stark, R. E., Rose, S. N., and McLagan, M., 1975, Features of infant sounds: the first eight weeks of life, *J. Child Lang.* 2:205-221.

Symmes, D. and Biben, M., 1985, Maternal recognition of individual infant squirrel monkeys from isolation call playback, *Amer. J. Primatol.* 9:39-46.

Symmes, D. and Biben, M., 1988, Conversational vocal exchanges in squirrel monkeys, *in*: "Primate Vocal Communication", D. Todt, P. Goedeking, and D. Symmes, eds., Springer-Verlag, Berlin.

Symmes, D. and Biben, M., 1992, Vocal development in nonhuman primates, *in*: "Nonverbal Vocal Communication: Comparative and Developmental Approaches", H. Papousek, U. Jürgens, and M. Papousek, eds., Cambridge Univ. Press, New York.

Symmes, D., Newman, J. D., Riggs, G., and Lieblich, A. K., 1979, Individuality and stability of isolation peeps in squirrel monkeys, *Anim. Behav.* 27: 1142-1152.

Winter, P., 1968, Social communication in the squirrel monkey, *in*: "The Squirrel Monkey", L. A. Rosenblum and R. W. Cooper, eds., Academic Press, New York.

Winter, P., Handley, P., Ploog, D., and Schott, D., 1973, Ontogeny of squirrel monkey calls under normal conditions and under acoustic isolation, *Behaviour* 47: 230-239.

Winter, P., Ploog, D., and Latta, J., 1966, Vocal repertoire of the squirrel monkey (*Saimiri sciureus*), its analysis and significance, *Exp. Brain Res.* 1: 359-384.

TEMPORAL AND ACOUSTIC FLEXIBILITY IN VOCAL EXCHANGES OF COO CALLS IN JAPANESE MACAQUES (*Macaca fuscata*)

Hideki Sugiura and Nobuo Masataka

Department of Anthropology
Faculty of Science
The University of Tokyo
Hongo, Bunkyo-ku, Tokyo 113, Japan

INTRODUCTION

A central issue in the study of primate vocalizations concerns the extent of control over vocal production. Information on the ontogeny of vocal communication of non-human primates is limited and conflicting. Until recently, our primary data were from squirrel monkeys (*Saimiri sciureus*), focusing on infant isolation peeps. The major finding was that call structures were inherited, not learned (Newman and Symmes, 1982). Failure to find evidence of learning is somewhat surprising, given the influence of learning on other behaviors, and the fact that there are some clear parallels in the development of bird song and the development of human speech (Marler, 1970).

The study of development of vocal production requires research on two interrelated areas: how species-specific calls develop and how animals come to temporally modify acoustic features of a given specific type of vocalization in particular circumstances. To answer the first question, Masataka and Fujita (1989) showed through cross-fostering experiments between Japanese macaques (*Macaca fuscata*) and rhesus macaques (*Macaca mulatta*) that production of one species-specific call (i. e., coo) is learned by each species. In their study, cross-fostering was initiated within 24 hours after the birth of subjects. The monkeys were tested when one year old since the vocalization chosen for study, labeled "coo calls" by Green (1975), are not made by Japanese or rhesus macaques until they were close to this age. Coo calls of Japanese and rhesus macaque infants reared by their biological mothers differed from each other in a single acoustic parameter. The calls of the Japanese macaque infant fostered by rhesus female at the age of 12 months differed from those of conspecifics reared by their biological mothers. Similarly, the vocalizations given by the two rhesus macaques fostered by Japanese macaque females were dissimilar to those of rhesus macaques reared by their biological mothers, but similar to those of Japanese macaques reared by their biological mothers. However, more recently, Owren et al. (1992) provide no evidence of vocal learning.

Concerning the second question, discriminative conditioning studies were conducted, as a first step to examine whether there was a voluntary component in non-human primate vocalizations. In such experiments, animals must learn to emit a call of given amplitude and duration in response to an arbitrary stimulus and withhold vocalizations in response to a second arbitrary stimulus in order to get food and, in some instances, to avoid punishment (Pierce, 1985). While some of the attempts to condition vocalizations failed (Myers, 1976), others, using the same species (rhesus macaques) have found that such discriminative vocal behavior is no more difficult to obtain under these training conditions than is a manual lever press response (Sutton et al., 1981).

Additional evidence has been presented from a series of studies of squirrel monkeys living in captive groups. Under naturalistic circumstances, animals of this species live in large social groups and inhabit forests with dense vegetation. To maintain contact with group members, individuals utter a variety of vocalizations. One of these vocalizations, the "chuck", was studied intensively to determine whether there are "question" and "answer" chucks which are used to gain information regarding the dynamics of particular social relationships. In this species, mutually preferred partners exchanged chucks more often and responded with shorter latencies than other pairs. The acoustic features of the vocalizations that followed other vocalizations closely were more similar than those of vocalizations that followed a period of silence (Biben et al., 1986). Masataka and Biben (1987) reported the presence of temporal "rules" that appeared to regulate vocal exchanges. If an animal responded to a call given by a group member, it did so within 0.5 sec after the sound was heard. If an animal uttered a call independently of the preceding call, the interval was greater than 0.5 sec. This temporal rule allowed the animals to determine whether the second vocalization was a response to the first or not, when two calls were heard consecutively. Squirrel monkeys therefore appear to use temporal "rules" to guide their vocal exchanges.

Research presented below has been conducted to extend this previous study of temporal flexibility of vocal behavior, with respect to the coo call of Japanese macaques (*Macaca fuscata*). Acoustically, the call has a basically tonal structure and a high degree of variability in the structure (Green, 1975). In this species, most members of a troop utter these vocalizations in calm or relaxed situations. In particular, in free-ranging situations, the calls are emitted most often during foraging and group progression, but regardless of any other ongoing activity (Itani, 1963). Animals appear to maintain within-group contact vocally, when visual contact is difficult to achieve. They also appear to recognize individuality of the vocalizations. Indeed, Pereira (1986) reported that Japanese macaque mothers were able to distinguish coos of their own juvenile from those of the other juveniles. Moreover, coos are often emitted consecutively by two or more group members. Mitani (1986) studied their "exchanges" and described a "network" of vocal exchange in a free-ranging troop. The vocalizations were found to be exchanged most frequently among females of the same kinship.

In this paper, the timing and acoustic properties of coo call exchanges are described and analyzed. It seems likely that often Japanese macaques respond vocally to a group member if two coos of different individuals follow each other with a short interval. Therefore, the temporal patterns of occurrence of consecutive coo calls were studied to determine whether or not following coos could be considered a response to preceding coos. We also examined whether the acoustic features of following coos were related to those of preceding coos. The observations suggested that part of the coos are uttered in response to preceding coos and that in those cases the acoustic features of following coos were similar to those of preceding coos. Therefore, a playback experiment with a group was performed to eliminate unobserved factors which could affect the result and to confirm the phenomenon experimentally. Moreover, a playback experiment with individuals was undertaken, to control the stimulus coo and response coo individually.

NATURALISTIC OBSERVATIONS OF VOCAL EXCHANGE OF COO CALLS

Methods

Subject Groups. Since 1990, we have been undertaking extensive vocal recordings and observations of behavior in two groups of Japanese macaques (Sugiura, 1993): the Yakushima P-group and the Ohirayama group. The two have been geographically separated (>700 km) and without any contact for the past 33 years. Animals of the Yakushima P-group range over approximately 40 ha of mountain forest of Yakushima Island, south of Kyushu. The group has been studied since 1973 without provisioning (Maruhashi, 1980). Age-sex structure of the two groups are shown in Table 1. During the study period, the group consisted of 5 adult females (≥6 yrs), 2 juvenile female (3-5yrs), 1 immature female (0-2 yrs), 6 adult males (≥6 yrs), 3 juvenile males (3-5 yrs), and 1 immature male (0-2 yrs).

The Ohirayama group was originally a translocated group. The original members of the group were captured together in 1957 on Yakushima Island and immediately flown to Mt. Ohirayama, Aichi Prefecture, located in the Japanese mainland. Thereafter, they have been provisioned under semifree-ranging conditions. A more detailed description of the group and its habitat was published by Kawai (1960). The group was made up of 15 adult females (≥6 yrs), 5 juvenile females (3-5 yrs), 9 immature females (0-2 yrs), 11 adult males (≥6 yrs), 9 juvenile males (3-5 yrs), and 15 immature males (0-2 yrs) during the observation period.

Table 1. Structure of the two groups by age-sex class. Number of target animals chosen are shown in parentheses. f: female; m: male.

	Yakushima				Ohirayama		
Class	Age	N	(Target)	Class	Age	N	(Target)
Adult f	≥6	5	(3)	Adult f	≥6	15	(4)
Juvenile f	3-5	2	(2)	Juvenile f	3-5	5	(4)
Immature f	0-2	1	(0)	Immature f	0-2	9	(0)
Adult m	≥6	6	(0)	Adult m	≥6	11	(0)
Juvenile m	3-5	3	(0)	Juvenile m	3-5	9	(0)
Immature m	0-2	1	(0)	Immature m	0-2	15	(0)
Total		18	(5)			64	(8)

Collection and Analysis of Data. Data presented here were drawn from observations of the Yakushima P-group conducted between August and October, 1990 (62 days), and those of the Ohirayama group between January and February, 1991 (49 days). For recording of data, we chose 5 females from the Yakushima P-group and 8 females from the Ohirayama group as target animals (all were 3 years or older, see Table 1). Mean ages were approximately 10 years (range: 3 to approximately 20 yrs) for the target females in the Yakushima P-group and 8.3 years (range: 3 to 17 yrs) for those in the Ohirayama group. The ages did not differ significantly between the two populations (U-test, U=17.5, N1=5, N2=8, p>0.72).

We collected data by the focal animal sampling method (Altmann, 1974), with a given observation session lasting for 60 min. In an observation session, the observer stood near the focal animal with a hand-held microphone (Sony ECM-672) aimed at it, and he recorded

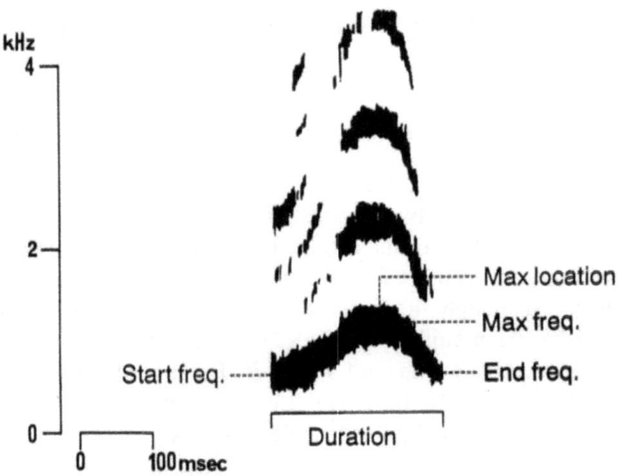

Figure 1. A representative sonagram of a Japanese macaque coo call, in which acoustic parameters measured for acoustic analyses are indicated.

all the vocalizations (target and others), identities of callers and behavioral correlates with a tape recorder (Sony WM-D6C). The observations were conducted when most of the group members were located within visible or audible range (1 to approx. 25 m). If an animal that emitted a vocalization could not be identified, the caller was treated as "unidentified" for the subsequent analysis. A total of 23 and 84 observation sessions were conducted with the Yakushima P-group and the Ohirayama group, respectively. Since, in general, the frequency of vocal emission varied considerably among individuals, data collection method was designed to provide a similar number of vocalizations for each individual per group. The observation sessions were distributed almost equally over the time between 8.00 and 17.00 h in the case of the Yakushima P-group and between 9.30 and 15.30 h in the case of the Ohirayama group. From this corpus of data, we extracted 5-min continuous recording segments where vocal exchange were successfully recorded, then chose 43 and 42 segments at random for spectrographic analysis for the Yakushima P-group and the Ohirayama group, respectively. All the vocalizations included in the segment were analyzed, using a Kay DSP Sonagraph 5500 with a wide band filter (128 pts: 234 Hz), and a frequency scale of 0-8000 Hz. Time and frequency parameters were measured with an accuracy of +/- 0.0032 sec and +/- 40 Hz, respectively.

We defined the coo call according to the classification by Green (1975). The calls are basically tonal and their F_0 elements are spectrographically distinct and are usually the most dominant frequency components (Fig. 1). Therefore, for each call, the duration, location of the maximum frequency in relation to the entire length of F_0 elements, start frequency, end frequency, minimum frequency, maximum frequency, and frequency ranges (maximum minus minimum frequency, maximum minus start frequency and maximum minus end frequency) were measured. Green divided coos into 7 "subtypes" from its acoustic structure. For example, he categorized coo calls discretely by duration of call ($\leq$0.19 sec and $\geq$0.20 sec) or maximum frequency location (<2/3 and $\geq$2/3). However, from our data, we could not find any discrete criteria for all parameters we measured. Therefore, we did not divide coos into subtypes but analyzed acoustic features of coos quantitatively. The contexts in which coos were recorded were feeding (69% and 30 %), moving (16% and 4%), grooming (1% and 32%), or resting (14% and 34%) for the Yakushima P-group and the Ohirayama groups, respectively. When all the vocalizations were categorized, more than 85% of the total vocal sample consisted of coos. Because of their low incidence, call types other than the coos were excluded from the subsequent analysis. Whenever two coos were uttered consecutively, being uninterrupted by other call types, the inter-call interval was defined as the time from the end of a call to the beginning of the next. If two or more coos occurred concurrently (e.g., the second coo overlapped the first), the inter-call interval was measured between the end of the *n*th call and the beginning of the (n+1)th call and given a negative value. When more than two coo calls occurred consecutively, we measured the interval between the first and the second call, and then between the second and the third. For example, if three calls occurred consecutively like A → B → C, we measured the interval between A and B, and that between B and C. In such case, B was treated as the second call to A, and was also treated as the first call to C.

Results

Sequences of two consecutive coos were classified into two categories as follows:
(1) sequence of two consecutive coos where the second call was uttered by a different caller from the first (*DC* sequence) and
(2) sequence of two consecutive coos uttered by the same caller (*SC* sequence).
A total of 244 and 140 intervals in *DC* sequence were analyzed for the Yakushima P-group and the Ohirayama group, respectively. Inter-call interval in *DC* sequence can be grouped (1) according to the first caller or (2) according to the second caller. Distribution of intervals of *DC* sequence grouped by the first caller are shown in Table 2. Median interval of *DC* sequence grouped by the first caller represents an interval between the coo of a target and the following coo by a group member when two coos were uttered consecutively. Table 3 shows the distribution of intervals grouped by the second caller. Median interval of *DC* sequence grouped by the second caller represents an interval between the coo of a group member and the following coo by a target animal. A total of 141 and 74 intervals in *SC* sequence were analyzed for the Yakushima P-group and the Ohirayama group, respectively. Distribution of intervals in *SC* sequences are shown in Table 4. These values mean the intervals between two coos by the same animal, not interrupted by another group member.
Among the 244 and 140 intervals in *DC* sequence, 182 (74.5 %) and 124 (86.1 %) second coos occurred between 0-4 sec in the Yakushima P-group and the Ohirayama group, respectively. Of the 141 and 74 intervals in *SC* sequence, 83 (58.8 %) and 64 (86.4 %) intervals were shorter than 4 sec in the Yakushima P-group and the Ohirayama group, respectively. Median intervals in *DC* and *SC* sequence between 0-4 sec were calculated by target females, to compare the duration of intervals in *DC* sequence with those in *SC* sequence individually. For each of 13 target animals in the two populations, the median intervals in *SC* sequence were plotted against the median intervals in *DC* sequence, grouped according to the first caller (Fig. 2) and against those in *DC* sequence, grouped according to the second caller (Fig. 3). Compared with the median intervals in *DC* sequence where the first callers were target animals, those in SC sequence were statistically significantly longer for all individuals (Wilcoxon's signed ranks test: $N=13$, $p<0.01$). Also, the median intervals in SC sequence were longer than those in *DC* sequence where the second callers were target animals for all individuals (Wilcoxon's signed ranks test: $N=13$, $p<0.01$). Because there were no significant individual differences in the intervals in *DC* sequence or in *SC* sequence within each population [Kruskal-Wallis test: the Yakushima P-group, $H=2.36$, $N=5$, $p=0.67$, (*DC* sequence where the first callers were target animals), $H=2.36$ $N=5$, $p=0.67$, (DC sequence where the second callers were target animals), $H=4.41$, $N=5$, $p=0.35$, (SC sequence); the Ohirayama group, $H=3.71$, $N=8$, $p=0.82$, (DC sequence where the first callers were target animals), $H=4.43$, $N=8$, $p=0.73$, (DC sequence where the second callers were target animals), $H=10.5$, $N=8$, $p=0.16$, (SC sequence)], pooling the data seems justified.

Table 2. Distribution of median inter-call intervals in *DC* sequence grouped by the first caller.

	Yakushima				Ohirayama		
1st caller	(Age)	Median (sec)	(N)	1st caller	(Age)	Median (sec)	(N)
Hp	(>15)	0.416	(56)	Bt	(17)	0.428	(12)
Ma	(>15)	0.377	(18)	Nk	(15)	0.453	(13)
On	(6)	0.569	(13)	Tr	(9)	0.466	(7)
Ni	(5)	0.534	(26)	Bb	(8)	0.534	(16)
Ti	(3)	0.512	(39)	Bk	(5)	0.550	(18)
				Nc	(5)	0.242	(6)
				Tn	(4)	0.509	(11)
				Ng	(3)	0.512	(9)
Other than target		0.661	(18)	Other than target		0.186	(11)
Unidentified		0.431	(74)	Unidentified		0.500	(37)
Overall		0.478	(244)	Overall		0.514	(140)

Table 3. Distribution of median inter-call intervals in *DC* sequence grouped by the second caller.

	Yakushima				Ohirayama		
2nd caller	(Age)	Median (sec)	(N)	2nd caller	(Age)	Median (sec)	(N)
Hp	(>15)	0.488	(54)	Bt	(17)	0.556	(12)
Ma	(>15)	0.947	(17)	Nk	(15)	0.516	(13)
On	(6)	0.678	(11)	Tr	(9)	0.397	(9)
Ni	(5)	0.494	(32)	Bb	(8)	0.539	(12)
Ti	(3)	0.373	(40)	Bk	(5)	0.312	(13)
				Nc	(5)	0.456	(5)
				Tn	(4)	0.594	(11)
				Ng	(3)	0.402	(10)
Other than target		0.569	(19)	Other than target		0.847	(10)
Unidentified		0.387	(71)	Unidentified		0.500	(45)
Overall		0.478	(244)	Overall		0.514	(140)

Table 4. Distribution of median inter-call intervals in *SC* sequence.

	Yakushima				Ohirayama		
Caller	(Age)	Median (sec)	(N)	Caller	(Age)	Median (sec)	(N)
Hp	(>15)	9.250	(27)	Bt	(17)	0.812	(7)
Ma	(>15)	1.204	(30)	Nk	(15)	1.420	(6)
On	(6)	12.000	(26)	Tr	(9)	1.107	(6)
Ni	(5)	1.825	(21)	Bb	(8)	1.278	(7)
Ti	(3)	2.252	(20)	Bk	(5)	1.539	(14)
				Nc	(5)	1.086	(6)
				Tn	(4)	0.688	(6)
				Ng	(3)	1.023	(10)
Other than target		1.603	(13)	Other than target		1.181	(7)
Unidentified		1.462	(4)	Unidentified		0.809	(5)
Overall		1.744	(141)	Overall		1.139	(74)

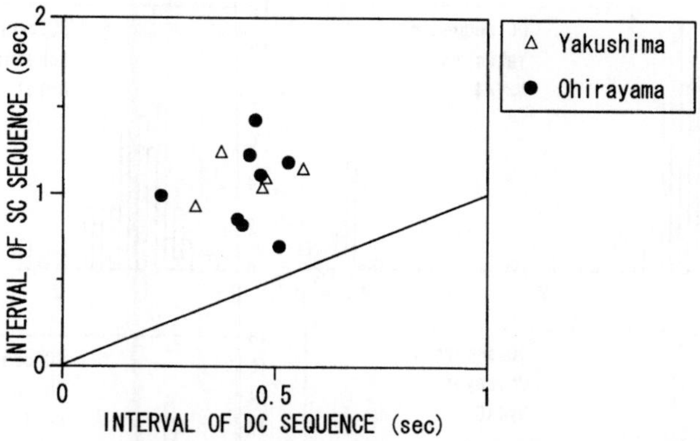

Figure 2. Distribution of 13 target animals with regard to median intervals in *DC* sequence between 0-4sec, grouped according to the first caller (target individuals — group members), and median intervals in *SC* sequence between 0-4 sec. If an animal is plotted on the line, it would mean that mean intervals in *DC* sequence and in *SC* sequence are the same for the animal.

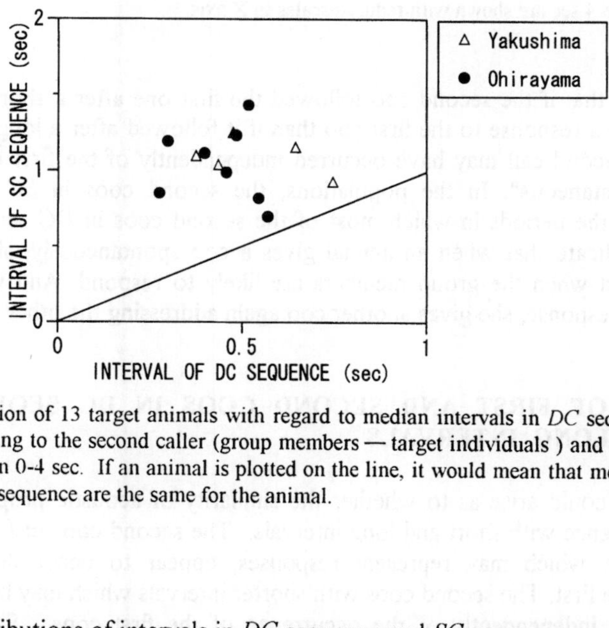

Figure 3. Distribution of 13 target animals with regard to median intervals in *DC* sequence between 0-4 sec, grouped according to the second caller (group members — target individuals) and median intervals in SC sequence between 0-4 sec. If an animal is plotted on the line, it would mean that mean intervals in *DC* sequence and in *SC* sequence are the same for the animal.

Overall distributions of intervals in *DC* sequence and *SC* sequence are shown in Fig. 4. Intervals of DC sequence were significantly shorter than those of *SC* sequence in the two populations (Mann-Whitney U-test: Yakushima P-group, N1=182, N2=83, U=1923, $p<0.001$; the Ohirayama group, N1=124, N2=64, U=1471, $p<0.001$). Median intervals of *SC* sequences between 0 and 4 sec were 1.1 sec and 1.0 sec in the Yakushima P-group and the Ohirayama group, respectively. In the Yakushima P-group, 88 % of second coos in *DC* sequence that were separated by intervals of 0-4 sec were distributed between 0 sec and 1.1 sec; the median intervals of *SC* sequence. While the 87 % of second coos were distributed between 0 and 1.0 sec (median interval of *SC* sequence) in the Ohirayama group.

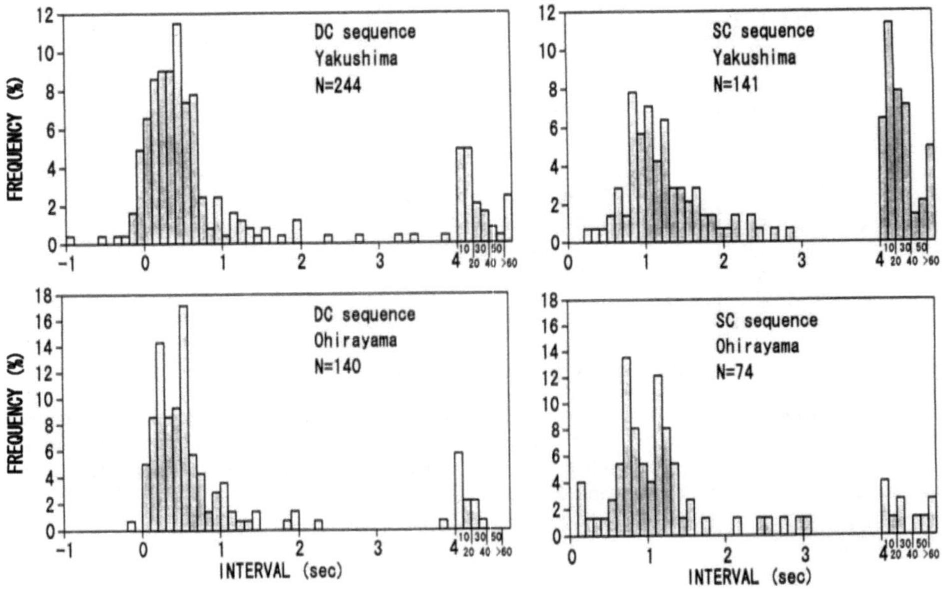

Figure 4. Overall distributions of intervals in *DC* sequences and *SC* sequences in the two populations. Intervals longer than 4 sec are shown with reduced scales in X axis.

We assume that if the second coo followed the first one after a short interval, it was more likely to be a response to the first coo than if it followed after a long interval. In the latter case the second call may have occurred independently of the first call, and thus be considered "spontaneous". In the populations, the second coos in *SC* sequence rarely occurred during the periods in which most of the second coos in *DC* sequence occurred. These results indicate that when an animal gives a coo spontaneously, she remains silent during the period when the group members are likely to respond. And when no animals make any vocal response, she gives another coo again addressing the other animals.

SIMILARITY OF FIRST AND SECOND COOS IN DC SEQUENCE WITH SHORT AND LONG INTERVALS

A question could arise as to whether the similarity of acoustic properties differs for calls of *DC* sequence with short and long intervals. The second coos in *DC* sequence with shorter intervals, which may represent responses, appear to occur depending on the occurrence of the first. The second coos with shorter intervals which may be "spontaneous", appear to occur independently of the occurrence of the first coos. Similarly, acoustic features of the second coos given by different callers with short intervals might be dependent on those of the first, second coos with long intervals might acoustically independent of the first coos. For example, one might hypothesize greater similarity between coos following short intervals than coos following longer intervals. Therefore, a correlation analysis was performed between the acoustic features of two consecutive coos following each other by different intervals. As the median intervals of *SC* sequence between 0-4 sec were 1.1 sec, when the data of the Yakushima P-group and that of the Ohirayama group were pooled, the entire data set for each population was divided into two subsets according to the following criteria:

(1) two consecutive coos that occurred at intervals of 0-1.1 sec and

Table 5. Results of single classification analyses of variance performed on the nine acoustic parameters of coos of target animals in each population recorded in naturalistic observations.

Parameter	Yakushima F(df=4)	Ohirayama F(df=7)
Duration	17.0***	5.9***
Max location	16.6***	6.1***
Min freq.	201.3***	47.9***
Start freq.	152.2***	53.7***
Max freq.	28.6***	7.9***
End freq.	35.4***	8.2***
Max-min freq.	17.0***	5.6***
Max-start freq.	18.7***	5.8***
Max-end freq.	9.7***	2.0

***:p<0.001

Table 6. Distribution of second coos in *DC* sequence by caller.

Yakushima				Ohirayama			
		Interval				Interval	
Caller	(Age)	0-1.1 sec	>1.1 sec	Caller	(Age)	0-1.1sec	>1.1 sec
Hp	(>15)	37	13	Bt	(17)	9	3
Ma	(>15)	9	8	Nk	(15)	12	1
On	(6)	6	4	Tr	(9)	9	0
Ni	(5)	26	5	Bb	(8)	10	2
Ti	(3)	26	10	Bk	(5)	10	3
				Nc	(5)	5	0
				Tn	(4)	9	2
				Ng	(3)	9	1
Other than target		11	7	Other than target		7	3
Unidentified		46	16	Unidentified		33	11
Overall		161	63	Overall		113	26

(2) two consecutive coos that occurred at intervals longer than 1.1 sec.

Prior to the correlation analysis, we examined whether interindividual differences were present in the acoustic parameters of target animals or not. As shown in Table 5, single classification analyses of variance revealed that there were significant individual differences for all parameters except one: the maximum minus end frequency in the Ohirayama group. To exclude the effect of these interindividual differences, data for the second callers were standardized by transformation to the z score according to the formula: $z=(x-x_i)/s_i$ (x, raw value; x_i, mean of individual i; s_i, standard deviation of individual i). The z score represents the relative deviation from the mean value for each individual. We carried out z score conversion only for the data obtained when the second caller was a target animal and when the standard deviation was larger than zero. The data obtained when the second callers were other than target or unidentified, or when the standard deviation was zero were eliminated from the following analyses (N=145). Because the sample size were not large enough, the data of the Yakushima P-group and the Ohirayama group were pooled. The distribution of second coos by caller is shown in Table 6.

When a correlation analysis was undertaken between two consecutive coos in *DC* sequence using these data (Table 7), five parameters (duration, maximum frequency,

Table 7. Correlation coefficients of nine acoustic parameters between two consecutive coos in *DC* sequence separated by intervals of 0-1.1 sec and between those separated by intervals longer than 1.1 sec, where the second callers were target animals; data for the second callers were standardized individually by z score transformation.

	Interval			
	0-1.1 sec		>1.1 sec	
Parameter	r	df	r	df
Duration	0.149*	(175)	0.343*	(50)
Max location	0.018	(175)	0.101	(50)
Min freq.	-0.088	(175)	-0.301*	(50)
Start freq.	-0.047	(175)	-0.358**	(50)
Max freq.	0.250***	(175)	-0.013	(50)
End freq.	0.101	(170)	-0.343*	(50)
Max-min freq.	0.317***	(175)	0.210	(50)
Max-start freq.	0.327***	(175)	0.206	(50)
Max-end freq.	0.247**	(170)	0.252	(50)

*:p<0.05, **:p<0.01, ***:p<0.001

Table 8. Correlation coefficients of nine acoustic parameters between two consecutive coos in *DC* sequence separated by intervals of 0-1.1 sec and between those separated by intervals longer than 1.1 sec, where the second callers were adult target animals (≥ 6 yrs) or juvenile target animals (3-5 years); data for the second callers were standardized individually by z score transformation.

	Age							
	≥6 yrs Interval				3-5 yrs Interval			
	0-1.1 sec		>1.1 sec		0-1.1 sec		>1.1 sec	
Parameter	r	df	r	df	r	df	r	df
Duration	0.028	(90)	-0.204	(29)	0.291**	(83)	0.420	(19)
Max location	0.145	(90)	0.187	(29)	-0.154	(83)	0.303	(19)
Min freq.	-0.163	(90)	0.171	(29)	-0.007	(83)	-0.389	(19)
Start freq.	-0.124	(90)	0.217	(29)	0.072	(83)	-0.400	(19)
Max freq.	0.375***	(90)	-0.117	(29)	0.126	(83)	-0.049	(19)
End freq.	0.161	(87)	0.096	(29)	0.063	(81)	-0.427	(19)
Max-min freq.	0.427***	(90)	-0.012	(29)	0.194	(83)	0.286	(19)
Max-start freq.	0.458***	(90)	-0.160	(29)	0.178	(83)	0.328	(19)
Max-end freq.	0.389***	(87)	-0.449*	(29)	0.109	(81)	0.388	(19)

*:p<0.05, **:p<0.01, ***:p<0.001

maximum minus minimum frequency, maximum minus start frequency and maximum minus end frequency) of the second coos in *DC* sequence that were separated by intervals of 0-1.1 sec showed significant positive correlations with those of the first coos. Among the 9 parameters of the second coos in *DC* sequence that were separated by intervals longer than 1.1 sec, one acoustic parameter (duration) of the second coos in *DC* sequence showed a significant positive correlation and 3 acoustic parameters (minimum frequency, start frequency and end frequency) showed significant negative correlations with that of the first coos. The second coos given by a different caller within 1.1 sec were likely to be similar in several acoustic parameters with those of first coos. This similarity was not seen in coos that occurred at intervals longer than 1.1 sec. The findings imply the possibility that Japanese macaques might alter their acoustic features of responding coo calls matching to those of preceding coos.

The results raise the question of whether such vocal behavior develops during ontogeny or not. To test this possibility, we examined whether characteristics of vocal behavior of coo calls in Japanese macaques described above were influenced by callers' age:
(1) two consecutive coo calls where the second caller was an adult target (≥ 6 yrs) and
(2) two consecutive coo calls where the second caller was a juvenile target (3-5 yrs).
Correlation analyses were performed between two consecutive coos in *DC* sequences (Table 8). Results on adult animals indicated significant positive correlations for four acoustic parameters (maximum frequency, maximum minus minimum frequency, maximum minus start frequency and maximum minus end frequency) between the second coo uttered within 0-1.1 sec and the first coo. One parameter (maximum minus end frequency) of second coos of adult animals with the intervals longer than 1.1 sec showed a significant negative correlation with those of the first. For animals younger than 6 years, only one acoustic parameter (duration) of the second coo within 0-1.1 sec was significantly positively correlated with the first coos. No parameter of the second coo uttered by young animals with intervals longer than 1.1 sec revealed any correlation with those of the first coo.

The second call given by adult animals with intervals of 0-1.1 sec was similar in several parameters to the first, and those second calls with intervals longer than 1.1 sec were dissimilar to the first. Nevertheless, the similarity between the second and the first coo given by young animals was not independent of the intervals. A possible explanation for this result is that the young animals do not alter the acoustic features of their coos. However, the standard deviations of all acoustic parameters of coos by young target animals were not significantly different from those of adult targets (Mann-Whitney U-test, $N1=8$, $N2=7$, for 9 acoustic parameters, $p>0.05$). Therefore, young animals are likely to be able to emit acoustically different coo calls as adult animals do. Therefore this explanation can be discarded. Another explanation is that the juvenile animals more often responded to coos whose acoustic features were dissimilar to their own calls. As our data is not large enough to control the first callers, this explanation is possible. However, an alternative explanation is that the young animals are able to utter various calls, but are not able to match the features of their coos to those of preceding coos. If this explanation is right, there is the possibility that the ability to match acoustic properties of coos might develop gradually.

PLAYBACK EXPERIMENTS WITH A GROUP

From the observations of vocal exchange of coo calls, response coos appear to be acoustically similar to the first coos. We conducted playback experiments of recorded coo calls, in an attempt to eliminate exogenous unobserved stimulation responsible for a similarity in the structure of two consecutive calls. We assumed that, if our hypothesis was correct, playback of coos would be followed by acoustically similar coos in cases where they occurred within 1.1 sec after the playback, but by dissimilar coos in cases where they occurred after intervals longer than 1.1 sec.

Methods

The playback experiment with a group was conducted with the Ohirayama group in September and October, 1991. A choice was made of 14 tape-recorded coos as playback stimuli (2 coos from each of 7 target animals). Each trial was conducted when most group members were within audible range of the experimenter. The experimenter stood about 5 meters away from the speaker and waited until all the animals within his audible range were vocally inactive, at least for a 5 sec period prior to the playback, to prevent any other vocalizations affecting the subsequent calls of the animals. Thereafter, a single coo call was produced from the tape recorder. In a given experimental trial, one of the 14 stimuli was broadcast from a tape recorder (Sony WM-F203) through a speaker (Sony SRS-77G). Playback stimuli and vocalizations from the animals during a 4-sec period after the playback were recorded through two microphones (Sony ECM-672 directional microphone and Sony ECM-261 non-directional microphone) with a tape recorder (Sony WM-D6C). Whenever possible, identities of the animals that made vocalizations during the period were recorded. If any vocalization occurred during the 4-sec period after the playback, the attempt was regarded as successful regardless of which animal made the vocalization. Intervals in the *DC* sequence were measured from the end of playback to the start of the first vocalization thereafter. If any vocalization occurred simultaneously with a stimulus or if the first vocalization was not a coo, the trial was ignored in the subsequent analysis (N=43). If the caller could not be identified, it was treated as "unidentified" for the subsequent analysis. The mean distance from the speaker to the animals that uttered coo calls following the stimuli was 8 meters. At that distance the mean sound pressure of the stimuli was 57 dB (range 56-58 dB; Quest Elemetrics sound level meter model 215, A weighting).

Results

Acoustic parameters of the stimuli are shown in Table 9. The number of entire and successful trials for each stimulus are shown in Table 10. Of the 329 broadcast coos, 57 (17%) were successfully followed by coo calls with intervals of 0-4 sec. Of these 57 following coos, 35 (61%) occurred within 1.1 sec after the presentation of the stimulus, and 22 (39%) occurred after intervals longer than 1.1 sec. In the 57 successful trials, callers were identified in 14 (15%) cases, and in 43 (75%) cases the callers could not be identified. We observed one case in which an animal followed her own stimuli with an interval of 1.8 sec. The overall distribution of inter-call intervals between the stimuli and the subsequent coo is shown in Fig. 5. Correlation analyses were made between the stimuli and subsequent coos for the following two data sets:
(1) coos that followed the stimuli at intervals of 0-1.1 sec and
(2) coos that followed the stimuli at intervals of 1.1-4 sec.
Results indicate significant positive correlations between the stimuli and two acoustic parameters (maximum minus minimum frequency and maximum minus start frequency) of the subsequent coos that occurred at intervals of 0-1.1 sec (Table 11). No parameters of the coos that followed the stimuli at intervals of 1.1-4 sec correlated significantly with those of the stimuli.
In the Ohirayama group, the second coos given by a different caller within 1.1 sec were similar in several acoustic parameters with those of the first coos. This similarity was not seen in coos that occurred at intervals of longer than 1.1 sec. The phenomenon described in the naturalistic observations was confirmed experimentally. For these results, two explanations are possible. One is that individuals whose acoustic features of coos were similar to the acoustic quality of the playback stimuli may have responded more often and more quickly. Alternatively, individuals might modify acoustic features of their coos under certain conditions, adjusting them to the acoustic features of the call they hear.
The question can be answered if we undertake a playback experiment by directing prepared coo calls to a predetermined target adult and are able to evoke, an overt response by the animal. In the Ohirayama troop, the number of potential respondents were so great that it appeared to be very difficult to conduct such an experiment on a single predetermined target animal. On the other hand, in an unprovisioned group such as the Yakushima

Table 9. Acoustic parameters of 14 stimuli used in the playback experiment at the group level.

Parameter		Mean	SD
Duration	(ms)	200.2	51.3
Max Location	(%)	69.5	26.1
Min freq.	(Hz)	470	66.0
Start freq.	(Hz)	480	64.0
Max freq.	(Hz)	840	430.0
End freq.	(Hz)	740	390.0
Max-min freq.	(Hz)	370	430.0
Max-start freq.	(Hz)	360	430.0
Max-end freq.	(Hz)	100	120.0

Table 10. The numbers of successful trials and total number of trials (in parentheses) with 14 different stimuli in the playback experiment with a group.

	Interval between stimulus presentation and response		
Stimulus	0-1.1 sec	1.1-4 sec	(Total number of trials)
1	0	1	(16)
2	2	1	(17)
3	3	4	(33)
4	2	1	(10)
5	3	2	(27)
6	2	0	(24)
7	3	3	(30)
8	1	0	(11)
9	2	1	(42)
10	2	2	(14)
11	2	2	(28)
12	7	1	(24)
13	4	0	(34)
14	2	4	(19)
Total	35	22	(329)

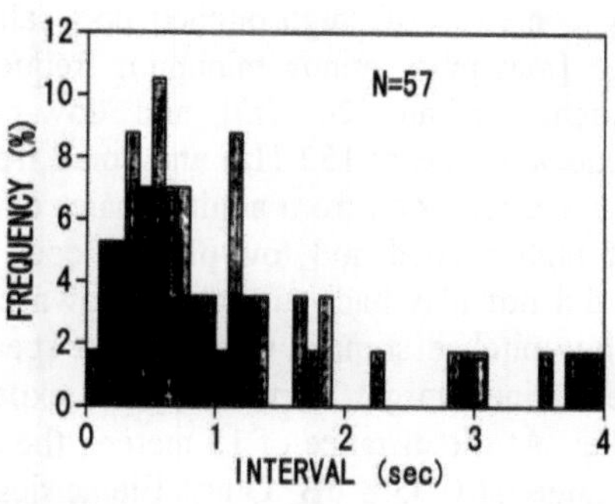

Figure 5. Overall distribution of intervals in the playback experiment with a group.

Table 11. Correlation coefficients of nine acoustic parameters between stimuli and following coos in the playback experiment at the group level.

	Interval between stimulus presentation and response	
	0-1.1 sec	1.1-4 sec
Parameter	r (df=28)	r (df=25)
Duration	0.247	0.083
Max location	-0.061	-0.268
Min freq.	-0.204	-0.033
Start freq.	-0.225	0.051
Max freq.	0.154	0.100
End freq.	0.210	0.043
Max-min freq.	0.491**	0.156
Max-start freq.	0.542***	-0.043
Max-end freq.	-0.195	-0.196

*:p<0.01

P-group, whose group size is relatively small and whose members are more likely to spread out over relatively large distances, the study should be feasible. The results of the playback experiment carried out at the group level suggest the possibility that animals might alter some acoustic features (maximum minus minimum frequency and maximum minus start frequency) of their vocalizations according to those of preceding coos. If the subject animals indeed match the acoustic features of their calls with those of playback vocalizations, the acoustic parameters of the subjects' coos should vary depending on the acoustic quality of the stimuli. To explore this possibility, we carried out the following playback experiment.

PLAYBACK EXPERIMENT WITH INDIVIDUALS

Methods

The following experiment was conducted with the Yakushima S-group between August 1992 and August 1993. The Yakushima S-group inhabits an approximately 40 ha area neighboring to the area where the Yakushima P-group lives. We started habituation of the group in April 1992, without provisioning, and individual identification of all adult animals was accomplished in May, 1992. At the beginning of our experimentation, the group consisted of 8 adult females, 8 adult males, at least 5 juveniles, and 1 infant.

Four adult females were chosen as target individuals. For each subject animal, a pair of stimuli was presented; they consisted of "high-pitched coo calls" with a relatively large frequency modulation range [maximum minus minimum frequency (mean=872 Hz) and maximum minus start frequency (mean=824 Hz)], and "low-pitched coo calls", with a relatively low maximum frequency (mean=152 Hz) and small frequency range (mean=120 Hz) (Figure 6). The stimuli were recorded from adult females of the same group. For each target animal, 1-2 pairs of high-pitched and low-pitched coos recorded from the same female were prepared. We did not play back an animal's own calls. A total of 5 coos (5 high-pitched stimuli and 5 low-pitched stimuli) was used. A speaker was hidden about 15 meters away from a predetermined target animal and the experimenter stood about 10 meters away from the speaker. At the distance of 15 meters, the mean sound pressure level of the stimuli was 42 dB (range 41.0-43.8 dB; Quest Elemetrics sound level meter model 215, A weighting). In a given experimental trial, one of the two stimuli was broadcast from a digital tape recorder (Sony NT-1) through a speaker. If any vocalization occurred simultaneously with a stimulus or if the first vocalization was other than a coo of the target animal, the trial was ignored in the subsequent analysis. Other experimental procedures and equipment used were the same as in the playback experiment within the group described above.

Results

Representative sonagrams of the playback experiment are shown in Figure 6. The numbers of entire trials and successful trials are shown in Table 12. A total of 147 high-

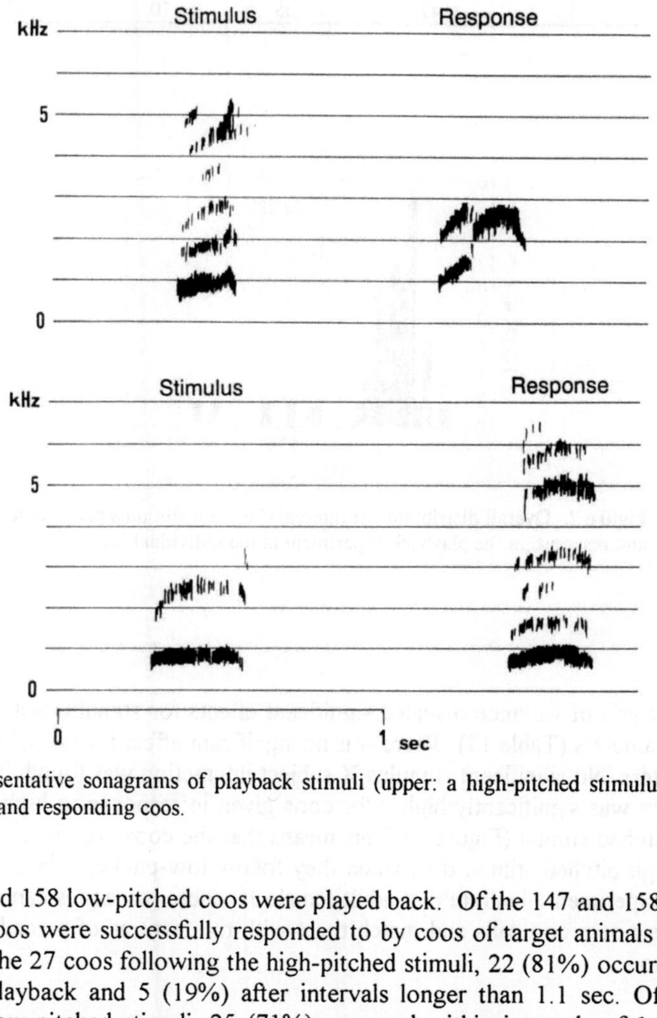

Figure 6. Representative sonagrams of playback stimuli (upper: a high-pitched stimulus, lower: a low-pitched stimulus) and responding coos.

pitched coos and 158 low-pitched coos were played back. Of the 147 and 158 trials, 27 and 35 broadcast coos were successfully responded to by coos of target animals with intervals of 0-4 sec. Of the 27 coos following the high-pitched stimuli, 22 (81%) occurred within 1.1 sec after the playback and 5 (19%) after intervals longer than 1.1 sec. Of the 35 coos following the low-pitched stimuli, 25 (71%) occurred within intervals of 1.1 sec, and 10 (29%) at intervals of 1.1-4 sec. Overall distribution of intervals in the playback experiment is shown in Figure 7. Acoustic parameters (maximum minus minimum frequency and maximum minus start frequency) of coos that occurred with intervals of 0-1.1 sec were analyzed by two-way analyses of variance. For the analysis, pitch of stimuli (high and low) and subject were chosen as two factors.

Table 12. The numbers of successful trials and total number of trials (in parentheses) with high-pitched and low-pitched stimuli by subject in the playback experiment at the individual level.

	High-pitched			Low-pitched		
	Interval between stimulus presented and response			Interval between stimulus presented and response		
Subject	0-1.1 sec	1.1-4 sec	(Total number of trials)	0-1.1 sec	1.1-4 sec	(Total number of trials)
Aoi	6	0	(26)	3	1	(23)
Tg	7	1	(18)	6	1	(18)
Ak	3	3	(48)	11	2	(51)
Nr	6	1	(55)	5	6	(66)
Total	22	5	(147)	25	10	(158)

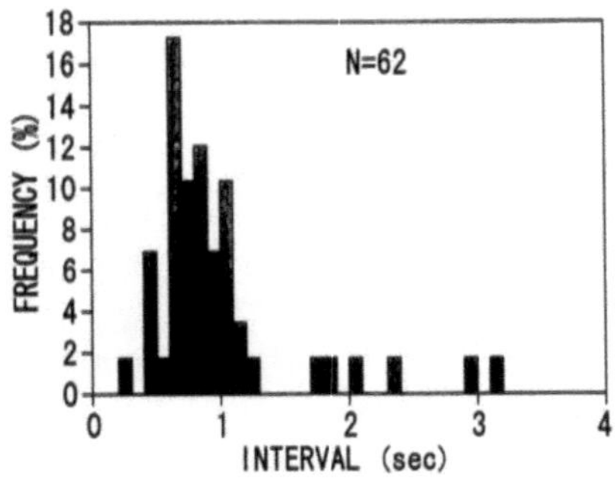

Figure 7. Overall distribution of intervals between stimulus presentation and response in the playback experiment at the individual level.

2 X 4 analysis of variance revealed significant effects for stimulus with regard to both of the two parameters (Table 13). There was no significant effect for the subject with regard to the parameters. No significant stimulus X subject interaction was found. The mean of the two parameters was significantly higher for coos given in response to high-pitched stimuli than to low-pitched stimuli (Figure 8). This means that the coos are more modulated when they follow high-pitched stimuli than when they follow low-pitched stimuli. The results of the playback experiment illustrate that adult female Japanese macaques can modify acoustic features of their own coo calls and match them with the features of preceding coos which they respond to.

PERSPECTIVES FOR FUTURE RESEARCH

Recently, one of us has published data on the cooing of 3- to 4-month-old human infants (Masataka, 1993), results that parallel those we report here for Japanese macaques. For human infants to achieve the skill of conversational turn-taking, parental speech

Table 13. Results of two-way analyses of variance performed on the acoustic properties of responding coos in the playback experiment at the individual level.

Parameter	Stimulus F (1,39)	Subject F (3,39)	Stimulus X Subject F (3,31)
Max-min freq.	6.59*	2.20	1.76
Max-start freq.	6.95*	2.10	1.83

*:$p<0.05$

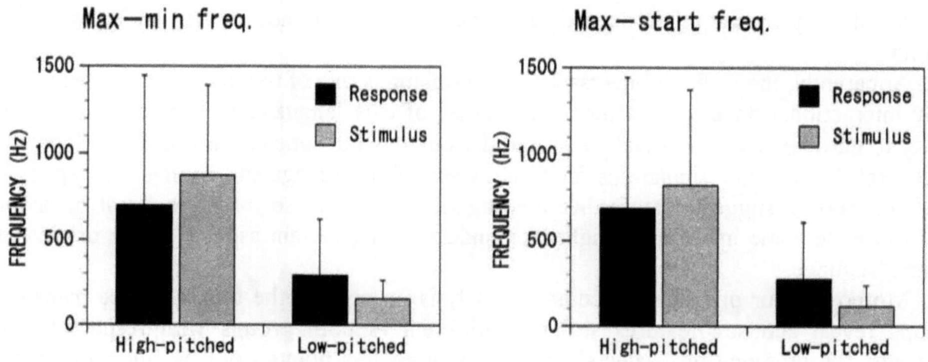

Figure 8. Means and standard deviations of the two parameters in response coos and stimuli.

apparently plays an important role. In very early conversations, adults play an active role in providing the temporal structure for taking turns by providing social stimulation contingent upon the infant's responses. We have demonstrated that Japanese infants as young as 3 months are capable of discriminating between adult behavior that occurs contingently and that which takes place randomly. In the study, a total of 48 male infants experienced either conversational turn-taking or random responsiveness when aged 3 months and 4 months. Throughout each testing session, the mother adopted a face-to-face position and maintained eye contact with the infant, who lay in a supine position in a crib. The session consisted of 2-min baseline period followed by a 6-min test period. In the 2-min baseline period, the mother took en face posture and maintained eye contact. However she remained unresponsive to the infant. In the following 6-min period, the mother actively interacted with the infant with a smile, light touch on the abdomen and the phrase "Hi" (baby's name) simultaneously.

When the first experiment was undertaken, the mother-infant dyads were randomly assigned to two groups — the contingent group and the random group. For infants in the contingent group, the mother responded only after an infant vocalization, thereby maintaining conversational turn-taking with her child. Infants in the random group, however, were interacted at a prearranged schedule and independently of the timing of their vocal utterances. The mother responded to her infant in the same manner and at the same time intervals as her counterpart in the contingent group. The responses were cued by audio-taped signals delivered through an earphone, which resulted in a violation of turn-taking in the random group infants. Second testing was conducted 30 days later (when the infants were 4 months old) in the same manner as that undertaken previously.

Results of the experiment revealed that in both periods, the infant's rate of vocalizing was not significantly influenced by the contingency of the mother's response, but that contingency altered the temporal parameters of the infant's vocal pattern. Infants tended to produce more bursts or packets of vocalizations when the mother talked to the infant in a random pattern. When the infants were 3 months old, such bursts occurred most often at intervals of 0.5-1.5 sec whereas at 4 month, they took place most frequently at significantly longer intervals of 1.0-2.0 sec. Interestingly, the difference corresponded to the difference between intervals with which mothers responded contingently to vocalizations of infants aged 3 months and 4 months. While the intervals (between the onset of the infant's vocalization and the onset of the mother's vocalization) rarely exceeded 0.5 sec when the infant was 3 months old, they were generally distributed between 0.5 and 1.0 sec at 4 months. The results imply that after vocalizing spontaneously, the infant tended to pause as if to listen for a possible vocal response from the mother. In the absence of a response, repeated vocalization occurred. The interval between the two consecutive vocalizations was changed flexibly by the infant according to the recent experience of turn-taking with the mother.

Apparently, the skill of conversational turn-taking is one of the major milestones in the early interactional development before the onset of true language in human infants. The ability to participate cooperatively in shared discourse is fundamental to social development in general. Remarkable similarities found between infant prelinguistic cooing and Japanese macaque cooing suggest that further investigation on Japanese macaque vocal behavior could provide some important insights for understanding certain aspects of the origins of human language.

Moreover, our preliminary acoustical analysis on coos in the two Japanese macaque groups reveal acoustic features strikingly different in both groups with respect to F_0 elements: the presence of "dialects" may be assumed. As pointed out by Snowdon et al. (1982), to study vocal ontogeny in non-human primates, that is, to determine how flexibly primate vocalizations are produced, it is most urgent to find an appropriate species. From the present study, we conclude that the vocal behavior of Japanese macaques is more flexible with regard to its temporal and acoustic structures than has commonly been supposed. We strongly suggest the Japanese macaque as a possible candidate.

CONCLUSION

Vocal exchanges of coo calls in female Japanese macaques were observed in two populations under natural condition. Temporal patterns of occurrence of these vocalizations during vocal interaction were studied by analyzing inter-call intervals between two consecutive coos. When the second call was uttered by another caller than the first (*DC* sequence), most of the second calls occurred at short intervals. These results indicate that second calls separated by short intervals may occur in response to the first call, whereas second calls separated by a relatively long interval may occur independently of the preceding call. When two consecutive calls were uttered by the same caller (*SC* sequence), the second coos rarely occurred during the period in which most of the second coos in *DC* sequence occurred. These results suggest that when an animal utters a coo spontaneously, it appears to remain silent for a short interval; when no response occurs, she is likely to give further coos addressing group members.

Acoustic analysis of two consecutive coos in *DC* sequence showed that the second coos given by a different adult caller and occurring within 1.1 sec significantly correlated with those of first coos with respect to several acoustic parameters. This correlation was not observed when coos occurred at intervals longer than 1.1 sec. Coos that followed different callers with short intervals were acoustically similar to the preceding coos. This similarity between two coos was observed when the second callers were adult animals. However, the similarity was not seen when the second callers were juvenile animals. These results suggest that Japanese macaques alter the acoustic features of their responding coos, matching them to the first coos. The results also suggest the possibility that this vocal matching behavior might develop during ontogeny. We conducted playback experiments with a group as well as individuals to confirm the results observed under naturalistic condition and to eliminate exogenous unobserved factors. The results of the experiments confirmed that the animals match the acoustic features of their coos in response to the first coos.

Our recent study of conversational turn-taking between three- to four-months-old human infants and their mothers showed that there are remarkable parallels between early stages of turn-taking ability of human infants and vocal behaviors during coo calling of Japanese macaques.

ACKNOWLEDGEMENTS

We thank Dr. S. Kodera of the Japan Monkey Center for permission to study in the Inuyama Monkey Park, and K. Sakurai, Y. Ishida, M. Takai and other members of the staff of the Japan Monkey Center for their assistance with the present research. We also thank the researchers and residents of Yakushima Island for their help during the field research. This study was supported by Grant-in-Aid from Ministry of Education, Science and Culture, Japan entitled "Emergence of human cognition and language (05206104)" to N. M., and Grant-in-Aid for JSPS Fellowships for Japan Junior Scientists and Cooperative Research Fund of the Primate Research Institute, Kyoto University to H. S.

REFERENCES

Altmann, J., 1974, Observational study of behavior: sampling methods, *Behaviour* 49:227-267.

Biben, M., Symmes, D., and Masataka, N., 1986, Temporal and structural analysis of affiliative vocal exchanges in squirrel monkeys (*Saimiri sciureus*), *Behaviour* 98:259-273.

Green, S., 1975, Variation of vocal pattern with social situation in the Japanese monkey (*Macaca fuscata*): a field study, *in*: "Primate Behavior, Vol. 4", L. A. Rosenblum, ed., Academic Press, New York.

Itani, J., 1963, Vocal communication of the wild Japanese monkey, *Primates* 4: 11-13.

Kawai, M., 1960, A field experiment on the process of group formation in the Japanese monkey (*Macaca fuscata*), and the releasing of the group at Ohirayama, *Primates* 2:181-253.

Marler, P., 1970, Bird song and speech development: could there be parallels? *Amer. Scientist* 58:669-673.

Maruhashi, T., 1980, Feeding behavior and diet of Japanese monkey (*Macaca fuscata yakui*) on Yakushima Island, Japan, *Primates* 21:141-160.

Masataka, N., 1993, Effects of contingent and noncontingent maternal stimulation on the vocal behaviour of three- to four-months-old Japanese infants, *J. Child Lang.* 20:303-312.

Masataka, N. and Biben, M., 1987, Temporal rules regulating affiliative vocal exchanges of squirrel monkeys, *Behaviour* 101:311-319.

Masataka, N. and Fujita, K., 1989, Vocal learning of Japanese and rhesus monkeys, *Behaviour* 109:191-199.

Mitani, M., 1986, Voiceprint identification and its application to sociological studies of wild Japanese monkeys (*Macaca fuscata yakui*), *Primates* 27:397-412.

Myers, R. E., 1976, Comparative neurology of vocalization and speech: Proof of a dichotomy. *in*: "Origins and Evolution of Language and Speech", S.R. Harnad, H.D. Steklis and J. Lancaster, eds., Ann. New York Acad. Sci. 280:745-757.

Newman, J.D. and Symmes, D., 1982, Inheritance and experience in the acquisition of primate acoustic behavior, *in*: "Primate Communication", C. T. Snowdon, C. H. Brown, and M. R. Petersen, eds., Cambridge University Press, Cambridge.

Owren, M. J., Dieter, J. A., Seyfarth, R. M. and Cheney, D. L., 1992, 'Food' calls produced by adult female rhesus (*Macaca mulatta*) and Japanese (*M. fuscata*) macaques, their normally-raised offspring, and offspring cross fostered between species, *Behaviour* 120: 218-231.

Pereira, M.E., 1986, Maternal recognition of juvenile offspring coo vocalizations in Japanese macaques, *Anim. Behav.* 34: 935-937.

Pierce, J.D., 1985, A review of attempts to condition operantly alloprimate vocalizations, *Primates* 26: 202-213.
Snowdon, C.T., Brown, C.H., and Petersen, M.R., 1982, "Primate Communication", Cambridge University Press, Cambridge.
Sugiura, H., 1993, Temporal and acoustic correlates in vocal exchange of coo calls in Japanese macaques, *Behaviour* 124: 207-225
Sutton, D., Trachy, R.E., and Lindeman, R.C., 1981, Vocal and nonvocal discriminative performance in monkeys, *Brain Lang.* 14:93-105.

THE VOCAL BEHAVIOR OF BARBARY MACAQUES (*MACACA SYLVANUS*): CALL FEATURES AND THEIR PERFORMANCE IN INFANTS AND ADULTS

Dietmar Todt, Kurt Hammerschmidt, Viveka Ansorge and Julia Fischer

Institut für Verhaltensbiologie
Freie Universität Berlin
Haderslebenerstrasse 9
D-12163 Berlin, Germany

INTRODUCTION

Listening to primate vocal signals entails witnessing episodes of social significance, and students of primate social behavior are well advised to investigate such episodes, but at same time also the features of signals escorting them. Today, the study of vocal communication profits tremendously from progress in technology, which in particular allows for a parametric analysis of signal patterns and their subsequent categorization by non-subjective criteria (Beeman, 1989; Gautier and Gautier-Hion, 1988; Hammerschmidt, 1990; Hauser et al., 1993; Newman and Goedeking, 1992; Schrader and Todt, 1993). These procedures confirmed, for instance, that the concept of 'multi-dimensional definition of call types' is a useful tool for addressing both the composition of signal repertoires and the rules of repertoire performance or use, respectively (Hammerschmidt et al., 1994b; Owren et al., 1992). Compared to a pragmatic view describing call types as 'vocal patterns invariantly appearing in the same uniform configuration of structural properties' (Robinson and Whitehead, 1988), the novel concept is more operational: It defines calls within a multi-dimensional parameter space and characterizes a particular type of call by a nucleus of significantly correlated parameter expressions within that space (Todt, 1986; Todt et al., 1992). This concept allows to investigate call type variations and also the conditions to which modifications of parameter expressions covary . Thereby, it provides a chance to address the interactional role of vocal patterns in a sophisticated manner.

Classically, signals are described as members of a repertoire that may be specific to a studied species, or to an individual's age and gender. Often, such descriptions do not consider the actual dynamic of signals given by their performance mode in a coherent communication process. The investigation of this performance, however, allows one to uncover fascinating rules of behavioral organization and dynamics (for citations see Todt and Hultsch 1992, 1994). The most basic rules that hold for a number of vertebrates in general concern the following aspects. First, evaluation of the time domain of vocalizations permits to distinguish two hierarchy levels of organization, namely calls and bouts of calls.

Current Topics in Primate Vocal Communication
Edited by E. Zimmermann *et al.*, Plenum Press, New York, 1995

Second, the duration of calls mirrors their communicative function, that is, they are long enough to provide an appropriate vehicle for a signaller's message, but not too long to delay his/her checking for a potential answer. Third, besides calls made up by a temporally coherent sound pattern there are others composed of a number of vocal elements. In primates, the latter are phonologically similar in most cases; thus, they form a sort of homotype series which invites study of the potential meaning of its structure and dynamics. Finally, instead of occurring as single acoustical events, many calls are given in a bout or a sequence, respectively. And such bouts can result from either a repetition of the same call type (label: homotype call sequences) or a sequential combination of different call types (label: heterotype sequences; see page 5).

In this chapter, we report findings of studies that have investigated these aspects in a paradigmatical manner. The studies concentrated on the Barbary macaque (*Macaca sylvanus*), a species known for its conspicuous vocal activities. Current research objectives have treated developmental issues, such as vocalizations of infants and their role during infant socialization, and in addition, interactional accomplishments of adults, such as rhythmical calls of females used during the mating season, or vocalizations of different individuals uttered when they were confronted with a sort of 'disturbing stimulus'. These issues were approached by investigating the covariation between a given signal pattern and (1) individual properties of the signaller, (2) social properties of the respective context, (3) immediate consequences of a signal apparent within that context, and (4) individual properties of signal addressees or recipients. Special efforts allowed us to substantially improve the analysis of signal patterns and to extend its application from single or solitary patterns to longer processes of signalling. This extension was sensible because the majority of Barbary macaque vocalizations do occur in a serial manner.

In line with the aim of this book our contribution is divided into the following sections: short introduction to the species and the methods of our approaches, an overview of the call type repertoire of infants and adults, then, a brief report about the role of serial signalling in the social life of Barbary macaques, and finally, a sketch of issues that will be part of our future studies on primate communication. The chapter provides a foundation for comparing Barbary macaques to other nonhuman primates, but is not designed as a comparative paper itself.

INTRODUCING THE SPECIES

The Barbary macaque (*Macaca sylvanus*) is the only representative of the genus *Macaca* in the Western section of the Eurasian hemisphere. Currently, it has been listed as an endangered species which can be found in a variety of habitat types of higher mountain areas in Morocco and Algeria (Fa, 1984; Ménard et al., 1985; 1990). The species is semi-terrestrial, and uses trees for sleep, protection against predators, and as a food supply (Deag 1974). As an adaptation to the Northern climates, the reproductive behavior of Barbary macaques is strictly seasonal. The mating season occurs in autumn and the breeding season in spring (Taub, 1980b; Küster and Paul, 1984). Barbary macaques become sexually active at the age of 3.5 years. Newborn infants are extremely attractive to all group members, and alloparental behavior plays a significant role in the socialization of infants (Deag and Crook, 1971; Taub, 1984; Small, 1990a; Todt, 1989).

Barbary macaques form multi-male groups of 15 to 20 individuals (Deag and Crook, 1971; Ménard et al., 1990) with a sex ratio of almost 1/1. Subadult males may leave their natal groups and join other ones (Paul and Küster, 1985; Mehlmann, 1986). For females, a clear maternal lineage and a linear dominance hierarchy has been described (Timme, 1989; Small, 1990b). In males, on the other hand, dominance relationships appear more variable and complex (Bartecki, 1986). A great deal of parental care is provided by adult males

(Taub, 1984; Küster and Paul, 1986; Riechelmann et al., 1994). Along with infant carrying, parental behavior includes a remarkable display which originally was called 'agonistic buffering' (Deag and Crook, 1971; Deag, 1980), but currently is termed 'triadic interaction'. It is observed primarily among adult males and involves one individual with an infant who then is the target of a number of vocal and non-vocal displays. Although this behavior has been subjected to a series of investigations, its function remained quite opaque (Deag, 1980; Taub, 1980a; Paul, 1984; Roshani et al., 1994).

Barbary macaques show a complex syndrome of vigilance behaviors which mainly serve to check and monitor social events (Quintilius and Hultsch, 1991). Group members regularly travel, feed and rest together. At dusk, they assemble near a communal roosting site, where they finally climb into the trees and form sleeping clusters (preferred cluster size: two and three). Their prenocturnal interactions and also their nocturnal behaviors turned out to be extremely interesting, especially with regard to both partner choice and cluster composition (Ansorge et al., 1992). The prenocturnal phase is characterized by a rise in vocal activity ('dusk calling') produced by infants and younger juveniles who experience difficulties in entering a sleeping group (Todt et al., 1985; Hammerschmidt et al., 1994b).

Most of the current knowledge on Barbary macaques originated from research conducted in a number of semi-free enclosures, where they have been successfully bred, for instance, at the Rock of Gibraltar, in France (Kintzheim and Rocamadour) and in Germany (Salem and Rheine). Mainly for two reasons, this knowledge can be regarded as biologically relevant. First, is because of site and population properties. The site of *La Forêt des Singes* at Rocamadour, for instance, makes an excellent monkey habitat that covers a semi-free area of about (now more) 15 ha and holds a prospering population of about 200 individuals. Second, is because the behavioral data recorded in these confinements do not appear to be significantly different from the data collected in the North African study sites (Fa, 1984; Fischer et al., submitted). Consequently, the confinements provide a promising chance for coordinated long-term studies combining ecological and behavioral issues with physiological ones (review in Todt et al., 1992). In addition to the fascinating social life of the Barbary macaques, this should render them a paradigmatic species for primate research.

INTRODUCING METHODOLOGICAL PROCEDURES

Vocalizations were studied as part of the Barbary macaques' behavioral repertoire, by the following procedures: (l) Multi-channel recording of a coherent stream of behavioral and contextual data. Channels referred to, e.g., individual primates, selected behaviors such as gestures and vocalizations and/or contextual events. (2) Formal analysis of data yielding to rules of the structural organization and the temporal dynamics of the behavior. (3) Listing of all results and, after this, examination of potential relationships among them. (4) Development of hypotheses that explain the findings and provide testable predictions about either the mechanisms underlying the investigated processes or functional aspects, respectively. (5) Examination of such predictions in a subsequent line of studies (for citations and further details see Todt et al., 1992).

Vocalizations were recorded with cassette recorders (Marantz cp 430, Sony WMDC6, Sony DAT DCD-D3) equipped with Sennheiser microphones (KN3/ME80, or ME88 plus windscreen). For visual inspection, vocalizations were transferred to a spectrum analyzer (NSC UA 500A) that, through an oscilloscope (Phillips PM 320) and a high-speed camera, delivered hard copies of their sound spectrograms (35mm positive film). In addition, vocalizations were examined using the real-time spectrograms provided by RTS (Engineering Design, Boston). Vocal patterns were digitized with a sample frequency of 37.5 kHz. Subsequently, a Fast Fourier Transformation (FFT) was conducted with the

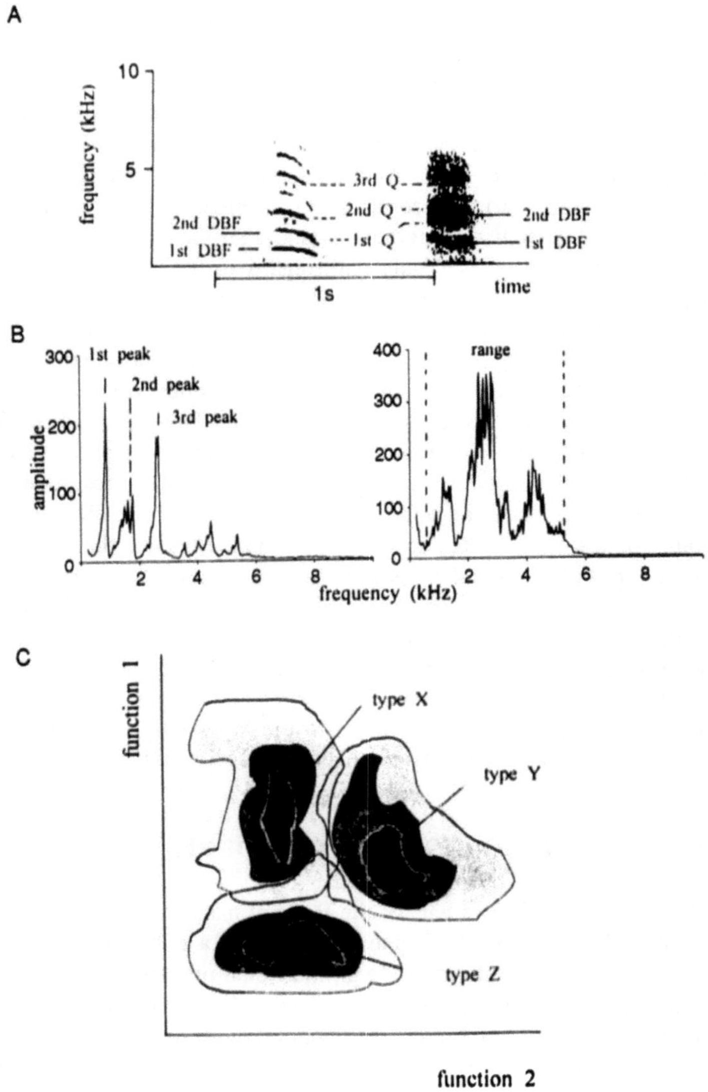

Figure 1. Illustration of acoustic patterns and their analysis. Examples of two acoustic patterns, a tonal and a noisy one (A), and their respective power spectra (B). The first and second dominant frequency band of the tonal call are marked (DFB).Their statistical measures provide suitable parameters for its description of tonal vocal patterns. In contrast, the first, second and third quartile of the frequency amplitude (marked by Q in both calls) are appropriate for describing noisy calls. The quartile denote the frequency up to which one, two or three quarters of the total amplitude are located. Marks in the graphs (B) indicate selected parameters calculated from the power spectra (amplitude peaks and frequency range). On bottom (part C) the figure shows schematically the categorization of acoustic patterns. In the depicted case, the frequency distribution of two acoustical parameters allows to distinguish the core area of three pattern types.

SIGNAL system (time resolution: 5 ms; frequency range: 15 kHz; frequency resolution: 37 Hz).

The call parameters, that were calculated subsequently, were classified as follows: (1) parameters characterizing structural features of acoustic patterns; (2) parameters characterizing the distribution of the frequency amplitudes; (3) parameters calculated from the average amplitude-frequency spectra (power spectra). For evaluation of parameters accepted for characterizing the acoustic structure of a given pattern, a statistical threshold filter was applied to emphasize dominant frequency bands. The threshold of the filter was set at the sum of the mean and the standard deviation of the amplitudes for each time segment (5 ms). This allowed reduction of pattern artefacts produced by head movement or locomotion of the signaller.

A measure that had proven useful, to characterize the frequently uttered noisy calls, was the distribution of energy within the frequency spectrum. The frequency amplitudes of each time segment were summarized according to their respective frequency value, and the first, the second and the third quartile was calculated. In order to calculate the power spectra, the values of the amplitude frequency spectra of all time segments were averaged. This resulted in a different signal-to-noise ratio compared to the parameter calculation mentioned in the first step. Out of the power spectra parameters were determined which described the spectral density, or the frequency range and the peak amplitude of those parts which represent the main call energies. On the basis of the filtered and unfiltered frequency-time-spectra and power spectra, finally, 85 parameters were calculated. Filtering and parameter calculations were run with a software program (LMA 3.0) developed by K. Hammerschmidt (Figure 1).

As a kind of preventive measure allowing for a compensation of problems in call type identification by estimating partially correlated parameters, our analyses included a calculation of large numbers of parameters. In particular the following reasons were relevant here. First, calls often differed considerably, for example, in their degree of tonality. Some parameters, which were useful for describing one type of call, turned out to be inappropriate for describing another one. Second, the fundamental frequency was often so low that it got masked by background noise. In contrast, young animals often gave calls with an extremely high and modulated fundamental, that also often lacked overtones. In those cases the fundamental frequency could not be assessed. Finally, as in other outdoor recordings (see Brown and Waser, 1988) the acoustical amplitude could not be assessed in a reliable manner, and therefore had to be compensated for by other parameters.

All parameters obtained in that way were analyzed with multivariate statistical techniques (multi- and univariate analyses of variance, and discriminant function analysis (SPSSWIN)). The discriminant function analysis provided a classification procedure which assigned each acoustic pattern to its appropriate group (correct assignment) or a different one (incorrect assignment). In addition, the discriminant function analysis calculated the rank order in which the parameters contributed to a correct assignment.

TERMINOLOGY: In line with other approaches our study in the Barbary macaques used the following labels. The term '*call*' refers to vocalizations that form a temporally coherent acoustic pattern. The term '*call type*' stands for a cluster of calls that are identical or at least similar in their acoustic shape. Call or signal categorization by the 'type concept' is done according to the distribution mode of signal parameter values within a multidimensional scale. This concept takes the joint occurrence of several parameter values for a definition of 'types', rather than taking the boundary between values of only *one* parameter as a categorical criterion. The term '*call sequence*' designates a bout of calls and indicates that this bout is investigated in terms of sequential properties. Depending on their organization, we distinguish between 'homotype' and 'heterotype' call sequences. The first ones originate from an immediate repetition of calls of the same type. Interestingly enough, homotype call sequences can occur as part of a heterotype call sequence (schema: A-A-A-

A-A-B-B-B-B-C-C-). Because of this, they are labeled separately by the term *'call series'*, and in addition, defined as compounds which from the perspective of behavioral organization are hierarchically inferior to such a call sequence, but superior to the calls composing them.

THE VOCAL REPERTOIRE OF INFANTS AND ADULTS

Signal repertoires have been described by considering (1) the acoustical features of signal patterns and individual properties of the signaller, (2) contextual variables of signalling, and (3) conclusions about possible signal functions (Rowell and Hinde, 1962; Green, 1975; Cleveland and Snowdon, 1982). This section concerns the first two matters only, and leaves the last issue for the next section. For comparative purposes, however, we would like to mention beforehand that the vocal repertoire of the Barbary macaques is primarily comprised of signals that have been documented for other Catarrhines also, including calls related to contact, distress, play, mating, threatening, warning and alarm.

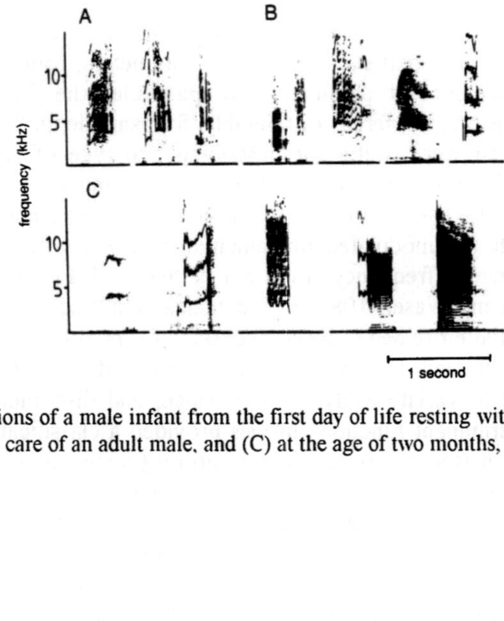

Figure 2. (A) Vocalizations of a male infant from the first day of life resting with the mother; (B) from the third day of life while in care of an adult male, and (C) at the age of two months, again during male care.

Our analysis of signal parameters revealed that, on the level of the entire vocal repertoire, the acoustical parameters of call types covaried with individuality, age, and sex of the signaller. Covariations were also found between call type and context. In some cases, clear relationships existed. Sometimes, however, many different call types within one apparently consistent context were observed. This leads to the question whether the contextual categories set up by the observer correspond to those of the subject (see page 16).

A great variety of vocal patterns exists from birth (Hammerschmidt and Ansorge,

1989). The vocalizations from the infants' first day of life resemble those from the third day of life, or after two months (Figure 2). Visual inspection of spectrograms of samples from infants of two to eleven weeks of age allowed us to distinguish 14 pattern types. And there is evidence that, interestingly enough, only a few of these types occur in a context-specific manner, and most of them are uttered in more than one social context (Wolff and Todt, 1985). One major change during early ontogeny is an increase in the stability of the amplitude of vocal patterns. This might be due to growth and improvement of motor control. The variety of vocal patterns uttered during the first day of life, the third day, or

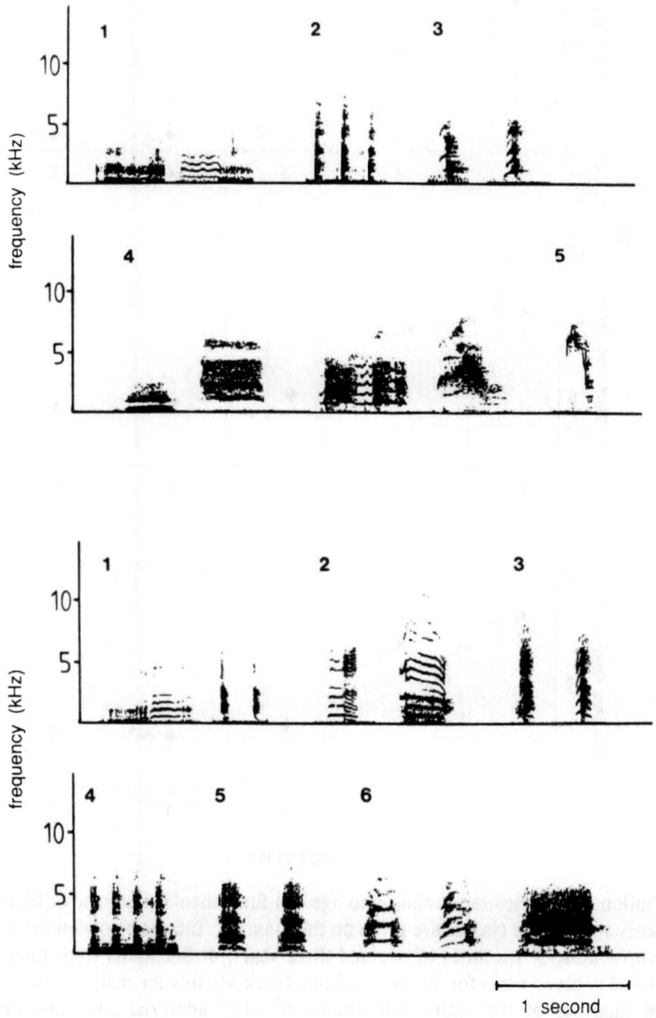

Figure 3. Spectrograms of vocal patterns of adult Barbary macaques uttered in different behavioral contexts. (A) adult males: 1. socio-positive vocalizations (growling), 2. threat barks, 3. alarm calls, 4. agonistic vocalizations. 5. escape (threatened); (B) adult females: 1. socio-positive vocalizations (growling, contact calls), 2. searching for infant. 3. alarm calls, 4. parts of a mating call, 5. mobbing, 6. agonistic vocalizations.

after two months does not differ fundamentally. Major developmental trajectories were an increasing use of similar vocal patterns in the same behavioral context. The increase in the variety of vocal patterns and of overall vocal activity during the first month of life correlates to the finding that infants spend an increasing amount of time without maternal body contact, e.g. in male care or in care of other group members. From 6 to 18 months during the period of weaning, juveniles show a further increase in vocal activity.

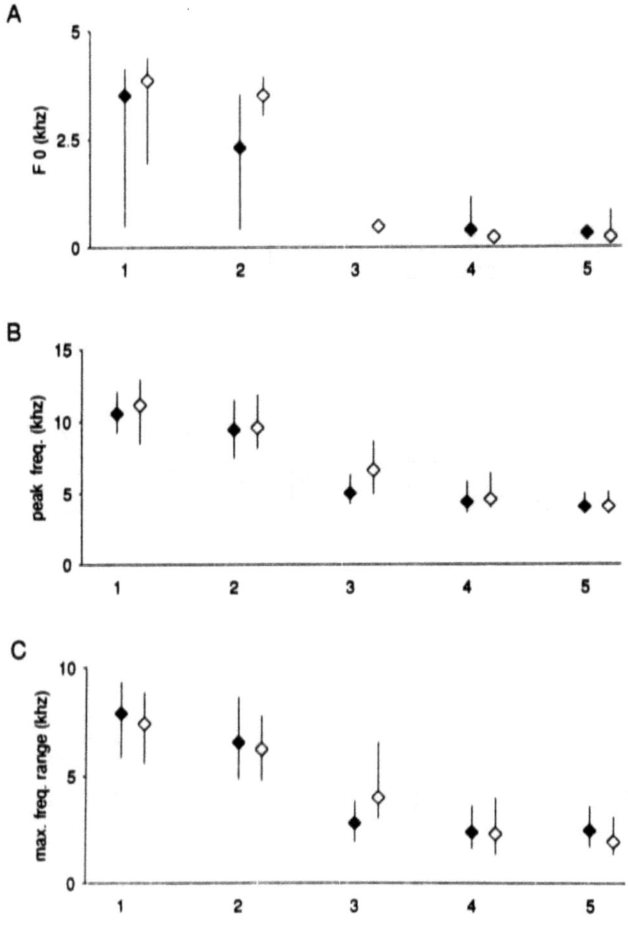

Figure 4. Distribution of parameters in relation to age (A) fundamental frequency, (B) peak frequency, (C) maximum frequency range. Age classes are given on the x-axis. 1: Infants below one year; 2: Juveniles from twelve to eighteen months; 3: Juveniles of two and three years; 4: Subadults from three and a half to five years for females and to seven years for males; 5: adults; black squares for males, white squares for females. The median and quartiles of the entire call sample of 5187 analyzed calls are shown (fundamental frequency: N=1999; the smaller number of analyzed calls compared to other analyses is due to the fact that for the majority of calls the fundamental frequency is hard to determine exactly. The value for males in age-class 3 was omitted because of an uneven distribution of individuals). In general, age related changes were: 1. Both the absolute value and the variability of fundamental frequency become lower. 2. The frequency range of calls becomes smaller and the peak frequency becomes lower.

During the day, infants and juveniles respond to maternal rejecting behavior with short or longer crying periods. At dusk, during the search for a sleeping partner, they utter protracted call sequences, composed of various series of cries (dusk calling ; Todt et al., 1985; Hammerschmidt, et al., 1994b). Later in development, the rate of vocal activity of Barbary macaques declines. Concurrently, the variability of their vocal patterns decreases. This is especially true for males. Beside nasal vocalizations and growling in friendly situations, their vocalizations are confined to agonistic and alarm contexts (Fig. 3A). Females, on the other hand, continue to vocalize in a greater number of different behavioral contexts. Along with the contexts mentioned for males, they vocalize in mating contexts or when they have lost their infant (Fig. 3B). Compared to adult males, females are more inclined to comment vocally on both affiliative and aversive situations.

As repeatedly reported, the stability of the pitch period increases with a primate's age (Hauser, 1989), and often young and old animals can be distinguished by the mean values of their fundamental frequency. In Barbary macaques, young animals on average have a higher fundamental frequency than adult animals. However, young animals can be found to sometimes produce vocalizations with a broad frequency range. The fundamental frequency may reach extreme values ranging between 250 Hz and 5000 Hz (Fig. 4A). This parameter, therefore, cannot be accepted as a reliable means for determining age. The two parameters found appropriate for assigning age classes are the peak frequency and the maximum frequency range (Fig. 4B,C). Despite these age-related trends in call parameters there are some exceptions. Adult animals in distressed situations like inter-group fights may utter extremely high pitched vocalizations, with a fundamental frequency up to 5000 Hz, which are similar to infant cries or screams (Hammerschmidt et al., 1994a).

SOCIAL ASPECTS OF SERIAL SIGNALLING

First paradigm : Crying in infants

As reported earlier in this article, alloparental care is a prominent feature of Barbary macaque society, and has a significant impact on the early socialization of infants. Alloparental behavior often begins a few days after birth and lasts for several weeks (Riechelmann et al., 1994). Usually, the number of different caretakers increases during this period of time. Concurrently, the duration of caretaking episodes decreases. Both developments covary with the following factors: (1) As in other primates, contacting or carrying a newborn is attractive for various group members. Mothers, however, are quite selective in handing their young to an alloparent. This selectivity seems to decline later on. (2) With growing age of a young infant, transfers among alloparents become more frequent. (3) In parallel, infants more often increase the spatial distance to their adult caretaker and thus raise the number of body contacts to other individuals (for review see Riechelmann et al., 1994). Group members acting as alloparents are either female members of the mother's matriline (Small, 1990a) or males that are not neccessarily related to the infant (Paul et al., 1992). Interestingly enough, male alloparents seem to prefer male infants. This preference is correlated with the recruitment of allies and the development of friendly relationships among males of different generations.

Various studies have shown that infant vocalizations are a significant feature of interactional contexts, and most of the calls are uttered in a serial manner. Up to an age of 18 months of life the major portion of these series can be labelled as 'crying' (synonymous term: screaming). In Barbary macaque infants, crying is a correlate of two developmental phases: The early weeks of life when alloparental episodes may last more than one hour and the infant may have problems with an individual caretaker. And the age span from 6 to 18 months of life, when infants may have problems gaining contact to a particular group, especially to a sleeping cluster. Crying behaviors occurring at dusk are often extremely protracted and are a correlate of weaning (Hammerschmidt et al., 1994b).

Crying has been defined as an intra-group display that indicates psychophysical discomfort and may comprise two different messages, namely a need for support and an announcement of disagreement. Normally, the first message is addressed to the infant's mother or her substitute, whereas the latter appears to be directed towards individuals who should be signified or even repelled (Todt, 1988, 1989). Whereas the function of crying as an appeal for help is widely considered in primate studies (Newman, 1985; Gouzoules et al., 1984; Hauser, 1993), its role as a signal of rejection and protest merits more attention than it has received in the past (Todt, 1994). This role is supported by our observations in Rocamadour and, in particular, by the specific acoustic features of crying behavior. Such features are, for instance, (i) a high acoustic density (relation between the durations of calls and silent intervals) within the protracted series, (ii) an often high proportion of atonal or even shrill process segments, (iii) a frequent and often extremely sudden variation of particular call parameters, such as the distribution of amplitude frequencies, frequency range or duration and amplitude of calls.

When discussing these features, Todt (1988) has emphasized four functional aspects: (1) The structural properties of crying are designed to induce specific nonacoustic responses. Especially intense crying that is characterized by a high acoustical density, given e.g. if the silent intervals between successive calls are very short, is inappropriate for a mutual exchange of vocal signals. (2) Regardless of the fact that specific information may be encoded in its various calls, the noise quality of at least parts of a crying process may influence addressees (Figure 5); namely by the following impact: Noisy cries can occupy the auditory channel of near group members and, for instance, by impeding their ability to auditorily monitor their environment, finally raise the probability that a group member will respond. (3) With regard to responses of a social ally, e. g. an infant's mother, it has been hypothesized that the sudden and extreme increases in cry intensity, which are quite typical for this display (Figure 6), can operate as a timer stimulus for caregiving responses. This has been exemplified by a dual component model. The model predicts that hearing a crying

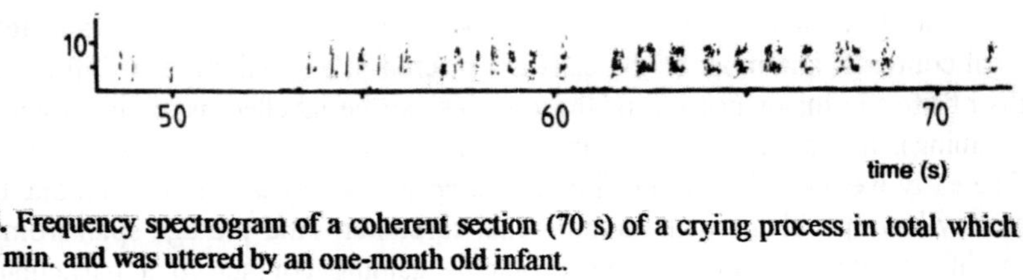

Figure 5. Frequency spectrogram of a coherent section (70 s) of a crying process in total which lasted for about 30 min. and was uttered by an one-month old infant.

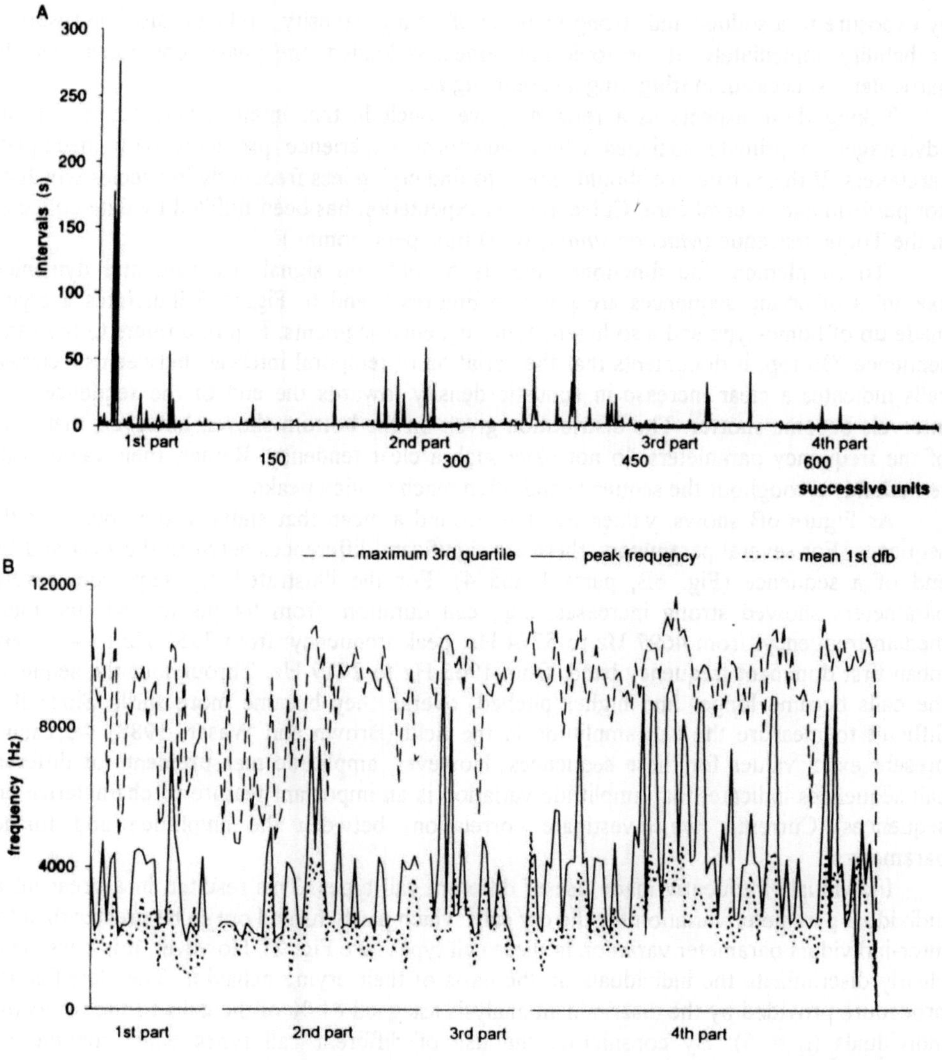

Figure 6. Time course of cry parameters (entire sequence; 30 minutes). (A) Duration of call intervals (time between onsets of successive calls). (B) Values of 'third quartile maximum'; 'peak frequency'; and 'mean first dominant frequency band'. The four parts in the lower graph correspond to the marked areas in the upper graph.

process influences the occurrence of a caregiving act by means of two different components. On the one hand, there is a tonic change in the potential of one component. This change which normally is an increase, follows from exposure to crying for a longer time and raises the response probability in a gradual manner. On the other hand, there are phasic changes in the potential of another component. Either of the phasic changes is caused by exposure to a sudden and strong increase of crying intensity, and can raise the response probability immediately. If the tonic component is loaded, the phasic component can be particularly successful in triggering a caregiving act.

Taking these aspects as a reference, we conclude that infant crying has significant advantages in primate societies where newborns experience problems with inadequate caretakers. If this is true one should expect to find crying less frequently in species which do not perform alloparental care. Currently this expectation has been fulfilled by data collected in the Toque macaque (*Macaca sinica*; W. Dittus, pers. comm.).

To supplement the functional aspects by data on signal structure and dynamics, examples of crying sequences are given in Figures 5 and 6. Figure 5 illustrates a crying made up of homo-type and also hetero-type sequence segments. Figure 6 refers to the same sequence. On top, it documents that the variation of temporal intervals between successive calls indicates a clear increase in acoustic density towards the end of the sequence, that intervals become shorter. The distribution given on the bottom shows, however, that most of the frequency parameters do not have such a clear tendency. Rather, their values vary remarkably throughout the sequence and often reach sudden peaks.

As Figure 6B shows, values oscillate around a mean that shifts in the course of the sequence. For several parameters, there are significant differences between the start and the end of a sequence (Fig. 6B, parts 1 and 4). For the illustrated cry sequence, several parameters showed strong increases, e.g., call duration: from 94 ms to 141 ms; mean median frequency: from 4697 Hz to 5774 Hz; peak frequency: from 3356 Hz to 4469 Hz; mean first dominant frequency band: from 1965 Hz to 2719 Hz. Throughout the sequence the calls became longer and higher pitched; overall they became more shrill. Since it is difficult to measure the call amplitude in the field (Brown and Waser, 1988) we cannot present exact values for these sequences. However, amplitude measurements of different call sequences indicate, that amplitude variation is an important feature to characterize cry sequences. Currently we investigate correlations between the amplitude and further parameters.

Individuals frequently made use of different call types. This resulted in a great intra-individual parameter variation for the cry calls which partly turned out to be greater than the inter-individual parameter variation in these call types (see Fig. 2). However, it is possible to clearly discriminate the individuals on the basis of their crying behavior. The classification procedure provided by the discriminant analysis assigned 81 % of the calls to the respective individuals (n = 5). By considering the use of different call types when running the discriminant analysis we could raise the rate of correct assignment up to 95 % (Hammerschmidt and Todt, 1994). The capacity of the vocal patterns to facilitate individual recognition was also shown with the aid of playback experiments (Hammerschmidt, 1990). Here, four Barbary macaque mothers were exposed to the simultaneous presentation ('two speaker choice design') of three cry calls of their own infant and three cry calls of a different infant. In all 25 playback settings, mothers never looked first to the loudspeaker of the other infant. In 16 cases they looked first to their own infant's loudspeaker, in 9 cases mothers looked to the position they presumed their infant to be.

There is evidence that the hetero-type organization of a crying process can amplify the interactional role of the variable dynamic of its acoustical parameters mentioned above. This effect which recently has been tested for human infants, is characterized as follows: (1) providing temporal triggers for the onset of caregiving in mothers or alloparents; (2) attracting the attention of adult listeners by, e.g., producing a kind of uncertainty about the meaning of perceived signals; (3) reducing the probability that signal addressees will habituate. In addition, however, one has to consider that both the variable dynamic and also the heterotype organization of crying may be a correlate of the ontogenetically early stage of vocalization and as such, a reflection of some incompetence in selective use of vocal signals (Todt, 1994).

Second paradigm : Calling during the mating season

Being incompatible with general implications of female selectivity (e.g., Trivers, 1972), the mating system of Barbary macaques was labeled 'promiscuous' (Taub, 1980 b). During the mating season (September to December) females often invite mounting of a selected male and also seem to control the course of a copulation (v. d. Bergh, 1989; Lehmann, 1989). Obviously, they do so by three sorts of gender-specific signals: A visual fixation directed backwards to the mounting male, termed *mating stare*, a use of one hand grasping one of the male's legs, termed *mating grasp*, and last but not least, a vocalization formed by a rhythmic succession of low frequency grunts, termed mating call (Todt and Pohl, 1984).

Because of the conspicuousness of the mating call it has been asked why females may produce this vocalization. Answers were proposed by three working hypotheses. The *trigger* hypothesis, stating that the calling, together with the other signals, provides cues which promote a successful copulation, for instance, by triggering a male's ejaculation. The *affiliation* hypothesis, saying that a female's signal patterns affect the sensory attention of a male which - taking copulation as a reinforcer - then may condition and enhance his affiliation with the particular mate. The *attraction* hypothesis, stating that a female's vocalizations, in contrast to mating stares and mating grasps, could stimulate not only the mounting individual, but also influence others by informing and attracting them via sound transmission, and so raise the probability of male-male competition (Todt and Pohl, 1984). To test these hypotheses, observers focussed on particular females (n=15) of a well known Barbary macaque population (Rocamadour site, see page 3) continuously for a longer period of time (three to six hours per day). Normally such observation times included several different episodes of mate contact, copulation and postcopulatory behaviors such as grooming. Concurrently, other observers were monitoring the area around a given consort pair and especially checking for approaches of other Barbary macaques. Analysis of data followed the normal procedures. The evaluation of behavioral data revealed results that can be summarized as follows (for further details see Todt et al., 1994).

First, there was a strong positive correlation between the occurrence rate of mating calls and the oestrus of females. Second, there was a positive correlation also between the mating calls given during copulation (here and in the following specified as *MC-c* calls) and the success of this copulation in terms of a male's ejaculation. In addition the analysis allowed us to identify a clear latency interval between the onset of a mating call and an ejaculation. Third, there was an even stronger correlation between *MC-c* calls and a postcopulatory maintenance of the consort pair. In almost all cases when this call was given, the male remained together with the female significantly longer than normally, and the majority of these episodes were filled by an extended mutual grooming. Finally, we did not find any evidence for an approach of another male when a female had given her mating call. These results allowed us to conclude, that Barbary macaque females address their *MC-c* calls predominatly to the mounting mates. In respect to the *trigger* hypothesis, we had to consider, however, that a direct stimulation of a male's ejaculation through a female could have been caused also by other behaviors, such as mating stare or mating grasp. The *affiliation* hypothesis, on the other hand, could neither be clearly accepted nor rejected to date. Nevertheless, the prolongation of postcopulatory behaviors observed after *MC-c* call utterance appears of interest in this concern. Currently, this phenomenon is interpreted by a novel working hypothesis saying that the postcopulatory behavior can be regarded as a sort of mate-guarding caused by the calls of the females. We assume that the male members of consort pairs may just not want to run any risk predicted by the *attraction* hypothesis (Todt et al., 1994). Although, in Barbary macaques, there is no finding yet, which meets such a prediction, this hypothesis merits further attention (Blaffer Hrdy and Whitten, 1987; Hauser, 1990; Todt et al., 1992).

During the mating season and related to female-male encounters, Barbary macaque females produce vocalizations which, just by ear, are difficult to distinguish from the *MC-c* calls given during copulation, but nevertheless do clearly differ from the latter ones in terms of two sets of characters: First, the fine-grained expression of specific acoustical parameters, and second, their relation to specific social contexts. To document these differences, this signal type received the label *MC-d* call (Figure 7, Table 1).

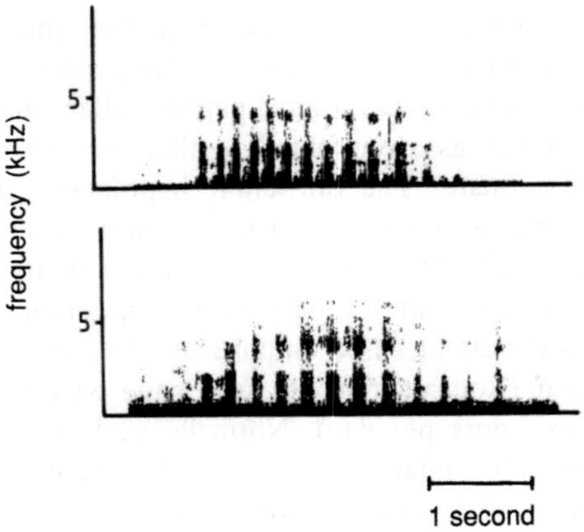

Figure 7. Spectrograms of the two types of mating call recorded from a given female. TOP: MC-c call uttered during copulation. BOTTOM: MC-d call vocalized during other sorts of female-male encounters in the mating season. Although the patterns look similar, the analysis of their parameters revealed significant differences between the two types (see Table 1).

Table 1. Differences between the two types of mating calls (MC-c, MC-d) ascertained by contrasting their acoustic parameters.

	MC-c	MC-d
First quartile (mean) [Hz]	725	1077
Second quartile (mean) [Hz]	1372	1677
max. peak frequency [Hz]	650	1132
frequency range (mean) [Hz]	1340	1546
First dominant frequency (mean) [Hz]	396	647

The *MC-d* calls are significantly correlated to the following three contexts: (1) The approached male is not ready to mate, either not showing any reaction to the female's approach and presenting, or threatens, slaps, and chases the female. The female typically runs away, turns around, and calls while looking at the male. (2) A consort pair is harassed by a second male. The female either calls when the first male is replaced or when the second male chases her. (3) A consort pair is disturbed by the approach of a competing female; the first female either calls immediately or sometimes threatens or chases the second female while calling. In all three situations females are unsuccessful in sexually soliciting a partner or in their attempt to mate. In order to test for differences between the two kinds of call series, a total of 1000 calls was submitted to a discriminant analysis. The analysis correctly classified 81.2 % of vocal patterns to the two call types (with a probability of correct assignment of 50 % due to chance).

In order to test the quality of assignment of MC-c and MC-D calls to the respective individuals a second discriminant analysis was conducted. 74 % of calls (n=500) were correctly assigned to the respective females (n=7). Since the classification results are based on parameter calculations of the single calls only, all sequential information which is contained in the series was disregarded. Taking into account that subjects can decode further information given in the entire series, we assume that the animals can recognize the calling individual and very likely can differentiate between MC-c and MC-d calls. In addition, we assume that both types of calls provide information about the female's readiness to mate, particularly because they most likely are individually recognizable.

Third paradigm : Calling in response to disturbances in the surroundings

Upon exposure to predators which pose an immediate danger, Barbary macaques produce alarm calls that normally lead to flight reactions or cryptic behavior in other group members. Sometimes, however, they utter long-lasting series of vocalizations which to the human ear sound like alarm calls. Also the spectrograms resemble those of alarm calls (Fig.3, A3 and B3). These call series usually do not evoke flight reactions. Therefore, we provisionally term these calls 'disturbance calls'. Both alarm calls eliciting escape reactions and long-lasting call series were also recorded in the field in Morocco in response to observers, shepherds and dogs.

To elicit such serial calling, subjects were repeatedly approached by a human at night. Acoustic analyses (Fischer et al., submitted) suggested that vocal patterns uttered in call series elicited by three different stimuli ('approached by a human at night', 'dog', and 'upon encountering a burrow in a new area') differed significantly from each other ($F=74.9$; $p<.001$). A total of 2679 calls from 19 different individuals were submitted to analysis. A discriminant analysis yielded a 83.7 % correct classification of vocal patterns to the three contexts. Figure 8A shows the three most significant parameters for discriminating between contexts. These parameters are median frequency, first dominant frequency width, and call duration. Whereas replies to the stimuli 'dog' and 'burrow' show similar values for the median frequency, they seem to differ in values of call duration. This indicates that for a correct assignment a detailed examination of more than one acoustic parameter is a crucial prerequisite. Moreover, a previous analysis (Hammerschmidt et al., 1994a) showed that the degree of correct assignment depends on the number of parameters available.

Our analysis also checked for differences within the call series mentioned above. Changes within one series were studied in those call series which were uttered after animals were approached by an observer at night. To determine within-context variation, 20 calls of each of the three positions '*start*', '*middle*', and '*end*' per series were examined. 56.3 % of the calls were correctly classified in regard to the different positions in a sequence. Here, the most significant parameters are mean amplitude, peak frequency, and first dominant frequency (Fig. 8B).

Besides differences in the amount of calling, we found significant differences for the following parameters, if these were calculated for the three serial positions *start, middle* and *end*: call duration, call density (call duration/interval; both $p<.001$), and for interval length (time between onsets of successive calls) ($p<.05$; Friedman Test) were assessed for the three positions. Whereas call duration and call density decreased, the interval length increased (Fig. 8C). Significant differences ($p<.05$) were revealed between *start* and *end* for the temporal parameters. In addition, calls sounded lower pitched along a call series.

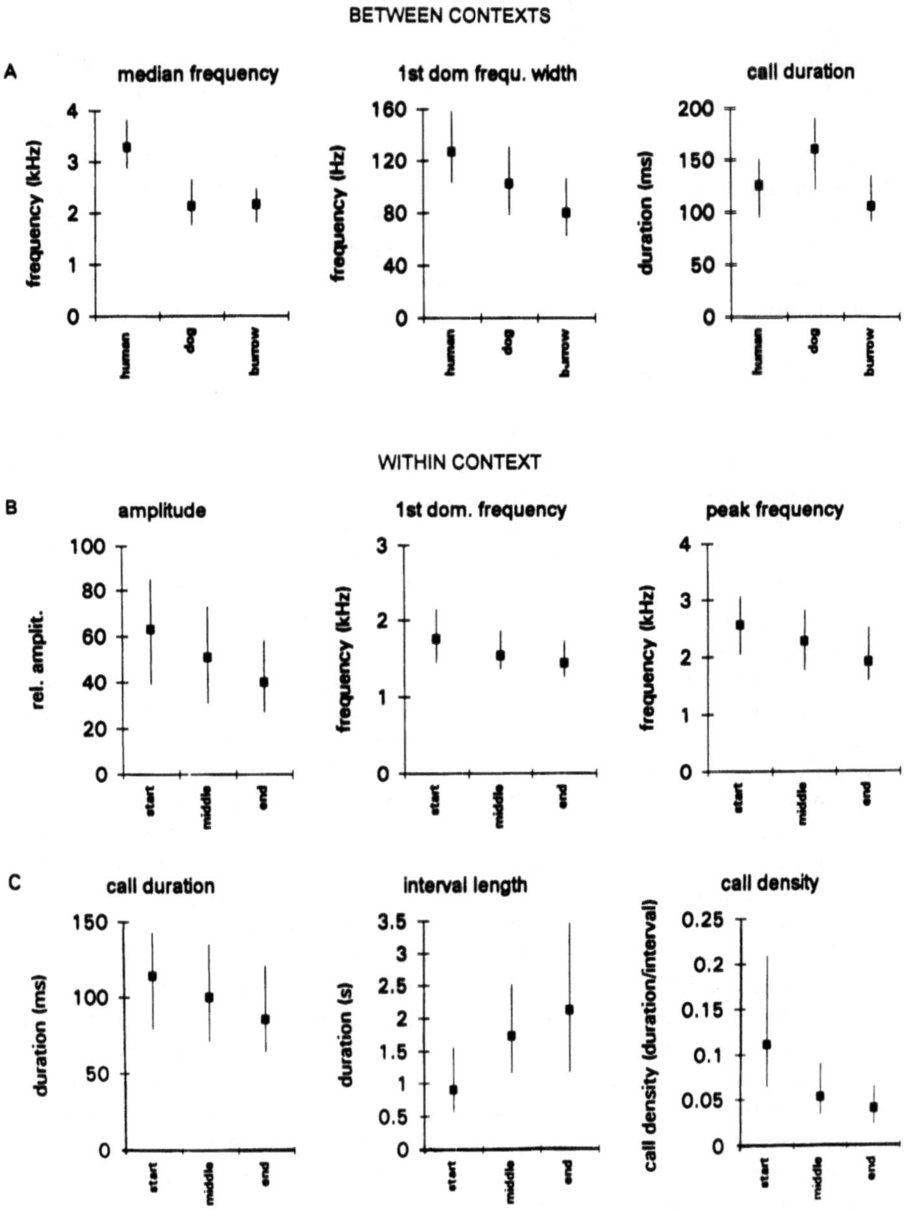

Figure 8. Median and 1st and 3rd quartile for (A) three parameters which were most crucial for a discrimination between contexts; (B) parameters crucial for the discrimination of positions in a series (within context); (C) temporal parameters for the three positions.

Finally, the calls given after experimental disturbance exhibited clear individual characteristics. The classification procedure for 15 individuals which were entered with 40-150 calls (total 1574 calls) into the analysis yielded a correct assignment of 86 %. It seems likely that the individual characteristics may be used for individual recognition of subjects.

Accordingly we conclude that Barbary macaques utter individually distinctive disturbance calls which, in terms of their acoustic features, vary with context. We assume that these variations are meaningful and may function in a referential manner (Marler et al., 1992). Within one apparently homogenous context, both temporal and frequency characteristics often exhibited gradual changes which probably point to properties of internal mechanisms underlying them.

ISSUES OF OUR FORTHCOMING RESEARCH

Our studies on the vocal behavior of Barbary macaques invite subsequent approaches which basically address four issues. First, is a parametric analysis of the complete vocal repertoire of this species, including a detailed elucidation of ontogenetic trajectories of signal development. Thereby, we expect to establish a basis for comparing the accomplishments of Barbary macaques to those of other nonhuman primates (Snowdon, 1988; Hauser, 1989; Symmes and Biben, 1992). Second, is an analysis of correlations between vocal and nonvocal displays of this species. Here, a special focus will be on the coordination of signals performed during playful interaction. Thereby, we expect to improve the understanding of strategies of signal choice and use in a widely neglected domain of behavior (Biben and Symmes, 1986; Goedeking, 1988). Third, is a further investigation of serial vocalizations on different hierarchy levels of signal organization, namely within and between calls. Thereby, we expect to achieve a behavioral access to properties of the mechanisms which underlie intra-serial variations of signal parameters. Finally, is a more extended study on correlates of individuality and also a more sophisticated inquiry into the relevance of external factors normally notated just as 'context'. The third and the last issue merit a brief explanation.

Serial parameter variation: Appropriate models to address rules of parameter variation make four alternative predictions, if the measurement is done unit-wise over a given succession of homo-type units, e. g. calls or call elements: (1) Solofluent model: Parameter values remain almost the same from the beginning to the end of a series. (2) Confluent model: Parameters vary in values at the beginning of a series but converge later on and show a specific value at a series' end. (3) Diffluent model: Parameters show a specific value at a series' beginning but diverge and vary in values towards its end. (4) Multifluent model: Parameter values vary during the entire series, often in an unpredictable manner (Thimm, 1973). Biological implications of these models have been discussed by Todt (1986). First results on the vocal behavior of Barbary macaques indicate that serial variation following the solofluent model is rare. As far as macaque vocalizations follow one of the other three models, parameters often do so without showing a fixed coupling. This points to mechanisms which control the retrieval and performance of vocal patterns by an intrinsic parallel proccessing (Todt and Hultsch, 1994).

Individuality and Context: Notations about animal communication should contain information about four items: the signals, their producers, the context and responses of recipients. In the case of acoustic signals their recording is not a problem, normally, and the same is true for specifying a signaller and a context. However, other variables require analytic treatment that has to be improved if a research is to progress to a sophisticated level of understanding. Our study on Barbary macaque vocalizations attempts to address this challenge.

A satisfying specification of signallers, and of signal addressees or recipients, for instance, may be a problem because there is more to a given primate than the properties that allow his/her classification. Studying this problem entails dealing with the diversity that individuals often show in their behavioral performances. In this respect, a novel approach is necessary which examines *individuality* as a complex of different variables. A concept suitable to guide such an approach is provided by a biological version of systems theory. This concept suggests that comparison of individual *systems* should be postponed until each of them has been studied separately and with sufficient rigor for predictions of its individual properties. Presently, we are applying methods of systems theory in order to better address rules of intra-individual variations in Barbary macaque signalling (for details see Todt et al., 1992).

Intra-individual variation in signalling is a particularly challenging issue when it occurs in situations which, according to conventional notation are classified as belonging to the '*same context*' (e. g. 'attacked by juvenile X, 'returned to mother', 'on nipple'). Notwithstanding the possibility that variable occurrences of vocalizations could be caused by short-term changes in the intrinsic state of a signaller, one has to take into account that the conventional classification of contexts may be too rough or even insufficient. At present, we are trying to assess by which criteria the Barbary macaques categorize contexts. In other words, and in borrowing from Cheney and Seyfarth (1990), we want to find out how monkeys see and judge their respective worlds. For this aim, the study of vocal expressions that mirror such judgements by, e.g., subtle modifications of specific signal parameters, seems to be an appropriate tool.

ACKNOWLEDGEMENT

We thank Ellen Merz and Gilbert de Turckheim for permission to study at 'La Forêt des Singes', and Lucien Ambiehl and M. Legrelle for cooperation in the park. In addition, we are grateful for financial support provided by the DFG (To 13/15-1, To 13/21-1).

REFERENCES

Ansorge, V., Hammerschmidt, K., and Todt, D., 1992, Communal roosting and formation of sleeping clusters in Barbary macaques (*Macaca sylvanus*), *Amer. J. Primatol.* 28: 271-280.

Bartecki, U., 1986, The social position of male Barbary macaques (*Macaca sylvanus*) in a semifree ranging population, *Primates* 27: 173-183.

Beeman, K., 1989, SIGNAL User's Guide. Belmont, Massachusetts: Engineering Design.

Bergh, U. v. d., 1989, Paarungsverhalten von Berberaffen (*Macaca sylvanus*): Analyse einzelner Verhaltensparameter im Paarungsablauf. Diplom-Thesis, Freie Universität, Berlin.

Biben, M. and Symmes, D., 1986, Play vocalizations of squirrel monkeys (*Saimiri sciureus*), *Folia Primatol.* 46: 173-182.

Blaffer Hrdy, S. and Whitten, P.L., 1987, Patterning of sexual activity, *in:* "Primate Societies", B.B. Smuts, D.L. Cheney, R.M. Seyfarth, R.W. Wrangham, and T.T. Struhsaker, eds., University of Chicago Press, Chicago.

Brown, C.H. and Waser, P.M, 1988, Environmental influences on the structure of primate vocalizations, *in:* "Primate Vocal Communication", D. Todt, P. Goedeking, and D. Symmes, eds., Springer, Berlin, New York.

Cheney, D.L. and Seyfarth, R.M., 1990, "How Monkeys see the World", University of Chicago Press, Chicago.

Cleveland, J. and Snowdon, C.T., 1982, The complex vocal repertoire of the adult cotton-top tamarin (*Sanguinus oedipus oedipus*), *Z. Tierpsychol.* 58: 231-270.

Deag, J.M., 1974, A study of the social behavior and ecology of the wild Barbary macaque, *Macaca sylvanus*, L. 1758. Dissertation, University of Bristol, Bristol.

Deag, J.M., 1980, Interactions between males and unweaned Barbary macaques: Testing the agonistic buffering hypothesis, *Behaviour.* 75: 54-81.
Deag, J.M. and Crook, J.H., 1971, Social behaviour and agonistic buffering in the wild Barbary macaque, *Macaca sylvanus*, *Folia Primatol.* 15: 183-200.
Green, S. ,1975, Variation of vocal pattern with social situation in the Japanese monkey (*Macaca fuscata*). *in*: "Primate behavior", Vol.4, L. A. Rosenblum, ed., Academic Press, New York.
Fa, J.F., 1984, "The Barbary Macaque: a Case Study in Conservation". Plenum Press, New York.
Fischer, J., Hammerschmidt, K. and Todt, D., submitted, Factors affecting acoustic variation in Barbary macaque (*Macaca sylvanus*): disturbance calls, *Ethology.*
Gautier, J.P. and Gautier-Hion, A., 1988, Vocal quavering: A basis for recognition in forest guenons. *in:* "Primate Vocal Communication", D. Todt, P. Goedeking, and D. Symmes, eds., Springer, Berlin.
Goedeking, P., 1988, Vocal play behavior in cotton-top tamarins. *in:* "Primate Vocal Communication", D. Todt, P. Goedeking, and D. Symmes, eds., Springer, Berlin.
Gouzoules, H., Gouzoules, S., and Marler, P., 1984, Rhesus monkey (*Macaca mulatta*) screams: representational signalling in the recruitment of agonistic aid, *Anim. Behav.* 32: 182-193.
Green, S., 1975, Variation of vocal pattern with social situation in the Japanese monkey (*Macaca fuscata*): a field study. *in:*"Primate behavior", L.A. Rosenblum, ed., Academic Press, New York.
Hammerschmidt, K., 1990, Individuelle Lautmuster bei Berberaffen (*Macaca sylvanus*): Ein Ansatz zum Verständnis ihrer vokalen Kommunikation. Dissertation, Freie Universität, Berlin.
Hammerschmidt, K. and Ansorge, V., 1989, Birth of a Barbary macaque (*Macaca sylvanus*): Acoustic and behavioural features, *Folia Primatol.* 52: 78-87.
Hammerschmidt, K. and Todt, D., submitted, Individual differences and recognition of vocal patterns in young Barbary macaques (*Macaca sylvanus*), *Behaviour.*
Hammerschmidt, K., Ansorge, V., and Fischer, J., 1994a, Age-related variations in the vocal repertoire of Barbary macaques, *Curr. Primatol.* 2: 57-68.
Hammerschmidt, K., Ansorge, V., Fischer, J., and Todt, D., 1994b, Dusk calling in Barbary macaques (*Macaca sylvanus*): Demand for social shelter, *Amer. J. Primatol.* 18: 324-335.
Hauser, M.D., 1989, Ontogenetic changes in the comprehension and production of vervet monkey (*Cercopithecus aethiops*) vocalizations, *J. Comp. Psychol.* 103: 149-158.
Hauser, M.D., 1990, Do chimpanzee copulatory calls incite male-male competition? *Anim. Behav.* 39: 596-597.
Hauser, M.D., 1993, Do infant vervet monkeys cry wolf? *Anim. Behav.* 45: 149-158.
Hauser, M.D., Evans, C. S., and Marler, P., 1993, The role of articulation in the production of rhesus monkey, *Macaca mulatta*, vocalizations, *Anim. Behav.* 45, 423-433.
Küster, J. and Paul, A., 1984, Female reproductive characteristics in semi-free ranging Barbary macaques (*Macaca sylvanus*), *Folia Primatol.* 43: 69-83.
Küster, J. and Paul, A., 1986, Male-infant relationships in semi-free ranging Barbary macaques (*Macaca sylvanus*) of Affenberg Salem/FRG: testing the 'male care' hypothesis, *Amer. J. Primatol.* 10: 315-327.
Lehmann, J., 1989, Akustische Parameter vokaler Muster weiblicher Berberaffen (Macaca sylvanus) im Kontext des Paarungsverhaltens. Diplomarbeit, Freie Universität, Berlin.
Marler, P., Evans, C.S., and Hauser, M.D., 1992, Animal signals: Motivational, referential, or both ? *in:* "Nonverbal vocal communication", H. Papousek, U. Jürgens, and M. Papousek, eds., Cambridge Univ. Press, Cambridge.
Mehlmann, P., 1986, Male intergroup mobility in a wild population of the Barbary macaque (*Macaca sylvanus*), Ghorman Rif/Morocco, *Amer. J. Primatol.* 10: 67-81.
Mènard, N., Vallet, D., and Gautier-Hion, A. 1985, Démographie et reproduction de *Macaca sylvanus* dans différents habitats en Algérie, *Folia Primatol.* 44: 65-81.
Mènard, N., Hecham, R., Vallet, D., Chikhi, H., and Gautier-Hion, A., 1990, Grouping patterns of a mountain population of *Macaca sylvanus* in Algeria - a fission-fusion system? *Folia Primatol.* 55: 166-175.
Newman, J.D., 1985, The infant cry of primates: an evolutionary perspective. *in:* "Infant Crying" B.M. Lester and C.F.Z. Boukydis, eds., Plenum Press, New York.
Newman, J.D. and Goedeking, P., 1992, Noncategorical vocal communication in primates: The example of common marmoset phee calls, *in:* "Nonverbal vocal communication", H. Papousek, U. Jürgens, and M. Papousek, eds., Cambridge Univ. Press, Cambridge.
Owren, M.J. , Seyfarth, R. M., and Hopp, S. L., 1992, Categorical vocal signaling in nonhuman primates, *in:* "Nonverbal vocal communication", H. Papousek, U. Jürgens, and M. Papousek, eds., Cambridge Univ. Press, Cambridge.

Paul, A., 1984, Zur Sozialstruktur und Sozialisation semi-freilebender Berberaffen (*Macaca sylvanus* L.1758). Dissertation, Universität Kiel.

Paul, A. and Küster, J., 1985, Intergroup transfer and incest avoidance in semi-free ranging Barbary macaques (*Macaca sylvanus*) at Salem (FRG), *Amer. J. Primatol.* 8: 317-322.

Paul, A., Küster, J., and Arnemann, J., 1992, DNA fingerprinting reveals that infant care by male Barbary macaques (*Macaca sylvanus*) is not paternal investment, *Folia Primatol.* 58: 93-98.

Quintilius, U. and Hultsch, H., 1991, Correlates of visual monitoring in Barbary macaques (*Macaca sylvanus* L.), *Primate Report* 31: 30.

Riechelmann, C., Todt, D., and Hultsch, H., 1994, Early development of social relationships in Barbary macaques (*Macaca sylvanus*): trajectories of alloparental behaviour during the infant's first three months of life, *Curr. Primatol.* 2, 31-38.

Robinson, J.G. and Whitehead, J.M., 1988, Glossary of terms, *in:* "Primate Vocal Communication", D. Todt, P. Goedeking, and D. Symmes, eds., Springer, Berlin.

Roshani, A., Todt, D., and Janik, V.M., 1994, Yawning in male Barbary macaques, *Curr. Primatol.* 2: 39-45.

Rowell, T.E. and Hinde, R.A., 1962, Vocal communication by the rhesus monkey (*Macaca mulatta*), *Proc. Zool. Soc. Lond.* 138: 279-294.

Schrader, L. and Todt, D., 1993, Contact call parameters covary with social context in common marmosets (*Callithrix j. jacchus*), *Anim Behav.* 46: 1026-1028.

Small, M.F., 1990a, Alloparental behaviour in Barbary macaques, *Macaca sylvanus*, *Anim. Behav.* 39: 297-306.

Small, M.F., 1990b, Promiscuity in Barbary macaques (*Macaca sylvanus*), *Amer. J. Primatol.* 20: 267-282.

Snowdon, C. T., 1988, Communication as social interaction: Its importance in ontogeny and adult behavior. *in:* "Primate Vocal Communication" D. Todt, P. Goedeking, and D. Symmes, eds., Springer, Berlin.

Symmes, D. and Biben, M., 1992, Vocal development in nonhuman primates. *in:* "Nonverbal vocal communication", H. Papousek, U. Jürgens, and M. Papousek, eds., Cambridge Univ. Press, Cambridge.

Taub, D.M., 1980a, Testing the 'agonistic buffering' hypothesis. 1. The dynamics of participation in the triadic interaction, *Behav. Ecol. Sociobiol.* 6: 187-197.

Taub, D.M., 1980b, Female choice and mating strategies among wild Barbary macaques (*Macaca sylvanus*). *in:* "The Macaques: Studies in Ecology, Behaviour and Evolution", D.M. Taub, ed., Van Nostrand Reinhold, New York.

Taub, D.M., 1984, Male caretaking behaviour among wild Barbary macaques. *in:* "Primate Paternalism", D.M. Taub, ed., Van Nostrand Reinhold, New York.

Thimm, F., 1973, Sequentielle und zeitliche Beziehungen im Reviergesang des Gartenrotschwanzes (*Phoenicurus phoenicurus* L.), *J. Comp. Physiol.* 84: 311-334.

Timme, A., 1989, Differentielle Aufzuchtmethoden semifrei lebender Berberaffenweibchen (*Macaca sylvanus*). Dissertation, Universität Göttingen.

Todt, D., 1986, Hinweis-Charakter und Mittler-Funktion von Verhalten, *Z. Semiotik* 8: 183-232.

Todt, D., 1988, Serial calling as a mediator of interaction processes: crying in primates, *in:* "Primate Vocal Communication", D. Todt, P. Goedeking, and D. Symmes, eds., Springer, Berlin.

Todt, D., 1989, Das Schreiverhalten von Primatenkindern: Basis für ein besseres Verständnis auch unserer eigenen Traglinge? *Fortschr. Sozialpädiatrie* 13: 150-170.

Todt, D., 1994, The dynamic of crying and its interactional role: phasic versus tonic components, *in:* "Social and Biological Aspects of Infant Crying", B. Lester, J.D. Newman, and F. Pedersen, eds., Plenum Press, New York.

Todt, D. and Hultsch, H., 1992, Variations that follow rules, *Behav. Brain Sci.* 15: 289-290.

Todt, D. and Hultsch, H., 1994, Decoding information from serial signal prossess. Proc. XVth Conf. Int. Primatol. Soc., Bali, p.272.

Todt, D. and Pohl, R., 1984, Communicative strategies in estrous Barbary ape females (Macaca sylvanus) during copulation behaviour: advertising, triggering, affiliating, *Verh. Dtsch. Zool. Ges.* 7: 225.

Todt, D., Hammerschmidt, K., and Hultsch, H., 1992, The behaviour of Barbary macaques (*Macaca sylvanus* L. 1758): Perspective and projects of a long-term study, *Primate Report* 32: 19-30.

Todt, D., Hammerschmidt, K., and Pohl, R., 1985, The dusk-chorus of the Barbary macaque (*Macaca sylvanus*): an age-specific performance, *Verh. Dtsch. Zool. Ges.* 78: 217.

Trivers, R.L., 1972, Parental investment and sexual selection. *in:* "Sexual selection and the descent of man", B. Campbell, ed., Aldine-Atherton, Chicago.

Turckheim, G., De and Merz, E., 1984, Breeding Barbary macaques in outdoor open enclosures, *in:* "The Barbary Macaque: a Case Study in Conservation", J.F. Fa, ed., Plenum Press, New York.

Wolff, A. and Todt, D., 1985, The effect of vocal signals performed by newborn monkeys (*Macaca sylvanus*) on infant-caretaker distance, *Verh. Dtsch. Zool. Ges.* 78: 219.

LOUD CALLS IN GREAT APES: SEX DIFFERENCES AND SOCIAL CORRELATES

Gottfried Hohmann and Barbara Fruth

Forschungsstelle für Humanethologie
der Max-Planck-Gesellschaft, 82346 Andechs, Germany
Zoologisches Institut der Universität München
80333 München, Germany

INTRODUCTION

If one were to scan the literature on primate behavior accumulated during the last two decades, one single taxon, the great apes, would likely dominate. Switching the key word to "communication" would already decrease the bias drastically, and when selecting for "vocal communication", the taxon would produce a rather marginal file (Marler, 1976; Marler and Tenaza, 1977). Behaving "almost human" in some aspects, there is no evidence that vocal communication by the great apes exceeds that of other non-human primates. This view is certainly biased. Until recently, we knew little about the structure and function of their natural vocal repertoires. Instead, researchers taught chimpanzees sign language and other artificial communication systems (for reviews see Sebeok and Umiker-Sebeok, 1980; Wallman, 1992). These studies may have quantified the cognitive abilities of the subjects but little was learned about the mode of natural communication among conspecifics. The picture is slowly changing. Long term field studies on all four species permit systematic research of vocal communication of focal individuals. Studies of animals in captivity use the advantages of more controlled conditions to evaluate the functional significance of particular calls. In the field, attention focused on elements of the vocal repertoire which are likely to facilitate distance communication. For several reasons this bias is not surprising. First, these *loud calls* are easy to detect by human researchers even against high levels of background noise. Second, because of their high amplitude and other physical adaptations for long-range transmission, loud calls are easy to record. Third, loud calls are of special interest because they are thought to encode and transmit information exclusively in an acoustic manner. Finally, field observations on various primate species have demonstrated that conspecifics' responses to loud calls are both detectable and predictable. This is in contrast to most of the other elements which either elicit no obvious behavioral response or a variety of different responses.

Current Topics in Primate Vocal Communication
Edited by E. Zimmermann *et al.*, Plenum Press, New York, 1995

Loud calls are used by many primate species and have been the subject of detailed studies (e.g., Gautier and Gautier, 1977; Zimmermann, this volume). Analyses of primate calls that may facilitate long distance communication have revealed that superior propagation depends on both the energy distribution in the spectrum of ambient noise ("sound windows") and the spectral features of the signals (Brown, 1989; Waser and Brown, 1984). Comparative data from two species (blue monkeys and mangabeys) have also demonstrated a positive correlation between the range of transmission of long distance calls and home range size (Brown, 1989). However, species-specific differences in signal structure have been attributed to differences in competition between males rather than to differences in the acoustic environment (Waser, 1982). The classification of a vocal signal as a "loud call" is normally based on several criteria including intensity (sound pressure level), structural originality and low variability in structural features (Gautier and Gautier, 1977). There are only few studies in which the sound pressure level of primate loud calls was measured (e.g., Brown, 1989; Whitehead, 1987) and in most species (including all the great apes) evaluations of the intensity were based on subjective judgements of a human listener. Evaluation of the extent of structural variability of a particular call requires samples from a large number of different subjects, and samples of the same individual collected during a longer period of time. For some species, these data are available (e.g., Gautier and Gautier, 1977, for forest guenons and Deputte, 1982, for gibbons). However, they are not available for the four species of great apes.

In spite of their high intensity, loud calls may not always facilitate long distance communication but have a relatively short audible range (Brown, 1989). The calls selected for this comparison are among the loudest produced by the given species and can be heard by observers on the ground several hundred meters away from the source. Nevertheless, it is important to emphasize that other calls produced by each species may carry further, and experience less degradation and less attenuation than the samples selected here. Comparing the audible range of vocal signals with home range size of mangabeys and forest guenons, Brown (1989) was able to distinguish "public calls" from "private calls". One could probably do the same with gorillas who live in coherent harem groups. Considering the fission-fusion society of chimpanzees and bonobos the distinction between the two categories would be more difficult. In orangutans, all vocalizations (except for calls used for communication between mother and dependent offspring) would be scored as "public" signals. In the absence of empirical data on both, the vocal signals and the acoustic environment we will use the terms "loud call" and "distance communication" in a conventional way.

The following paper reviews results of current studies on loud calls in the four great ape species. In addition to reviewing written work, we present the first data from bonobos collected by the authors during an ongoing field study at Lomako (Zaire). Special attention will be given to the various hypotheses concerning the potential function of these calls in the different species. We also examine sex differences in the use of loud calls, focusing on the question of why females of some species refrain from loud calling while females of other species participate in this activity. To complement the great ape data, we also address this question using data from simian primates. The hypothesis we propose can be tested more systematically in future projects. Provided this happens, we will have achieved the major goal of this paper, that is, to stimulate greater interest in a previously neglected subject, the significance of female loud calls.

For description of the acoustic properties of the calls, we adopted part of the terminology introduced by Struhsaker (1967) which contains the following elements: unit (tonal, non-tonal), phrase and bout. Tonal units consist of discrete frequency bands, while non-tonal units are characterised by their diffuse spectral distribution. Units may be simple (one element) or compound (two elements follow one another). Phrases consist of several units with the same structure (uniform) or different structural features (mixed) produced by the same individual. Bouts are clusters of phrases given by one and the same individual. In contrast, choruses consist of a number of phrases from different individuals uttered more or less simultaneously.

LOUD CALLS OF GREAT APES

Orangutan

There is little information on vocal communication in orangutans and the only vocalization which has been subject to more detailed studies is the *long call* (Galdikas, 1983; Galdikas and Insley, 1988; MacKinnon, 1974; Mitani, 1985).

Structure (Table 1): The term *long call* refers to phrases of units with different physical features. It starts with pulsed grumbles, continues with bellows, and trails off with grumbles and sighs (Galdikas and Insley, 1988; MacKinnon, 1974). There might be individual markers because human observers, familiar with the subjects, were able to identify the different males by their long calls (Galdikas, 1983; Mitani, 1985). Galdikas and Insley (1988) distinguish a second type of distance call, the *fast call*. Compared with long calls, fast calls were uttered less frequently (1.7% of all long calls), structural variation within fast calls seemed to be less pronounced, phrases were of longer duration and units were emitted at a higher rate.

Site Specificity: As a prelude to their long calls, males from Tanjung Puting (Central Kalimantan) push over large branchless trees (Galdikas, 1983). Originally, it was proposed that this peculiarity is unique to this site. However, in Kutai Game Reserve (East Kalimantan) the noise of falling branches was always followed by long calls (Mitani, 1985), indicating that at least the connection between this external stimulus and long calls is not restricted to the population of Tanjung Puting.

Context: Empirical data on the context of long calling are available from studies at Tanjung Puting (Galdikas, 1983; Galdikas and Insley, 1988) and from Kutai (Mitani, 1985). At both sites the vast majority of samples (93% at Tanjung Puting and 82% at Kutai) were given spontaneously. Loud noises caused by falling trees or long calls of conspecifics evoked corresponding calls, too. Encounters between adult males were the most reliable releaser for long calls. With one exception, all encounters were characterized by aggressive interactions, and it was the dominant individual who gave long calls. Males not visible but calling in close proximity also increased the rate of vocalization by the dominant male. Mating seems to be another context for long calling. Most copulations observed (87% at Tanjung Puting, 67% at Kutai) were preceded or accompanied by long calls. Galdikas (1983) observed that long calls by adult males often (43%) evoked proceptive behavior in adolescent females and that receptive adolescent females were locating (and approaching) adult males by their long calls. She suggested that initiation of contact with other individuals was the major stimulus for long call vocalization. Conversely, Mitani (1985) found that males called most frequently following associations between males and females. Since the fast calls given by male orangutans at Tanjung Puting have been characterized as a "post-conflict" vocalization (Galdikas and Insley, 1988), this disparity could be a matter of definition rather than a site-specific pattern. Both studies have demonstrated that the response to long calls varies with the distance between sender and receiver. In Tanjung Puting, focal animals showed visible responses most often when the distance of the vocalizer was below 400 m. In Kutai females associated with adult or subadult males moved away from the loudspeaker even at distances above 400 m. Observations from both sites suggest that the response of males does not depend solely on the distance between sender and receiver but is also an expression of the dominance relationship between the two individuals. The motoric response (approach-avoidance) of three adult males from Tanjung Puting towards long calls given within the spatial limits of 400 m indicated a linear hierarchy.

Function: It has been hypothesized that male long calls have two different functions: spacing among males and mate attraction. Both functions are not mutually exclusive but may in fact complement each other. To test both hypotheses, Mitani (1985) conducted a series of playback experiments. Tests of the male-spacing hypothesis included the following: (a) playback of a long call from the dominant (resident) male to 4 subordinate (non-resident) adults and 4 subadults, and (b) playback of a long call given by a subordinate (non-resident) adult to the dominant male. In response to playbacks of the long call from a subordinate adult, the counter-calling dominant male approached the playback site. Hearing the long call of the dominant male, subordinate adults and subadults silently moved away from the location of playback and moved with a faster speed than before the experiment. To test the mate-attraction hypothesis, a long call of the dominant male was presented to (a) 3 adult females with infants and (b) 2 adult females without infants but accompanied by subadult males. Following the playback, females remained stationary or avoided the adult male. These data do not support the mate-attraction hypothesis but are not sufficient to reject it completely (Mitani, 1985).

Gorilla

Recent studies on vocal communication of gorillas have focused on calls used for communication over short distances (Harcourt et al., 1986, 1993). Information on distance communication has come from studies on mountain gorillas by Schaller (1963) and Fossey (1972). Out of 22 different vocal patterns distinguished by Schaller (1963), four may serve communication over large distances. Fossey (1972) described 16 types of vocalization including the *roar*, *wraagh* and *hoot series*. Both studies agree that roars and hoot series are given almost exclusively by silverback males and are usually followed by aggressive displays. Wraaghs in contrast are given by members of all age/sex classes (except infants) and are not accompanied by aggressive displays. Hoot series are that gorilla loud call with the highest similarity to loud calls of the other great apes.

Structure (Table 1): Hoots are tonal or mixed units usually given in phrases (hoot series) consisting of up to 20 units. Considering differences in duration of units and intervals, Schaller (1963) distinguished two different sections (hooting and growling) within a phrase. Hoot series are frequently terminated by other acoustic cues like chest beating, ground thumps and branch braking.

Context: Hoots were given exclusively during encounters with other groups or lone males (Fossey, 1972; Elliot, 1976). During close encounters (600 m or less) hoot series were more likely to be terminated by displays (chest beat, parallel run). The most common releaser for hoot series were corresponding vocalizations from other males. Group members responded to hoots with dispersion.

Table 1. Some structural features of loud calls of great apes (references are given in the text).

Species	Type/ section	Fundam. frequency (Hz)	Duration of unit (sec)	Units per phrase	Phrase duration (sec)	Transmiss. range (km)
Pongo pygmaeus	long call			18	14.1	0.8
	- grumbles	100-500	0.01-0.05			
	- bellows	100-1200	0.13-1.63			
	- sighs	100-600	0.13-1.30			
	fast call	?	?	34	54.7	?
Gorilla gorilla	hoot	100-700	0.03-0.20			
	series	1400-1800		>20	6.8	?
Pan troglodytes	pant hoot			23	2.0	0.6
	- introduction		?			
	- build up	250-350	0.15-0.25			
	- climax	600-1400	0.40-1.00			
	- led down					
Pan paniscus	low hoot	500	0.10-0.27	5-17	1.2-4.5	0.3
	high hoot	2300	0.15-0.70	4-41	1.5-5.8	0.5

Function: Observations on the context of emission of hoot series have indicated their significance for intergroup communication. Field observations gave the impression that the members of a given group responded differently (retreat or approach) to hoot series from different silverbacks (Akles, cited in Fossey, 1972) suggesting that conspecifics may be able to recognize the identity of the vocalizer by voice. Changes in spacing might be the most important consequence of hoot series, but the data available are not sufficient to determine if the function is primarily distance-increasing, distance-maintaining, or both.

Chimpanzee

The vocal repertoire of chimpanzees includes a number of calls for long range communication (Goodall, 1968, 1986). Among those, *pant hoots* are one of the most frequently uttered calls of adult chimpanzees. They are given by males and females of all age classes and can be heard during all hours of the day and from nesting sites at night (Goodall, 1986).

Structure (Table 1)**:** Single units consist of a voiced inhalation and exhalation phase and are given in phrases. Within a phrase, different sections can be distinguished (Marler and Hobbet, 1975): (1) the *introduction* with relatively long, unmodulated units, (2) the *build-up* characterised by an increase in both amplitude and range of frequency modulation,

(3) the *climax* consisting of comparatively long units with extensive modulation, and (4) the *let-down* which mirrors the build-up. Phrases can incorporate units of other vocal and non-vocal cues (e.g., shrieks and drumming) and may vary within a wide range according to the identity, age and sex of the vocalizer, and perhaps also with the context of emittance.

Individuality and Rank: Most field-workers agree that pant hoots of different individuals can be distinguished by humans even at large distances. Marler and Hobbet (1975) provided the first detailed description of structural differences suggesting various dimensions for individual specificity. However, recent studies suggest that at least some of the individual-specific features are surprisingly flexible. Disappearance of four males from a habituated community observed by Boesch (1991) caused major structural changes in the pant hoots of three other males. One male adopted the style of pant hooting formerly used by one of the missing males. At the same time the alpha male (Brutus) had modified his pant hoots and another male (Falstaff) "...started to produce exactly the long series of rapid hoots that Brutus had used previously" (Boesch, 1991, p. 83). This suggests that certain features of pant hoots could be related to social rank and/or dominance status. More evidence for rank-related differences comes from a study conducted by Mitani and Nishida (1993) at Mahale. In this study, seven adult focal males were ranked according to the outcome of agonistic interactions and the direction of pant grunts. High ranking males gave more pant hoots than low ranking males. While the calls given by the alpha male received the highest rate of vocal responses from other males, the alpha male responded least to the calls of others. However, none of the other six focal males showed the same pattern of vocal activity.

Age and Sex: Individually known chimpanzees of habituated communities have been studied for more than three decades. However, information about the ontogeny of vocal communication is virtually non-existent. Pusey (1990) reported that the rate of pant hooting increases with age and that in the case of females this development seems to be connected to the first estrous. Recordings from Gombe suggest that phrases of older males were shorter compared to phrases given by younger males (Marler and Hobbet, 1975). However, older males were also higher in rank and the structural variation could therefore equally express rank differences.

It is unfortunate that even in the more detailed studies, the pant hoots given by females were - for whatever reason - excluded from analyses (e.g., Mitani and Nishida, 1993; Mitani et al., 1992) or pooled with male calls (Clark, 1991). Hence, the existence of sex differences remains to be demonstrated. One difference between males and females might be the rate of pant hooting. Juvenile females produced pant hoots less frequently and joined choruses less often than male peers (Pusey, 1990). Figures on vocal rates from several other studies indicate a similar tendency (e.g., Clark, 1991; Marler, 1976; Mitani and Nishida, 1993). However, because of the consistent rank differences between the sexes (males are always dominant over females), it is not yet possible to distinguish precisely between the effects of rank and sex.

Site Specificity: Samples of male pant hoots recorded at Gombe and Mahale, respectively, have revealed differences in temporal and spectral features (Mitani et al.,1992). Individuals from Mahale, emitted the second section (build-up) of their pant hoots at a faster rate and the climax with a higher pitch than chimpanzees at Gombe. However, any interpretation of these site-specific differences is difficult because the two data sets are not really compatible. In Gombe, recordings were made during a short period of time and only from animals visiting the artificial feeding site. In Mahale, the period of sampling spans 6 years and recordings were collected under natural conditions (Mitani et al., 1992). Both sites are isolated and genetic differences between the two populations cannot be excluded. Genetic variation could have an impact on anatomical and/or morphological traits which in turn could cause differences in sound production. Alternatively, chimpanzees could acquire their vocal performance through learning and develop site-specific dialects. However, until compatible data sets are available the problem of site specificity remains an open question.

Context: Pant hoots are given in a wide range of diverse and partly disparate contexts (Goodall, 1965, 1986): arrival at a rich feeding site, capture of animal prey, joining of other community members, response to voices of strange conspecifics, and during agonistic displays. However, until recently evidence for context-related structural variation was lacking. According to Wrangham (1977) pant hoots given on arrival at large feeding sites sounded different from pant hoots given in other contexts. The structural attribute that appeared to be related to this specific context is the presence/absence of the last section (let-down) of a phrase. In a study at Kibale Forest (Uganda), Clark (1991) and Clark and Wrangham (1993) analysed recordings from adult males and females visiting fruit-bearing fig trees. They found that a high proportion (43%) of the pant hoots given within 5 minutes after arrival included the let-down section. In samples given more than five minutes after arrival, the proportion of pant hoots with let-downs was much lower (11%) but still higher than in any other context category. Recordings from Mahale did not confirm the findings of Kibale, and it was concluded that the results did not "support the claim that chimpanzees have a distinct vocalization that they give uniquely upon arrival at food trees" (Clark, 1991, p. 91). Nevertheless, the study has demonstrated that the utterance of pant hoots with a let-down section is at least closely connected to the feeding context.

Function: Field observations on the context of pant hooting have raised various propositions about its possible function including (1) intergroup spacing, (2) establishment and maintainance of social parties, (3) attraction of allies, (4) attraction of mates, and (5) food availability (Goodall, 1986; Wrangham, 1977; Wrangham and Smuts, 1980). All of them involve spatial regulation and each single hypothesis is compatible with one or several others. Are there empirical data supporting or rejecting one or the other hypothesis?

(1) Goodall (1986) reported that upon hearing choruses of distant individuals from another community, chimpanzees move either away from the source of sound or engage in vigorous displays. Apparently, assessments of the number of males present in distant parties were based on pant hoots and drumming. Although the effect of pant hoots for inter-community spacing should be tested more rigorously (e.g., with playback experiments), there is at least some indirect evidence in support of the first hypothesis.

(2) Evidence for the proposed function of coordination between parties of the same community comes from the Tai forest where Boesch (1991) collected data on distance communication from the alpha male of a habituated community. The study focused on sequences of pant hoots, drumming and climax screams. The observations suggested that the response of other community members depended on the repetition of sequences within 2 minutes or less, and the location of repeated sequences. Depending on the spatio-temporal design of the performance, the behavior of the alpha male following his own performance, and the behavioral response of other members, human observers distinguished three different types of messages: (a) change of travel direction, (b) announcement of a resting period, (c) a combination of both. Boesch (1991) concluded that some of the pant-hoot-drumming sequences of Tai chimpanzees transmitted precise information to other community members and that this communication system can partly function in a symbolic way. These data provide support for the second hypothesis.

(3 and 4) Recently, Mitani and Nishida (1993) have investigated the influence of several social factors on the production of pant hoots by adult males. In this study, two

aspects were considered: the size of the party to which the focal males belonged, and the proximity of focal males and particular social partners (*allies*, *associates* and *estrous females*). Vocal activity of the different males was measured for three different conditions (*within*, *nearby* or *far away* from the focal male). The results suggested the following: First, there was no correlation between party size and vocal activity of the focal males. Second, focal subjects called more often when allies and association partners were nearby than when these individuals were absent. Third, the two highest ranking males and one middle ranking male called more frequently when estrous females were nearby than when absent. Finally, the three low ranking males called less often when estrous females were nearby.

These findings support the third hypothesis, indicating that pant hoots of adult males may facilitate communication with allies and associates in order to maintain contact and/or recruit aid. However, the mate attraction hypothesis remains to be tested again (Mitani and Nishida, 1993); the authors considered the possibility that the conflicting results "...may reflect the different mating tactics employed by males of varying ranks" (p. 744).

(5) Food experiments conducted in a group of captive chimpanzees have revealed that with one exception, pant hoots were always given by adult males and primarily by subjects who discovered large amounts of food (Hauser et al., 1993). If the aim of food calling is to attract other individuals to a food source, one would expect that animals who vocalize are joined by a larger number of individuals than animals who keep quiet. Studies by Wrangham (1977) in Gombe and Ghiglieri (1984) at Kibale (Ngogo community) revealed that individuals giving food calls were joined more often (33% and 59%, respectively) than individuals not vocalizing (6% and 34%, respectively). Another study in the Kibale forest (Kanyawara community) failed to reproduce these results (Clark, 1991). Hence, the fifth hypothesis (food attraction) can be neither accepted nor rejected.

Bonobo

Until recently, information about communication among bonobos was restricted to brief descriptions of some frequent vocalizations (e.g., Jordan, 1977; Kuroda, 1979; Savage-Rumbaugh and Wilkerson, 1978). Recent studies on captive colonies in San Diego (de Waal, 1988) and Frankfurt (Tröster, 1988) revealed two groups of long distance calls: *low hooting* and *high hooting*. Distance communication among wild bonobos has been described to some extent by Mori (1983) and is now one of the major topics of an ongoing field project at Lomako (Hohmann and Fruth, in press).

Structure (Table 1): Low hoots consist of two phases (inhalation and exhalation) and are always given in phrases (de Waal, 1988). According to spectrographic analyses of high hoots from Lomako (Hohmann and Fruth, in press) and San Diego (de Waal, 1988) several types of high hoots can be distinguished. However, the current classification is still tentative and more data are needed for a useful subdivision of this category.

Sex: Analyses of high hoots from adult males and females collected at Lomako indicated sex differences in some physical features. First, in calls of adult females the average fundamental frequency was significantly lower than in calls from adult males. However, when analysing the samples of males separately, it became apparent that they also produced high hoots within the range of females and that the samples of male hoots with a low fundamental were recorded from lone individuals. In addition, the intervals in phrases of females were significantly shorter than in males (Hohmann and Fruth, in press).

Context: In captivity low hoots were not directed at a particular individual but given in response to environmental changes and disturbances (de Waal, 1988). In Lomako, low hoots were either directed towards the human observer or elicited by vocalizations of distant community members and were often accompanied by agonistic displays (e.g., stamping on the ground or tree drumming). In Lomako emission of high hoots coincided with corresponding calls of conspecifics (48%), loud calls of mangabeys (16%) and other less specific noises (3%). In the remaining cases, high hoots were given spontaneously (Hohmann and Fruth, in press). In captivity, increased rates of high hooting coincided with the regular feeding times, suggesting that expectation or discovery of food stimulates vocalization. However, measurements of the vocal activity during food experiments suggest that the quantity of food presented to the subjects does not affect the number of high hoots (van Krunkelsven, 1993). In Lomako, high rates of hooting marked the capture of prey, the beginning of food-sharing episodes or invasion of feeding trees. The highest rates of high hooting were scored when different parties gathered at nesting sites.

Function: Although more empirical data on the function of distance communication in bonobos is required, observations from Lomako confirm previous proposals that high hoots are part of a system of signals facilitating communication between members of different parties. The most common response to choruses of distant conspecifics was high hooting (more than 50%) or locomotion (40%). Together with simultaneous observations on locomotion and vocal activity of different parties, this suggests that the basic function of high hoots is the maintainance and/or reduction of distance between different parties. Moreover, data from Lomako revealed that the number of individuals gathering at the same nesting site for the night was consistently higher than in day parties (Fruth and Hohmann, in press). In the absence of any evidence for competition between individuals or parties for nesting sites, it is assumed that one specific function of high hoots is the assemblage of a large number of community members at nesting sites.

DISCUSSION

Reviewing loud calls of the great apes reveals correspondence in a number of patterns. Loud calls always occur in phrases consisting of more or less distinct sections, and the vocalizations were often combined with acoustical cues. In all species, the calls were given by adult males and in those species where relevant data are available, dominant individuals vocalized more often than subdominants. The majority of loud calls were uttered spontaneously or during encounters with conspecifics.

However, comparisons between some of the species revealed great differences. In orangutans and gorillas, loud calls were given exclusively by adult, dominant males and showed a rather stereotype structure (Fossey, 1972; Galdikas, 1983; Mitani, 1985). Behavioral observations and field experiments suggest that in both species spacing among single males or members of different groups is the major function of loud calls.

In chimpanzees and bonobos, mature and immature individuals of both sexes gave loud calls in a variety of contexts, ranging from agonistic to affiliative (Marler, 1976; de Waal, 1988). The bulk of the data consisted of calls used for communication between members of the same community. Spectrographic analyses have demonstrated a high degree of variation in structural features. Some of these features appear to be related to the rank of the vocalizer, whereas others appear to encode specific messages. The sudden structural changes of the pant hoots of three adult males from Tai forest invite speculations about the possibility of imitation of individual-specific and perhaps rank-related features. Studies on bonobos also indicate covariation of certain structural features of high hoots and context of emission. Analyses of sound recordings further suggest that bonobos may be able to adjust spectral parameters of high hoots according to corresponding calls of conspecifics (for similar observations see Sugiura and Masataka, this volume).

Interspecific differences in vocal behavior may reflect differences in the complexity of social organization or be related to a shift from inter-group communication to intra-group communication. This is precisely the concept adopted by Marler (1976) when comparing the vocal repertoire and vocal behavior of chimpanzees and gorillas, respectively. While comparison of structural features indicated a high degree of correspondence between the two species, striking disparities became apparent when the use of equivalent calls by males and females was considered. Although chimpanzees of different age and sex contributed to the overall production of a given call with different rates, all vocalizations were given by males and females. In contrast, three calls of the gorilla repertoire were heard only from adult males and a number of other calls were also produced more frequently by males than by females. Marler (1976) proposed a correlation between the contrasting patterns of social organization and the related differences in spatial distribution on one hand, and the differences in vocal behavior on the other. Accordingly, the low degree of age and sex differences in vocal behavior in chimpanzees was in line with both the spatial dispersion of community members and the relatively balanced participation of all age and sex classes in social affairs. In gorillas, superiority of vocal behavior by one age and sex class (adult silverback males) corresponded well with the strong cohesion between group members and the social dominance of adult silverbacks. Though the differences in vocal behavior between the two African apes may indeed be related to certain facets of their social organization, the influence of other factors should be considered as well. First, in contrast to the conditions with the chimpanzees at Gombe, the data from gorillas have been collected at the beginning of a long term study when subjects were poorly habituated, and it is possible that the true capacity of distance communication among gorillas has been heavily underestimated. Second, there is good evidence that the patterns of social organization may be flexible within and between populations of the same species (see Byrne et al., 1990, for baboons; Milton, 1984, and Strier et al., 1993, for woolly spider monkeys; Watanabe, 1981, for pigtailed langurs) but corresponding modifications of vocal communication have yet to be demonstrated. Third, chimpanzees and gorillas differ in a number of other attributes (e.g., size of home range, migration pattern, group size, adult sex ratio, sexual dimorphism in body size and weight) which may equally be responsible for the differences in vocal behavior.

The aim of the following chapter is twofold: First, we will evaluate the relationship between vocal behavior and social organization by considering a larger spectrum of primate species with grouping patterns similar to those of gorillas *(harem groups)* and chimpanzees *(fission-fusion,)* respectively. Second, one particular aspect of sex differences in vocal behavior - the use of loud calls - will be examined in more detail.

VOCAL BEHAVIOR AND ITS SOCIAL CORRELATES

Harem Groups

The term harem group refers to cohesive one-male breeding units with a small number of adult females (Vogel, 1975). Leader males may tolerate other males born within their group but are usually extremely aggressive towards extra-group males. Out of the 172 species listed by Smuts et al. (1986, Table A-1), 33 are living in one-male groups or show a

Table 2. Comparative data from primate species living in harem groups.

Species	Size of group	Adult sex ratio (M:F)	Migration pattern	Loud calls	Home range (km^2)	Sex dimorphism index of body weight (F/M)	References
Callimico							
goeldii	6-9	1:2	?	yes	0.3-0.6	0.82	1, 1, -, 2, 1, 3
Cebus							
capucinus	11-20	1:5	M&F	yes	0.5-0.8	0.71	4, 4, 4, 5, 4, 3
Alouatta							
seniculus	4-10	1:1.6	M&F	M&F	0.04-0.25	0.79	6, 6 , 7, 8, 6, 3
caraya	7-9	1:2	?	yes	?	0.85	6, 6, -, 9, -, 3
Cercopithecus							
nictitans	13-20	?	M	M	0.6-0.8	0.63	10, - ,11,12,13, 3
mitis	10-45	?	M	M	0.4-0.6	0.57	10, - ,11,12,13, 3
diana	14-40	1:7	M	M	1	0.62	10,14,11,12,13,15
cephus	8-15	?	M	M	0.1-2.8	0.70	10, - ,11,12,13, 3
campbelli	8-13	?	M	M	0.4	1.00	10, - ,11,12,13, 3
ascanius	17-40	?	M	M	0.2	0.69	10, - ,11,12,13, 3
Erythrocebus							
patas	5-35	1:10	M	?	20-30	0.56	10,16,11, - ,16, 3
Papio							
sphinx	20-25	1:5	M	yes	5	0.46	17,17,11,18,19, 3
hamadryas	2-23	1:3	F	?	?	0.43	17,17,11, - , - , 3
Theropithecus							
gelada	10	1:3	M	?	?	0.66	17,17,11, - , - , 3
Colobus							
guereza	12	1:1.3	M	M	0.02	0.78	20,20,11,21,20, 3
verus	8	?	?	M	0.26	0.94	20, - , - ,14,14, 3
satanas	15	1:2.6	M	M	?	0.79	22,22,11,21,22, 3
polykomos	16	?	?	M	0.17	0.81	23,23, - ,21,15, 3
Nasalis							
larvatus	9	1:3.6	M&F	M	0.8	0.48	24,24,24,25,24, 3
concolor	3-5	1:1.6	?	M	0.2-0.3	?	26,26, - ,27,26, 3
Presbytis							
thomasi	3-12	1:3	M	M	0.1-0.2	?	28,28,28,28,28, 3
senex	7-9	1:4	M&F	M	0.2-0.7	0.91	29,29,30,31,29, 3
pileata	4-13	1:4	M&F	M	0.14-0.22	0.90	32,32,32,32,32,33
johnii	8-16	1:2.8	M&F	M	0.06	0.81	34,34,35,36,37, 3
christata	32	1:12	M	M	0.2	0.94	38,38,11,38,38, 3
aygula	3-12	1:2.7	?	M	0.1-0.4	0.98	39,39, - ,39,39, 3
melalophus	9-15	1:8	M	M	?	0.98	40,40,11,41, - , 3
Rhinopithecus							
roxellanae	?	?	?	M&F	?	0.49	- , - , - ,42, - ,42
Gorilla							
gorilla	5-37	1:2.4	M&F	M	5-35	0.58	43,43,44,45,43, 3

grouping pattern similar to that of gorillas (small groups with a single or few adult males). The search for features common to all or most of these species demonstrates a remarkable incoherence (Table 2): In 20 species, loud calls are thought to be given exclusively by adult males but in 2 other species loud calls are given by both sexes. In the species grouped in this category all different types of migration patterns are practiced. Group size and home range size varies greatly and the degree of sexual dimorphism ranges from moderate (e.g., *Cercopithecus cephus, Presbytis aygula)* to extensive (e.g., *Papio hamadryas*). A previous review (Gautier and Gautier, 1977) identified 18 species in which loud calls (type 1) are given exclusively by adult males. Nine of these species are reported to live in one-male groups, seven in multi-male groups and one is probably monogamous (grouping patterns are from Smuts et al., 1986, Table A-1). In sum, there is little evidence that sex differences in loud call behaviour can be related to a particular type of social organization or grouping pattern. Instead, species showing similarities in vocal behavior differ with respect to other aspects of their social life and vice versa.

Fission-Fusion Society

The fission-fusion society is a rare social pattern among primates. Besides chimpanzees, the only other species with a similar social design are bonobos and the spider monkeys *Brachyteles* and *Ateles* (Table 3).

Pan: The grouping patterns of the two *Pan* species are more or less identical (Wrangham, 1986). Communities split into temporarily stable parties of differing size and composition. In contrast to chimpanzees, bonobos more often form bisexual parties and the number of adult males per party is higher. Female bonobos form unisexual coalitions, male bonding is comparatively weak and dominance relationships between the sexes are biased in favor of females (Nishida and Hiraiwa-Hasegawa, 1986; Parish, in press; White, 1988). Females participate actively during community encounters and occasionally gang up against adult males of their own community (personal observation). Mating occurs during the entire sexual cycle and involves most or all males. However, observations from Wamba and Lomako suggest that males of the same community mate at different rates (Kano, 1992; personal observation). Chimpanzees and bonobos are thought to vocally advertise the discovery of an abundant food patch (Wrangham, 1977; personal observation). Sex differences in vocal behavior appear minor. In chimpanzees, differences are most prominent in the rate animals of different age and sex use the different elements of the vocal repertoire (Marler, 1976). Preliminary analyses of vocalizations of bonobos recorded at Lomako

---

Table 2: M, male; F, female; yes, species has been reported to give loud calls but information on sex differences is not available; ?, no information available. The index of sex dimorphism was calculated by dividing the weight of females (in kg) by the weight of males using the data in the according references. References for each respective column are cited in the final column.

References of Table 2: 1, Goldizen (1986); 2, Masataka (1982); 3, Harvey et al. (1986); 4, Robinson and Janson (1986); 5, Oppenheimer (1977); 6, Crockett and Eisenberg (1986); 7, Sekulic (1982b); 8, Sekulic (1983); 9, Thorington et al. (1984); 10, Cords (1986); 11, Smuts et al. (1986,Tab. A-I); 12, Gautier (1988); 13, Gautier-Hion (1988); 14, Oates and Whitesides (1990); 15, Galat and Galat-Luong (1982); 16, Chism and Rowell (1988); 17, Stammbach (1986); 18, Kudo (1987); 19, Hoshino et al. (1984); 20, Dunbar (1986); 21, Oates and Trocco (1983); 22, McKey and Waterman (1982); 23, Galat and Galat-Luong (1985); 24, Bennet and Sebastian (1988); 25, Kawabe and Mano (1972); 26, Watanabe (1981); 27, Tenaza (1989); 28, Gurmaya (1986); 29, Rudran (1973); 30, Manley (1986); 31, Hohmann (1990); 32, Stanford (1991); 33, Napier and Napier (1970); 34, Hohmann (1989a); 35, Moore and Ali (1984); 36, Horwich (1976); 37, Horwich (1972); 38, Bernstein (1968); 39, Ruhiyat (1983); 40, Struhsaker and Leyland (1986); 41, Wilson and Wilson (1975); 42, Tenaza et al. (1988); 43, Stewart and Harcourt (1986); 44, Harcourt (1978); 45, Schaller (1963).

suggest sex-related (and perhaps context-related) spectral differences in high hoots (Hohmann and Fruth, in press).

Brachyteles: Data obtained from habituated groups at two study sites have revealed contradictory results about the nature of social organization. Cross-sectional data from one site covering a period of nine years demonstrate an ample degree of variation (Strier et al., 1993): During the first six years, the study group remained comparatively cohesive. Occasional fission into smaller subgroups occurred and was correlated with increasing rates of contact calls between the dispersed group members. However, these fissions were restricted to short periods of time. It was only later that the group showed a regular pattern of fission-fusion into two subgroups of variable size and composition. At the other study site, fission-fusion occurs regularly and resembles that reported from chimpanzees (Milton, 1984). Here, reunions between group members involved antagonistic displays, loud calls and ritualized greeting (embracing). The latter occurred among females and among males but not between the sexes. Males were apparently dominant over females and formed alliances during agonistic interactions. Observations of mating behavior indicated that females mate with several male group members within short succession and even with males from a neighboring group (Milton, 1985). A repertoire of eight different calls has been described and the only reported sex difference in vocal behavior concerns the use of the *mating twitter*, a call given exclusively by females to attract males (Milton, 1984).

Ateles: Several members of this genus (e.g., *A. belzebuth, A. paniscus*) have been reported to live in fission-fusion societies (Klein and Klein, 1977; Symington, 1990). Groups split regularly into subgroups of flexible size and composition. Data on agonistic interactions indicate that both sexes engage in unisexual coalitions and males always dominate females. Agonistic intra-group interactions from males against females are rare or absent and male-male aggression involves predominantly or exclusively individuals from different groups. Females participate in agonistic interactions during inter-group encounters but are less active if male group mates are involved. As in *Brachyteles* tolerance among male group members is very high and in *A. paniscus* average association indices were highest for male-male dyads (Symington, 1990). Estrous females may be consorted by a single male but have also been seen to copulate repeatedly with a larger number of mates. In one species *(A. geoffroyi)* males and females have been reported to give food calls in order to attract distant community members and manipulate the size of feeding parties (Chapman and Lefebvre, 1990). Males and females of *A. geoffroyi* use the same repertoire of agonistic *(ook ook, growling)* and non-agonistic *(whoop, heavy panting)* distance calls (Klein, 1974). Eisenberg (1976) who studied vocal communication in two other species *(A. belzebuth, A. fusciceps)* has reported a very low degree of sex differences in vocal behavior. Two types of loud calls were distinguished and from the corresponding descriptions it is inferred that one was preferably used for intergroup communication while the other was given during interactions among group members.

There are a number of striking parallels among all those species living in a fission-fusion society. Most important for the present paper, loud calls are habitually given by females and males alike, some species having different signals for distance communication, and loud calls are used for both intra- and inter-group communication. Correspondence among the species living in fission-fusion societies is not restricted to the spatial distribution of group members but also includes other factors such as the adult sex ratio, the mode of migration and the degree of sexual dimorphism (Table 3). Thus, one could hypothesize that instead of being related to the fission-fusion society, the existence (or absence) of female loud calls is perhaps related to one of these other traits.

Table 3. Comparative data from primates species living in fission-fusion societies

Species	Size of group / community	Adult sex ratio M:F	Mean party size	Migration pattern	Loud calls	Size of home range (km^2)	Index of sex dimorphism in body weight (F/M)	References
Pan								
troglodytes	19-106	1:1.4	2.6-8.6	F	F&M	5-300	0.74	1,2,1,2,3,2,1,4
paniscus	50-120	1:2.1	7.6-16.8	F	F&M	22-60	0.74	1,5,1,1,3,6,1,7
Brachyteles								
arachnoides	22-25	1:1.2	?	F	F&M	0.7-1.7	1.00	8,8, - ,9,3,9,10,4
Ateles								
belzebuth	18	1:2.9	3.5	F	F&M	0.3-0.4	0.93	8,8,11,12,3,12,13,4
paniscus	18	1:2.6	3.7-5.0	F	F&M	1.5-3.3	0.87	8,8,14,14,3,15,13,4
geoffroyi	15	?	?	F	F&M	0.1-1.2	0.93	8, -, -, -,3,15,13,4

References: 1, Wrangham (1986); 2, Goodall (1986); 3, Smuts et al. (1986,Tab.A-I); 4, Harvey et al. (1986); 5, Fruth and Hohmann (in press); 6, Hohmann and Fruth (in press); 7, Jungers and Susman (1984); 8, Robinson and Janson (1986); 9, Milton (1984); 10, Strier (1987); 11, Klein and Klein (1975); 12, Klein (1974);13, Symington (1990); 14, Symington (1988); 15, Eisenberg (1976).

While the pattern of fission-fusion is confined to a few species from two different taxons, the trait in question (female loud calls) is not. Although the number of species in which females give loud calls on a regular basis is restricted, it includes members of very different taxonomical groups (Table 4). Group structures and mating patterns are variable and include monogamous pairs, polygynous one-male groups as well as promiscuous multi-male groups and fission-fusion societies.

Rather than forming a uniform category, the samples of female loud calls pooled in Table 4 differ along several lines. (1) Physical structure: Loud calls of males and females may be monomorphic or dimorphic. (2) Mode of emission: Individuals of both sexes may vocalize simultaneously or independently of each other. If both sexes vocalize in unison, the calls given by different individuals may be superimposed on each other (chorusing) or males and females may alternate their vocalizations (duetting). (3) Time of utterance: Loud calls may be given at particular times (usually at dawn) or at any time of the day (or night). Along with field reports about the context of vocalization, these differences suggest that female loud calls may serve various functions. Behavioral observations and playback experiments have shown that female gibbons and howler monkeys give loud calls in order to aid exclusion of extragroup females (Raemaekers and Raemaekers, 1985; Sekulic, 1982a, b).

Female howlers interfere with loud calls when female group mates try to approach the dominant male and threaten subdominant females who try to howl together with the male (Sekulic, 1983). Parties of female spider monkeys encountering strange males respond with vocal bouts (ook-barking, whooping) lasting for periods as long as one hour (Klein, 1974). These and other examples (for corresponding data from *Aotus*, *Callicebus* and *Pithecia* see Robinson et al., 1986) suggest that female loud calls may facilitate defense of communal resources and intrasexual competition.

Vocal responses of female howlers towards the roaring of male group mates have been found to vary in relation to their reproductive status. It was suggested that howling by females in unison with their mates may enforce pair bonding and that affiliation with females in turn may strongly affect the male's ability to remain dominant (Sekulic, 1982a, 1983; Sekulic and Chivers, 1986). In chimpanzees the rate of pant hooting by females increases at the time of estrous (Pusey, 1990) and when travelling together with adult males in mixed parties (Clark, 1991). Thus, female loud calls may also advertise the reproductive state of the vocalizer and attract potential mates.

In bonobos, members of different female parties give high hoots spontaneously before or at the onset of locomotion. Individual parties may stay apart but remain in auditory contact for the entire day until they fuse at the nest site at night. The coincidence between onset of locomotion and distance vocalization indicates that high hoots maintain contact and coordinate travel routes between members of different parties (personal observation).

In conclusion, the use of loud calls by females is neither related to a particular type of social organization nor to a single function. In general, females seem to give loud calls for the same reasons as males (e.g., spatial regulation within and between groups, mate attraction, self-advertisement, intrasexual competition). Regarding function, loud calls are thought to facilitate communication between members of different groups. In addition, it is conceivable that loud calls are directed towards those individuals who are most likely to transfer between groups. If this is correct, one would expect to find a correlation between the mode of migration and sex differences in the use of loud calls. In most of the species (72%) listed in the according section (loud call type-1) by Gautier and Gautier (1977) adult males are the primary producers of loud calls, migration is male-biased or restricted to males. In contrast, Table 4 shows that in all species where females are known to give loud calls, migration is female-biased or both sexes migrate. It is therefore hypothesized that the

Table 4. Species in which loud calls are given by males and females.

Species	Migration pattern	Sex dimorphism (F/M)	Grouping pattern	Reference for vocalizations
Alouatta				
palliata	F&M	0.79	Sm-V	Baldwin & Baldwin, 1976
seniculus	F&M	0.77	Lg-M	Sekulic, 1982a
pigra	?	?	Sm-V	Horwich & Gebhard, 1983
Ateles				
belzebuth	F	0.94	C-i	Klein, 1974
fusciceps	?	1.02	C-i	Eisenberg, 1976
geoffroyi	F	0.94	C-i	Eisenberg, 1976
Brachyteles				
arachnoides	F	1.0	C-i / Lg-M	Milton, 1984
Lagothrix				
lagotricha	?	0.85	Lg-M	Williams, 1967
Aotus				
trivirgatus	F&M	1.09	Mon	Robinson et al., 1986
Callicebus				
moloch	F&M	0.95	Mon	Robinson et al., 1986
torquatus	F&M	1.00	Mon	Robinson et al., 1986
Pithecia				
monachus	F&M	?	Mon	Robinson et al., 1986
pithecia	F&M	0.88	Mon	Robinson et al., 1986
Presbytis				
potenziani	F&M	0.98	Mon	Tilson & Tenaza, 1976
Hylobates				
agilis	F&M	0.95	Mon	Gittins, 1980
concolor	F&M	1.04	Mon	Deputte, 1982
hoolock	F&M	0.94	Mon	Tilson, 1979
klossii	F&M	1.04	Mon	Tenaza, 1976
lar	F&M	0.93	Mon	Carpenter, 1940
moloch	F&M	0.95	Mon	Kappeler, 1984
muelleri	F&M	?	Mon	Mitani, 1984
pileatus	F&M	?	Mon	Srikosamatara, 1980
syndactylus	F&M	0.97	Mon	Chivers, 1974
Pan				
troglodytes	F	0.75	C-i	Goodall, 1968
paniscus	F	0.74	C-i	Hohmann & Fruth, in press

Grouping patterns and data on migration for species living in groups (communities) with more than one female were taken from Smuts et al., 1986, Tab. A-I. The index of sex dimorphism was calculated by using the data on body weight (kg) of males and females published by Harvey et al., 1986, and Jungers and Susman, 1984 (for *Pan paniscus*), respectively.

positive correlation between the two traits, female-biased migration and female loud calling, indicates a functional linkage.

We are aware that there are limitations to the concept developed above. First, it is not clear why in some species with female-biased migration, neither sex uses loud calls (e.g., *Macaca radiata*, Hohmann, 1989b; *Colobus badius*, Struhsaker, 1975). Second, members of the phylopatric sex may not always refrain from giving loud calls (e.g., male chimpanzees and bonobos). Third, there is apparently no species where loud calls are given exclusively by females. The only species coming close to that is the silver gibbon *(Hylobates moloch)* where females rather than males give loud calls (Kappeler, 1984). Finally, there are species where males give loud calls but females, in spite of their tendency towards exogamy, do not (e.g., *Nasalis larvatus*, *Gorilla gorilla*).

In gorillas, both sexes migrate but loud calling is thought to be restricted to adult males. However, Fossey (1972) indicates that there are exceptions from this rule. She reports two cases where hoot series were given once by an adult female and once by a juvenile of unknown sex. No contextual information is provided except that at the time of observation the female was a member of an all male band. Though these exceptional cases may justify changing the status of the behavior from "absent" to "rare", they do not diminish the overwhelming sex bias in loud calling of gorillas. Are there any alternative explanations why female gorillas refrain from producing loud calls?

Gorillas subsist on an evenly dispersed folivorous diet, they are not territorial and females may gain little (if any) nutritional benefits from migration (Watts, 1985). Instead, female gorillas seem to be associated with silverback males in order to receive protection against predators (Harcourt et al., 1981). There is probably little difference for the females whether a silverback extends his protection towards two or towards three of them. In contrast to nutritional resources, male protection could be considered a divisible resource. One could speculate that female gorillas refrain from loud calling because of the low degree of intrasexual competition. However, recent analyses of the distribution of aggressive behavior between female group members indicate high rates of social conflicts between long-term residents and recent immigrants (Watts, 1994).

Another factor not yet considered here is sexual dimorphism in body size and weight. One of the driving forces for sexual dimorphism is intra-sexual selection (Darwin, 1871). There is evidence that certain acoustical parameters of primate calls vary with body size (Gautier-Hion and Gautier, 1985; Hauser, 1993; Inoue, 1988). Assuming that judgements of the fighting abilities of potential opponents are based on vocal cues, weaker males would have no advantage from loud calling. Vocalizing in unison with other group members would be an alternative strategy because chorusing of males and females would probably prevent distant listeners from evaluating body size (and fighting ability) of opponents. With the support of many females, smaller males could be more successful than large males. It is obvious that such a trend would either inflate the value of the signal or decrease sexual dimorphism. Table 4 shows that in all species where females give loud calls and migrate, sexual dimorphism is indeed relatively low. In contrast, for those species where this correlation does not apply (e.g., gorilla, proboscis monkey, hamadryas) the indices for sexual dimorphism are exceptionally high (Harvey et al., 1986). However, Table 2 shows that in some species (e.g., *Colobus*, *Presbytis*) where both attributes, loud calling and migration, are thought to be a domain of males, sexual dimorphism can be very low. Hence, the relationship between sexual dimorphism and sex differences in the use of loud calls is inconsistent.

Suggestions for future studies: We are just beginning to understand vocal communication in the great apes and some of the noted examples indicate an unexpected capacity for natural communication. However, there are still large gaps in our knowledge.

For example, mountain gorillas have been studied for several decades but even the most basic aspects of their distance communication remain unexplored. Similarly, individual chimpanzees were monitored by researchers for their entire lives but the ontogeny of vocal communication is unknown. Finally, field studies on bonobos are less comprehensive than those on chimpanzees. If we are to understand the development of vocal communication, we must be able to compare patterns of all the great apes in light of their respective physical, ecological and social constraints. There are a number of recommendations for future research in this area (e.g., Boehm, 1989; Seyfarth, 1986) and the recent studies on chimpanzees can also serve as examples for corresponding studies on other species. However, what is required are more comprehensive field studies. For example, one of the issues raised in this chapter - the function of sex differences in loud call behavior - can be approached in different ways:

(1) Playback experiments: In some species (e.g., gibbons) the degree of sex differences of loud calls has been well documented. In other species (e.g., chimpanzees, howlers, spider monkeys) the degree of sex differences is not known and analyses of the structural features are required. Playback experiments (similar to those conducted by Raemaeker and Raemaeker, 1985) could be used to evaluate the response of males and females to loud calls from both sexes.

(2) Analysis of the context of loud calling: At which age do females and males begin to give loud calls and what social conditions lead to habitual use of this behavior ? For example, there may be a link between sexual maturity and loud call production (e.g., Gautier and Gautier, 1977). However, observations on some species indicate that this is not simply an age phenomenon but a social one. Male gibbons and langurs refrain from loud calling until they have established their own territory and are again accompanied by females (Hohmann and Vogl, 1991; Tenaza, 1976). Therefore, to evaluate the significance on female loud call performance, the vocal activity of individuals must be known in a number of contexts such as before, during and after migration.

(3) Relationship between vocalization and group dynamics: If loud call behavior of resident group members transmits information on the size and composition of the group and can act to attract or repel migrating conspecifics, one would expect the rate of vocalization, group size, rate of visits of potential migrants and actual rate of immigration to be correlated. Long term studies on well habituated groups or communities are required to collect this kind of data.

(4) Effect of group transfer: If loud calls of males and females are used to regulate migration (e.g., attract members of the opposite sex and/or repel individuals of the same sex), how is migration regulated in species in which the migrating sex (or both sexes) does not produce loud calls ? Differences in vocal behavior (presence/absence of loud calls) could be related to the total density of primates in a particular area, the degree of interspecific food competition or the presence of predators.

Finally, there is a very simple way to enhance our understanding of sex differences in loud call behavior: When recording calls, do not automatically score vocalizations as a "male loud call" but try to determine the sex and identity of the caller. As additional data on sex differences in the use of loud calls become available it will be then possible to examine in greater detail the social and ecological conditions under which female primates raised their voices.

CONCLUSIONS

All species of great apes utter loud calls which correspond in a number of structural and behavioral patterns. They are always given in phrases consisting of more or less distinct sections, and are often combined with acoustical cues. Loud calls are regularly given by adult males and in those species where relevant data are available, dominant individuals vocalized more often than subdominants. Concerning the use of these calls, species-specific differences are more prominent than common features. In two species (orangutan and gorilla), vocalization is restricted to adult males and behavioral observations and field experiments suggest that spacing among single males or members of different groups is the major function.

In chimpanzees and bonobos, loud calls are used by mature and immature individuals of both sexes and play a major role in communication between members of the same community. Structural features vary with differences in rank, context of emission, and specific functions. Sudden changes of individual-specific features indicate a high degree of flexibility in physical traits.

Comparison of a larger set of species does not support previous propositions concerning an interrelation between type of social organization (grouping pattern) and sex differences in utilization of loud calls. Instead, cross-species comparison suggests a positive correlation between migration patterns and use of loud calls: In species in which migration is male-biased or restricted to males, loud calls are given exclusively (or predominantly) by males. In species in which migration is female-biased or both sexes migrate, loud calls are given by both sexes. These findings are in general agreement with previous hypotheses concerning the functional significance of loud calls. However, the fact that some species do not follow this trend demonstrates the need for additional research.

ACKNOWLEDGEMENT

The authors thank Irenäus Eibl-Eibesfeldt, Gerhard Neuweiler and Detlev Ploog for continuous support. Excellent comments on an earlier version by Elke Zimmermann, Marc Hauser and Bertrande Deputte are gratefully acknowledged. Special thanks are due to Christopher Roberts and Tom Van't Hof for valuable recommendations, critical reading and corrections of the English text. Financial support for the field-work in Zaire was provided by the Max-Planck-Society, the University of Munich, the German Science Foundation, the German Academic Exchange Service, and a private donor.

REFERENCES

Baldwin, J.D. and Baldwin, J.I., 1976, Vocalizations of howler monkeys (*Alouatta palliata*) in Southwestern Panama, *Folia Primatol.* 26:81-108.

Bennet, E. and Sebastian, A., 1988, Social organization and ecology of proboscis monkeys (*Nasalis larvatus*) in mixed coastal forest in Sarawak, *Int. J. Primatol.* 9:233-254.

Bernstein, I., 1968, The lutong of Kuala Selangor, *Behaviour* 32:1-16.

Boehm, C., 1989, Methods for isolating chimpanzee vocal communication, *in:* "Understanding Chimpanzees", P. Heltne and L. Marquardt, eds., Harvard University Press, Cambridge.

Boesch, C., 1991, Symbolic communication in wild chimpanzees ?, *Human Evol.* 6: 81-89.

Brown, C.H., 1989, The acoustic ecology of East African primates and the perception of vocal signals by grey-cheeked mangabeys and blue monkeys,*in:* "The Comparative Psychology of Audition: Perceiving Complex Sounds". R. Dooling and S. Hulse. eds., Lawrence Erlbaum, Hillsdale.

Byrne, R., Whitten, A., and Henzi, S., 1990, Social relationships of mountain baboons: leadership and affiliation in a non-female-bonded monkey, *Amer. J. Primatol.* 20: 313-329.

Carpenter, C., 1940, A field study of the behavior and social relations of the gibbon (*Hylobates lar*), *Comp. Psychol. Monogr.* 10:1-212.

Chapman, C. and Levebvre, L., 1990, Manipulating foraging group size: spider monkey food calls at fruiting trees, *Anim. Behav.* 39:891-896.

Chism, J. and Rowell, T., 1988, The natural history of patas monkeys, *in:* "A Primate Radiation: Evolutionary Biology of the African Guenons", A. Gautier-Hion, F. Bourlier, J-P. Gautier and J. Kingdon, eds., Cambridge University Press, Cambridge.

Chivers, D., 1974, The siamang in Malaya, *Contrib. Primatol.* 4.

Clark, A., 1991, The socioecology of wild chimpanzee vocal behavior in the Kibale forest, Uganda, Ph.D. thesis, University of Michigan.

Clark, A. and Wrangham, R., 1993, Acoustic analysis of wild chimpanzee pant hoots: do Kibale forest chimpanzees have an acoustically distinct food arrival pant hoot ?, *Amer. J. Primatol.* 31: 99-109.

Cords, M., 1986, Forest guenons and patas monkeys: male-male competition in one-male groups,*in:* "Primate Societies", B. Smuts, D. Cheney, R. Seyfarth, R. Wrangham and T. Struhsaker, eds., University of Chicago Press, Chicago.

Crockett, C. and Eisenberg, J., 1986, Howlers: variations in group size and demography, *in:* "Primate Societies", B. Smuts, D. Cheney, R. Seyfarth, R. Wrangham and T. Struhsaker, eds., University of Chicago Press, Chicago.

Darwin, C., 1871, "The Descent of Man and Selection in Relation to Sex", John Murray, London.

Deputte, B., 1982, Duetting in male and female songs of the white-cheeked gibbon (*Hylobates concolor leucogenys*), *in:* "Primate communication", C. Snowdon, C. Brown, and M. Petersen, eds., Cambridge University Press, Cambridge.

Dunbar, R., 1986, The social ecology of gelada baboons, *in:* "Ecological Aspects of Social Evolution", D. Rubenstein and R. Wrangham, eds., Princeton University Press, Princeton.

Eisenberg, J., 1976, Communication mechanisms and social integration in the black spider monkey, *Ateles fusciceps robustus*, and related species, *Smithson. Contr. Zool.* 213.

Elliott, R., 1976, Observations on a small group of mountain gorillas (*Gorilla gorilla beringei*), *Folia Primatol.* 25:12-24.

Fossey, D., 1972, Vocalizations of the mountain gorilla (*Gorilla gorilla beringei*), *Anim. Behav.* 20:36-53.

Fruth, B. and Hohmann, G., in press, Comparative analyses of nestbuilding behavior in bonobos (*Pan paniscus*) and chimpanzees (*Pan troglodytes*), *in:* "Behavioral Diversity in Chimpanzees", P. Heltne and L. Marquardt, eds., Harvard University Press, Mass.

Galat, G. and Galat-Luong, A., 1982, Domaines vitaux, densités et stratification de sept Cercopithecines forestiers d'un même site du Parc National de Tai, Côte d'Ivoire: Premiers résultats, unpublished report.

Galat, G. and Galat-Luong, A., 1985, La Communauté de Primates Diurnes de la Fôret de Tai, Côte d'Ivoire, *Rev. Ecol. (Terre Vie)* 40:3-32.

Galdikas, B., 1983, The orangutan long call and snag crashing at Tanjung Puting Reserve, *Primates* 24:371-384.

Galdikas, B. and Insley, S., 1988, The fast call of the adult male orangutan, *J. Mamm.* 69:371-375.

Gautier, J-P., 1988, Interspecific affinities among guenons as deduced from vocalizations, *in:* "A Primate Radiation: Evolutionary Biology of the African Guenons", A. Gautier-Hion, F. Bourlier, J-P. Gautier and J. Kingdon, eds., Cambridge University Press, Cambridge.

Gautier, J.-P. and Gautier, A., 1977, Vocal communication in Old World monkeys, *in:* "How Animals Communicate", T. Sebeok, ed., Indiana University Press, Bloomington.

Gautier-Hion, A., 1988, Polyspecific associations among forest guenons: ecological, behavioral and evolutionary aspects,*in:* "A Primate Radiation: Evolutionary Biology of the African Guenons", A. Gautier-Hion, F. Bourlier, J.-P. Gautier and J. Kingdon, eds., Cambridge University Press, Cambridge.

Gautier-Hion, A. and Gautier, J-P., 1985, Sexual dimorphism, social units and ecology among sympatric forest guenons, *in:* "Human Sexual Dimorphism", J. Ghesquiere, R. Martin and F. Newcombe, eds., Taylor and Francis, London.

Ghiglieri, M., 1984, "The Chimpanzees of Kibale Forest: A Field Study of Ecology and Social Structure", Columbia University Press, New York.

Gittins, S., 1980, Territorial behavior in the agile gibbon, *Int. J. Primatol.* 1:381-399.

Goldizen, A., 1986, Tamarins and marmosets: communal care of offspring, *in:* "Primate Societies", B. Smuts, D. Cheney, R. Seyfarth, R. Wrangham and T. Struhsaker, eds., University of Chicago Press, Chicago.

Goodall, J., 1965, Chimpanzees of the Gombe Stream Reserve, *in:* "Field Studies of Monkeys and Apes", I. DeVore, ed., Holt, Rinehart and Winston, New York.

Goodall, J.v.L., 1968, A preliminary report on expressive movements and communications in the Gombe Stream chimpanzees, *in:* "Primates: Studies in Adaptation and Variability", P. Jay, ed., Holt, Rinehart and Winston, New York.

Goodall, J., 1986, "The Chimpanzees of Gombe", The Belknap Press of Harvard University Press, Cambridge.

Gurmaya, K., 1986, Ecology and behavior of *Presbytis thomasi* in Northern Sumatra, *Primates* 27:151-172.

Hauser, M., 1993, The evolution of non-human primate vocalizations: effects of physiology, body weight, and social context, *Amer. Nat*. 142:528-542.

Hauser, M., Teixidor, P., Field, L., and Flaherty, R., 1993, Food-elicited calls in chimpanzees: effects of food quantity and divisibility, *Anim. Behav.* 45: 817-819.

Harcourt, A., 1978, Strategies of emigration and transfer by primates, with particular reference to gorillas, *Z. Tierpsychol.* 48:401-420.

Harcourt, A., Stewart, K., and Fossey, D., 1981, Gorilla reproduction in the wild, *in:* "Reproductive Biology of Great Apes", C. Graham, ed., Academic Press, New York.

Harcourt, A., Stewart, K., and Harcourt, D., 1986, Vocalizations and social relationships of wild gorillas: a preliminary analysis, *in:* "Current Perspectives in Primate Social Dynamics", D. Taub and F. King, eds., Van Nostrand Reinold, New York.

Harcourt, A., Steward, K., and Hauser, M., 1993, Function of wild gorilla "close" calls. I. Repertoire, context, and interspecific comparison, *Behaviour* 124:89-122.

Harvey, P., Martin, R., and Clutton-Brock, T., 1986, Life histories in comparative perspective, *in:* "Primate Societies", B. Smuts, D. Cheney, R. Seyfarth, R. Wrangham, and T. Struhsaker, eds., University of Chicago Press, Chicago.

Hohmann, G., 1989a, Comparative study of vocal communication in two Asian leaf monkeys, *Presbytis johnii* and *Presbytis entellus, Folia Primatol.* 52:27-57.

Hohmann, G., 1989b, Vocal communication of wild bonnet macaques (*Macaca radiata*), *Primates* 30:325-345.

Hohmann, G, 1990, Loud calls of male purple-faced langurs (*Presbytis senex*), *Folia Primatol.* 55:200-206.

Hohmann, G. and Fruth, B., in press, Structure and use of distance calls in wild bonobos (*Pan paniscus*), *Int. J. Primatol.*

Hohmann, G. and Vogl, L., 1991, Loud calls of male Nilgiri langurs (*Presbytis johnii*): age-, individual-, and population-specific differences, *Int. J. Primatol.* 12:503-524.

Horwich, R., 1972, Home range and food habits of the Nilgiri langur, *Presbytis johnii*, *J. Bomb. Nat. Hist. Soc.* 69:255-267.

Horwich, R., 1976, The whooping display in Nilgiri langurs: an example of daily fluctuations superimposed on a serial trend, *Primates* 17:419-431.

Horwich, R., and Gebhard, K., 1983, Roaring rhythms in black howler monkeys (*Alouatta pigra*) of Belize, *Primates* 24:290-296.

Hoshino, J., Mori, A., Kudo, H., and Kawai, M., 1984, Preliminary report on the grouping of mandrills (*Mandrillus sphinx*) in Cameroon, *Primates* 25:295-307.

Inoue, M., 1988, Age gradations in vocalization and body weight in Japanese monkeys (*Macaca fuscata*), *Folia Primatol.* 51:76-86.

Jordan, C., 1977, "Das Verhalten zoolebender Zwergschimpansen", Dissertation, Universität Frankfurt a.M.

Jungers, W. and Susman, R., 1984, Body size and skeletal allometry in African apes, *in:* "The Pygmy Chimpanzee: Evolutionary Biology and Behavior", R. Susman, ed., Plenum Press, New York.

Kano, T., 1992, "The Last Ape: Pygmy Chimpanzee Behavior and Ecology", Stanford University Press, Stanford.

Kappeler, M., 1984, Vocal bouts and territorial maintainance in the moloch gibbon, *in:* "The Lesser Apes: Evolutionary and Behavioral Ecology", H. Preuschoft, D. Chivers, W. Brockelman and N. Creel, eds., Edinburgh University Press, Edinburgh.

Kawabe, M. and Mano, T., 1972, Ecology and behavior of the wild proboscis monkey, *Nasalis larvatus* (Wurmb), in Sabah, Malaysia, *Primates* 13:213-228.

Klein, L., 1974, Agonistic behavior in neotropical primates, *in*: "Primate Aggression, Territoriality, and Xenophobia", R. Holloway, ed., Academic Press, New York.

Klein, L. and Klein, D., 1977, Social and ecological contrasts between four taxa of neotropical primates, *in:* "Socioecology and Psychology of Primates", R. Tuttle, ed., Mouton, Paris.

Krunkelsven, E. v., 1993, Vocale Communicatie bij Bonobo's *Pan paniscus*, in Gevangenschap, Ms. thesis, University of Antwerp.

Kudo, H., 1987, The study of vocal communication of wild mandrills in Cameroon in relation to their social structure, *Primates* 28:289-308.

Kuroda, S., 1979, Grouping of the pygmy chimpanzees, *Primates* 20:161-183.

MacKinnon, J., 1974, The behaviour and ecology of wild orang-utans (*Pongo pygmaeus*), *Anim. Behav.* 22:3-74.

Manley, G., 1986, Through the territorial barrier: harem accretion in *Presbytis senex*, *in:* "Primate Ontogeny, Cognition and Social Behavior", J. Else and P. Lee, eds., Cambridge University Press, New York.
Marler, P., 1976, Social organization, communication and graded signals: the chimpanzee and the gorilla, *in:* "Growing Points in Ethology", P. Bateson and R. Hinde, eds., Cambridge University Press, London.
Marler, P. and Hobbet, L., 1975, Individuality in a long-range vocalization of wild chimpanzees, *Z. Tierpsychol.* 38: 97-109.
Marler, P. and Tenaza, R., 1977, Signalling behavior of apes with special reference to vocalization, *in:* "How Animals Communicate", T. Sebeok, ed., Indiana University, Bloomington.
Masataka, N., 1982, A field study on the vocalizations of Goeldi's monkeys (*Callimico goeldii*), *Primates* 23:206-219.
McKey, D. and Waterman, P., 1982, Ranging behavior of a group of black colobus (*Colobus satanas*) in the Douala-Edea Reserve, Cameroon, *Folia Primatol.* 39:264-304.
Milton, K., 1984, Habitat, diet, and activity patterns of free-ranging woolly spider *monkeys (Brachyteles arachnoides*, E. Geoffrey, 1806), *Int. J. Primatol.* 5: 491-513.
Milton, K., 1985, Mating patterns of woolly spider monkeys, *Brachyteles arachnoides*: implications for female choice, *Behav. Ecol. Sociobiol.* 7:53-59.
Mitani, J., 1984, The behavioral regulation of monogamy in gibbons (*Hylobates muelleri*), *Behav. Ecol. Sociobiol.* 15:225-229.
Mitani, J., 1985, Sexual selection and adult male orangutan long calls, *Anim. Behav.* 33:272-283.
Mitani, J. and Nishida, T., 1993, Contexts and social correlates of long-distance calling by male chimpanzees, *Anim. Behav.* 45: 735-746.
Mitani, J., Hasegawa, T., Gros-Louis, J., Marler, P., and Byrne, R., 1992, Dialects in wild chimpanzees ?, *Amer. J. Primatol.* 27: 233-243.
Moore, J. and Ali, R., 1984, Are dispersal and inbreeding avoidance related ?, *Anim. Behav.* 32:94-1122.
Mori, A., 1983, Comparison of the communicative vocalizations and behaviors of group ranging in Eastern gorillas, chimpanzees and pygmy chimpanzees, *Primates* 24: 486-500.
Napier, J. and Napier, P., 1970, "A Handbook of Living Primates", Academic Press, London.
Nishida, T. and Hiraiwa-Hasegawa, M., 1986, Chimpanzees and bonobos: cooperative relationship among males, *in:* "Primate Societies", B. Smuts, D. Cheney, R. Seyfarth, R. Wrangham and T. Struhsaker, eds., University of Chicago Press, Chicago.
Oates, J. and Trocco, T., 1983, Taxonomy and phylogeny of black-and-white colobus monkeys, *Folia Primatol.* 40:83-113.
Oates, J. and Whitesides, G., 1990, Associations between olive colobus (*Colobus verus*), diana guenons (*Cercopithecus diana*), and other forest monkeys in Sierra Leone, *Amer. J. Primatol.* 21:129-146.
Oppenheimer, J., 1977, Communication in New World monkeys, *in:* "How Animals Communicate", T. Sebeok, ed., Indiana University, Bloomington.
Parish, A., in press, Sex and food control in the "uncommon chimpanzee": how bonobo females overcome a phylogenetic legacy of male dominance, *Ethol. Sociobiol.*
Pusey, A., 1990, Behavioural changes at adolescence in chimpanzees, *Behaviour* 115:203-246.
Raemaekers, J. and Raemaekers, P., 1985, Field playback of loud calls to gibbons (*Hylobates lar*): territorial, sex-specific and species-specific responses, *Anim. Behav.* 33:481-493.
Robinson, J. and Janson, C., 1986, Capuchins, squirrel monkeys, and atelines: socioecological convergence with Old World primates, *in:* "Primate Societies", B. Smuts, D. Cheney, R. Seyfarth, R. Wrangham and T. Struhsaker, eds., University of Chicago Press, Chicago.
Robinson, J., Wright, P., and Kinzey, W., 1986, Monogamous cebids and their relatives: intergroup calls and spacing, *in:* "Primate Societies", B. Smuts, D. Cheney, R. Seyfarth, R. Wrangham and T. Struhsaker, eds., University of Chicago Press, Chicago.
Rudran, R., 1973, Adult male replacement in one-male troops of purple-faced langurs (*Presbytis senex senex*) and its effect on population structures, *Folia Primatol.* 19:166-192.
Ruhiyat, Y., 1983, Socio-ecological study of *Presbytis aygula* in West Java, *Primates* 24:344-359.
Savage-Rumbaugh, R. and Wilkerson, B., 1978, Socio-sexual behavior in *Pan paniscus* and *Pan troglodytes*: a comparative study, *J. Human Evol.* 7: 327-344.
Schaller, G., 1963, "The Mountain Gorilla: Ecology and Behavior", University of Chicago Press, Chicago.
Sebeok, T. and Umiker-Sebeok, J., 1980, "Speaking of Apes", Plenum Press, New York.
Sekulic, R., 1982a, The function of howling in red howler monkeys (*Alouatta seniculus*), *Behaviour* 81: 291-305.

Sekulic, R., 1982b, Behavior and ranging patterns of solitary female red howler (*Alouatta seniculus*), *Folia Primatol.* 38:217-232.
Sekulic, R., 1983, The effect of female call on male howling in red howler monkeys (*Alouatta seniculus*), *Int. J. Primatol.* 4:291-305.
Sekulic, R. and Chivers, D., 1986, The significance of call duration in howler monkeys, *Int. J. Primatol.* 7:183-190.
Seyfarth, R., 1986, Vocal communication and its relation to language, *in:* "Primate Societies", B. Smuts, D. Cheney, R. Seyfarth, R. Wrangham and T. Struhsaker, eds., University of Chicago Press, Chicago.
Smuts, B., Cheney, D., Seyfarth, R., Wrangham R., and Struhsaker, T., 1986, "Primate Societies", University of Chicago Press, Chicago.
Srikosamatara, S., 1980, Ecology of pileated gibbons in South-East Thailand, *in:* "The Lesser Apes: Evolutionary and Behavioral Ecology", H. Preuschoft, D. Chivers, W. Brockelman and N. Creel, eds., Edinburgh University Press, Edinburgh.
Stammbach, E., 1986, Desert, forest, and montane baboons: multilevel societies, *in:* "Primate Societies", B. Smuts, D. Cheney, R. Seyfarth, R. Wrangham and T. Struhsaker, eds., University of Chicago Press, Chicago.
Stanford, C., 1991, The capped langur in Bangladesh: behavioral ecology and reproductive tactics, *Contrib. Primatol.* 26.
Stewart, K. and Harcourt, A., 1986, Gorillas: variation in female relationships, *in:* "Primate Societies", B. Smuts, D. Cheney, R. Seyfarth, R. Wrangham and T. Struhsaker, eds., University of Chicago Press, Chicago.
Strier, K., 1987, Ranging behavior of woolly spider monkey, or muriquis, *Brachyteles arachnoides*, *Int. J. Primatol.* 8:575-591.
Strier, K., Mendes, F., Rimoli, J., and Rimoli, A., 1993, Demography and social structure of one group of muriquis (*Brachyteles arachnoides*), *Int. J. Primatol.* 14: 513-526.
Struhsaker, T., 1967, Auditory comunication among vervet monkeys (*Cercopithecus aethiops*), *in:* "Social Communication among Primates", S.T. Altman, ed., University of Chicago Press, Chicago.
Struhsaker, T., 1975, "The Red Colobus Monkey", University of Chicago Press, Chicago.
Struhsaker, T. and Leyland, L., 1986, Colobines: infanticide by adult males, *in:* "Primate Societies", B. Smuts, D. Cheney, R. Seyfarth, R. Wrangham and T. Struhsaker, eds., University of Chicago Press, Chicago.
Symington, M., 1988, Food competition and foraging party size in the black spider monkey (*Ateles paniscus chamek*), *Behaviour* 105:117-134.
Symington, M., 1990, Fission-fusion social organization in *Ateles* and *Pan*, *Int. J. Primatol.* 11:47-61.
Tenaza, R., 1976, Songs, choruses and countersinging of Kloss' gibbons (*Hylobates klossi*) in Siberut Island, Indonesia, *Z. Tierpsychol.* 40: 37-52.
Tenaza, R., 1989, Intergroup calls of male pig-tailed langurs (*Simias concolor*), *Primates* 30:199-206.
Tenaza, R., Fitch, H., and Lindburg, D., 1988, Vocal behavior of captive Sichuan golden monkeys (*Rhinopithecus r. roxellana*), *Amer. J. Primatol.* 14:1-9.
Thorington, R., Ruiz, J., and Eisenberg, J., 1984, A study of a black howling monkey (*Alouatta caraya*) population in northern Argentina, *Amer J. Primatol.* 6:357-366.
Tilson, R., 1977, Social organization of Simakobu monkeys (*Nasalis concolor*) in Siberut Island, Indonesia, *J. Mammal.* 58:202-212.
Tilson, R., 1979, Behavior of hoolock gibbons (*Hylobates hoolock*) during different seasons in Assam, India, *J. Bomb. Nat Hist. Soc.* 76:1-16.
Tilson, R. and Tenaza, R., 1976, Monogamy and duetting in an Old World monkey, *Nature* 263:320-321.
Tröster, P., 1988, Kommunikationsverhalten beim Bonobo (*Pan paniscus*, Schwarz 1929) unter besonderer Berücksichtigung des akustischen Bereichs, Diplomarbeit, Universität Heidelberg.
Vogel, C., 1975, Soziale Organisationsformen bei catarrhinen Primaten, *in:* "Hominisation and Behavior", G. Kurth and I. Eibl-Eibesfeldt, eds., Fischer, Stuttgart.
Waal, F. de, 1988, The communicative repertoire of captive bonobos (*Pan paniscus*), compared to that of chimpanzees, *Behaviour* 106: 183-251.
Wallman, J., 1992, "Aping Language", Cambridge University Press, Cambridge.
Waser, P., 1982, The evolution of male loud calls among mangabeys and baboons, *in:* "Primate Communication", C. Snowdon, C. Brown and M. Petersen, eds., Cambridge University Press, Cambridge.
Waser, P.M. and Brown, C.H., 1984, Is there a "sound window" for primate communication ?, *Behav. Ecol. Sociobiol.* 15: 73-76.

Watanabe, K., 1981, Variation in group composition and population density of the two sympatric Mentawaian leaf monkeys, *Primates* 22:145-160.
Watts, D., 1985, Relations between group size and composition and feeding competition in mountain gorilla groups, *Anim. Behav.* 33:72-85.
Watts, D., 1994, Social relationships of immigrant and resident female mountain gorillas. II: Relatedness, residence, and relationships between females, *Amer. J. Primatol.* 32:13-30.
White, F., 1988, Party composition and dynamics in *Pan paniscus*, *Int. J. Primatol.* 9:179-193.
Whitehead, J., 1987, Vocally mediated reciprocity between neighbouring groups of mantled howler monkeys, *Alouatta palliata palliata*, *Anim. Behav.* 35:1615-1627.
Williams, L., 1967, Breeding Humboldt's woolly monkey at Murrayton Woolly Monkey Sanctuary, *Int. Zoo Yearb.* 7:86-89.
Wilson, W. and Wilson, C., 1975, Species-specific vocalizations and the determination of phylogenetic affinities of the *Presbytis aygula-melalophus* group in Sumatra, *in:* "Contemporary Primatology", S. Kondo, M. Kawai and A. Ehara, eds., Karger, Basel.
Wrangham, R., 1977, Feeding behaviour of chimpanzees in Gombe National Park, Tanzania, *in:* "Primate Ecology", T.H. Clutton-Brock, ed., Academic Press, New York.
Wrangham, R., 1986, Ecology and social relations in two species of chimpanzee, *in:* "Ecological Aspects of Social Evolution", D. Rubenstein and R. Wrangham, eds., Princeton University Press, Princeton.
Wrangham, R. and Smuts, B., 1980, Sex differences in the behavioural ecology of chimpanzees in the Gombe National Park, Tanzania, *J. Reprod. Fert., Suppl.* 28:13-31.

A COMPARATIVE APPROACH TO THE NON-HUMAN PRIMATE VOCAL TRACT: IMPLICATIONS FOR SOUND PRODUCTION

Miguel A. Schön Ybarra

Department of Anatomy
School of Medicine
University of Puerto Rico
San Juan, Puerto Rico, 00936-5067

INTRODUCTION

Despite the fact that it is the vocal tract morphology what ultimately shapes the acoustic structure of the sounds used in vocal communication (see Lieberman, 1984), recent state-of-the-art publications on the subject of non-human primate communication have not been concerned with that morphology (e.g., Snowdon, Brown, and Petersen, 1982; Todt, Goedeking, and Symmes, 1988). To help fill the gap thus left in our comprehension of the biological bases of vocal communication in non-human primates, this chapter reviews current knowledge pertaining to the comparative and functional morphology of non-human primate vocal tract features that, because of their shape, structure, and location, can influence sound structure. Conceptually, the chapter is based on (1) known relationships among the human vocal tract anatomy, the movements of its walls, air flow patterns, and the production of speech sounds (see Lieberman and Blumstein, 1988; Pickett, 1980); (2) Hirano's structural model for vocal fold vibrator capabilities (Hirano, 1974, 1991; Gray et al., 1993); (3) the source-filter theory of speech production as summarized by Lieberman and Blumstein (1988).

State of the Art

Comparative morphological investigations have focused individually on the larynx, pharynx, mouth and nasal fossa of non-human primates (Brandes, 1931; Cave, 1967; Doyle et al., 1990; Flügel and Rohen, 1991; Geist, 1933; Hill, 1952, 1953-70; Hill and Booth, 1957; Hilloowala, 1975; Huber, 1931, Jordan, 1971; Kelemen, 1948, 1969; Kelemen and Sade, 1960; Lampert, 1926; Laitman et al., 1977; Negus, 1962; Nèmai, 1926; Schneider, 1958; Schön, 1964, 1970, 1971; Starck and Schneider, 1960). Most of these investigations have served primarily to clarify systematic relations. Some have explored the functional morphology of selected parts of the tract in relation to food ingestion, lung ventilation, or sound output. Schön Ybarra (1986, 1988) studied the vocal tract of howling monkeys as the peripheral effector organ for their deep, loud, and raucous barks and roars. Only

Lieberman (1969) has studied the luminal dimensions of the supralaryngeal part of the tract of some mature and immature monkeys and apes in relation to sound production, and Taylor et al. (1976) the tract dimensions in immature monkeys to identify growth trends. Neurological and histological data related to the control of vocal production have been produced (Bowman, 1968; Garret and Larson, 1991; Jürgens, 1988; Jürgens and Lu, 1993; Larson, 1985; Newman, 1988; Ploog, 1988; Raman and Darmanadan, 1989; Satoda et al., 1990; Strutz et al., 1988; Sutton et al., 1977; West and Larson, 1993; Yoshida et al., 1985).

Functional morphological perspectives, such as those used for elucidating sound output capabilities in the human vocal tract, have not been applied to the non-human primates. Thus, Hirano's (1974, 1991) characterization of the human vocal folds as consisting of a cover layer (epithelium and lamina propria) and a body layer (the thyroarytenoid muscle), each with different vibrator capabilities, has not been extended to non-human primates, except for Hirano (1991)'s characterization of those features in not identified monkey species. No extensive investigation of the molecular and cellular structure of the non-human primate vocal fold has been done (see Gray et al., 1993) despite the importance of the subject for the identification of animal models that can be used in experimental work related to the solution of human voice disorders (see Hirano, 1991). Studies on the rotary capabilities of the human cricoarytenoid articulation (Frable, 1961; Maue and Dickson, 1971; Sonesson 1959; von Leden and Moore, 1961) have not been applied to non-human primates albeit the importance of this joint in glottal shaping mechanisms (see Woodson, 1992). The morphologically intrinsic capability of the supra-, infrahyoid, and digastric muscles to regulate the position of the hyo-laryngeal skeleton and the tract's luminal configuration, hence, its acoustic response, has not been investigated in non-human primates notwithstanding the known effect of these muscles on the sound output of *Macaca mulatta* (Sapir et al., 1981) and humans (Sonninen, 1956). The arrangement of the extrinsic muscles of the tongue in the body of this organ, important as it is for understanding lingual maneuverability (Du Brul, 1976), has not been investigated in non-human primates.

Current views about the sound producing capabilities of the non-human primate vocal tract are that it (1) generates an unstable glottal source; (2) can produce some of the human phonetic contrasts, and, given its laryngeal air sacs, very loud vocalizations (Lieberman, 1969, 1975, 1977; Lieberman et al., 1969). Important and seminal as these views have been for understanding the morphological bases of speech sound production differences between modern humans and non-human primates, they have not been accompanied by an equivalent comprehension of the sound output faculties specific to the vocal tract of non-human primates. Recent studies suggest that these faculties have been underestimated (see Bauer, 1986; Marler and Tenaza, 1977; Mitani and Marler, 1989; Seyfarth and Cheney, 1986), and that the articulatory gestures associated with the non-human primate vocal output are likely to be more complicated than originally thought (Bauer, 1986; Hauser, 1989, 1993; Hauser et al., 1993; Hauser and Schön Ybarra, 1993). There is evidence that *Macaca mulatta* can produce, in a remarkable contrast with humans, spike-like Fo patterns characterized by a very rapid, steep and pronounced frequency upsweep that peaks very high and downsweeps just as precipitously (Hauser and Marler, 1993). Similar patterns are detectable in squirrel monkeys (Winter et al., 1966) and gibbons (Mitani and Marler, 1989); and some callithricids, *Galago senegalensis, crassicaudatus*, and *garnettii*, and *Microcebus murinus*, all of them small-bodied primates, can produce very high frequency calls (Epple, 1975; Masters, 1991; Snowdon, 1989; Zimmermann, 1981; Zimmermann and Lerch, 1993). The morphologies that can subserve this variety of sound outputs in non-human primates remain to be identified.

In this chapter, I shall (1) present data gathered in my laboratory and summarize information from available references, all of which demonstrates that the vocal tract of non-human primates has morphologically inherent sound output capabilities more diversified than has so far been realized; (2) outline lines of research that can lead to a better comprehension than is currently available about the non-human primate vocal tract morphology and its intrinsic sound output capabilities.

Data Gathering Procedures

I have procured comparative information about the following morphological features in non-human primates: (1) vocal fold shape and structure viewed as constraints on the generation of glottal excitations; (2) air sacs, as sound amplifiers and modulators; (3) the luminal configuration and cartilaginous framework of both the laryngeal vestibule and inlet as features capable of influencing laryngeal sound output quality and amplitude; (4) the thyrohyoid connections and the velo-pharyngeal morphology as constraints for the buccal vs. nasal emission of sounds; (5) the shape, muscular structure, and location of the tongue, as constraints for oral articulations; (6) the circumlabial musculature, as a factor on modifying buccal dimensions, hence, formant and even Fo patterns; (7) the length of the glotto-labial tract as constraint for basic formant patterning. I restricted my search of data to mature subjects to avoid problems attending to morphological differences in age.

I dissected and examined one adult female *Cebus albifrons* and six adult male and six adult female *Macaca mulatta* obtained from the Caribbean Primate Research Center; seven female and eight male *Alouatta seniculus* of my personal collection; the larynx of one adult male *Ateles spec.*, from the Smithsonian Institution, and of one *Symphalangus spec.*, made available to me by Dr. E. L. Simons. I use binomial names according to Fiedler (1956), and common names according to Napier and Napier (1967). From the authors listed in the following paragraph, I obtained information about the prosimian and simian species they studied. I present the names of these taxa as given by the sources consulted, and add that numbers refer to that of the specimens studied by the source, M = male, F = female, N = sex of specimen(s) not specified.

Brandes (1931): orang-utans, 9M, 4F. Hill (1952): *Colobus b. badius,* 1M, 4F; *C. b. preussi,* parts of several N specimens; *C. polykomos polykomos*, 1M; *C. abyssinicus*, 1M; *Semnopithecus priam,* 1M; *Trachypithecus obscurus flavicauda*, 1M; *Presbytis melalophus*, 2F. Jordan (1971): *Troglodytes niger*, 2M, 4F. Kelemen (1948): juvenile chimpanzee, 1M. Kelemen and Sade (1960): *Alouatta palliata*, 1M, 2F. Lampert (1926): *Ateles ater*, 1F; *Cebus macrocephalus* Spix, 1M; *Chrysothrix nigrivittatus* Wag., 1M; *Hapale spec.*, 1M; *Lagothrix castelnaui*, 1M; *Mycetes auratus*, 1M; *Mycetes belzebul*, 1 M; *Nictipithecus azarae*, 1F. Nèmai (1926): *Cebus capucinus*, 2N; Cercocebus collaris, 1N; *Hapale penicillata*, 1N; *Mycetes seniculus*, 2N. Schneider (1958): *Aotes trivirgatus*, 1N; *Callithrix argentata*, 1N; *Cebus variegatus*, 1N; *Cercopithecus aethiops*, 1N; *Cercopithecus talapoin*, 2N; *Colobus polykomos abyssinicus*, 1N; *Daubentonia madagascarensis*, 1N; *Galago crassicaudatus*, 2N; *Gorilla gorilla*, 2N; *Hylobates lar*, 1N; *Loris tardigradus*, 1N; *Macaca mulatta*, 4N; *Papio doguera*, 1N; *Papio hamadryas*, 1N; *Presbytis entellus*, 1N; *Pan troglodytes*, 1N; *Pongo pygmaeus*, 3N; *Saimiri sciureus*, 1N; *Tarsius syrichta*, 1N; *Theropithecus gelada*, 1N. Starck and Schneider (1960): *Aotes trivirgatus*, 1N; *Alouatta caraya*, 1M; *Alouatta seniculus*, 2M; *Ateles spec.*, 3N; *Callithrix jacchus*, 1N; *Cebus albifrons*, 2N; *Cebus variegatus*, 1N; *Cercopithecus aethiops*, 3M; *Cercopithecus cephus*, 1M; *Cercopithecus talapoin*, 1M; *Colobus polykomos abyssinicus*, 2M; *Daubentonia madagascarensis*, 1N; *Galago senegalensis*, 1N; *Galago crassicaudatus*, 1N; *Gorilla gorilla*, 2M; *Hylobates*, 1M; *Lemur catta*, 1F; *Leontocebus rosalia*, 1M; *Loris tardigradus*, 1F; *Macaca mulatta*, 3N; *Macaca nemestrina*, 1F; *Nycticebus coucang*, 1N; *Papio comatus*, 1F; *Papio doguera*, 2M; *Papio hamadryas*, 2M; *Papio papio*, 1N; *Perodicticus potto*, 1M; *Pan troglodytes*, 1F; *Pongo pygmaeus*, 3N; *Saimiri sciureus*, 1M; *Semnopithecus*, 2N; *Symphalangus syndactylus*, 1M; *Tarsius spectrum*, 1M.

I have utilized the following directional terms in anatomical descriptions: cranial and superiorly, when in the direction of the cranium of the animal; caudally, posteriorly, and inferiorly, when in that of its caudal end; ventral and anteriorly, when toward the ventral surface of its neck; posteriorly and dorsal, when toward its back or dorsum; other directional terms I use as employed in human anatomy.

RESULTS

(1) The shape and structure of the vocal folds. On the basis of my observations and the information gathered from references, I recognize three different patterns of vocal folds. (1) In the platyrrhines that I examined, in those studied by Starck and Schneider (1960), and in *Lemur catta* and *Pongo pygmaeus* (Starck and Schneider, 1960), the cover layer (epithelium and lamina propria) has two portions structurally distinct. The inferior appears

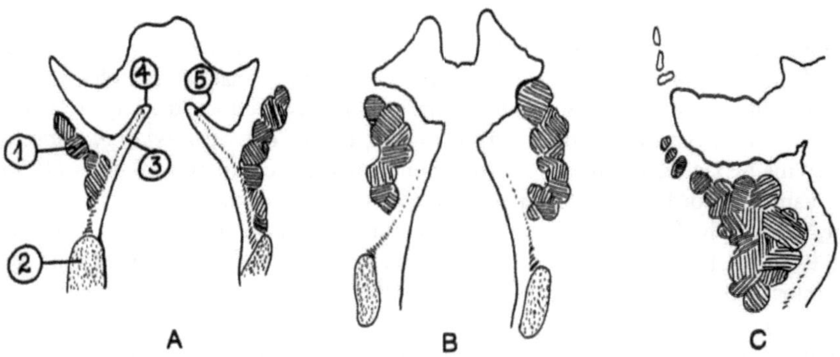

Figure 1. Tracings of coronal sections of the larynx showing: 1, thyroarytenoid muscle; 2, cricoid cartilage; 3, conus elasticus; 4, vocal lip; 5, ligamentum vocale. A, Pattern 1, *Saimiri sciureus*; B, Pattern 2, *Galago senegalensis* (modified from Starck and Schneider, 1960); C, Pattern 3, *Pan spec.* (modified from Kelemen, 1960).

to be firmly fastened through the conus elasticus to the underlying thyroarytenoid muscles. The superior is built as a thin upward expansion of the free edge of the membranous part of the fold: the vocal lip (Fig. 1A) that is said to contain a vocal ligament and a conus elasticus (Kelemen and Sade, 1960; Starck and Schneider, 1960). (2) In chimpanzees the free edge of the fold is thickened by an accumulation of connective tissue within the lamina propria at the edge of the fold, apparently in Reinke's space (Fig. 1C) (see Reinke, 1898, cited in Starck and Schneider, 1960). This edge is capped by a stub-like, vestigial lip with no conus elasticus or vocal ligament inside (Jordan, 1971; Kelemen, 1948). Something similar is reported to be the case for the vocal folds of gorillas, orangs, rhesus monkeys, *Papio hamadryas*, *Colobus*, bushbabies (Fig. 1B) and tarsiers, but in the latter two there seems to be no lip at all (see Brandes, 1931; Kirchner, 1993; Starck and Schneider, 1960). Hirano (1991) identifies a vocal fold configuration in "the monkey" that is similar to what I have called "pattern 2," and he indicates that the arrangement of tissues in the cover layer of "the monkey" is different from what it is in other mammals and adult man. According to him, the arrangement of elastic and collagenous fibers immediately under the epithelium makes the mucosa "stiffer in monkeys than in humans or dog" (Hirano, 1991, p. 34). (3) The vocal folds of hylobatids have a lip, and, in the basal part of the cover layer, a bulging strand that extends from the anterior commissure to the cricoid cartilage (Starck and Schneider, 1960). In patterns 1-3 the superior part of the cover layer appears to be unleashed from the tethering effect that is presumably created by the more intimate binding between the cover and body layer in the inferior part of the fold. Information available to me indicates that the cadaveric non-human primate glottis is lanceolate anteroposteriorly. Its dorsal extreme remains patent as a hiatus intervocalis (see Kelemen, 1948; Kelemen and Sade, 1960; Nèmai, 1926), except in gibbons (Starck and Schneider, 1960). The longitudinal and area proportions between the membranous and cartilaginous glotisses has not been investigated in non-human primates.

(2) Air sacs. I identify four kinds (Fig. 2). (1) *Symphalangus* (Starck and Schneider, 1960), the great apes (Brandes, 1931; Jordan, 1971; Starck and Schneider, 1960), and alouattines (except *A. palliata*, Kelemen and Sade, 1960) have large lateral air sacs that develop as outgrowths of the ventricular epithelium; their size and extent is quite large in gorillas and orangs (Fick, cited by Jordan, 1971, reported in 1929 a capacity of 6 liters for the air sacs of an orangutan). (2) Callithricids appear to be the only possessors of a pair of median sacs that originate just supraglottally from a single midline opening, then balloon forward between the cricoid and thyroid cartilages (Starck and Schneider, 1960); it is not

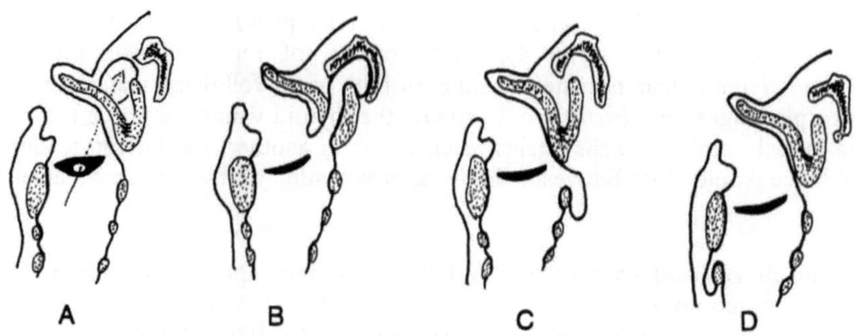

Figure 2. Four kinds of air sacs: A, lateral; B, superomedian; C, inferomedian; D, dorsal.

established that all callithricids have such sacs. (3) Old World monkeys (Starck and Schneider, 1960) and *Alouatta* (Kelemen and Sade, 1960; Lampert, 1926; Nèmai, 1926; Schön, 1970, 1971; Schön Ybarra, 1986, 1988) have a superomedian sac that, at its unpaired origin just above the glottis, splits the stem of the epiglottic cartilage and then expands cranially to fill the concavity of the hyoid bone. The latter is very large and bulla-like (size variations of the hyoid in *Alouatta* have been qualitatively assessed by Hershkovitz, 1949). My findings show that there is in *A. seniculus* a large hyo-epiglottic fatty pad behind both the ostium of the bulla and the membranous thyrohyoid canal that connects the vestibule with the part of the superomedian sac that fills the hyoid bulla. (4) *Indri brevicaudatus* (Milne Edwards, 1874, cited by Starck and Schneider, 1960), *Lemur variegatus* (Meyer, 1901, cited by Starck and Schneider, 1960), and spider monkeys possess an infraglottal posterior sac which passes dorsally between the cricoid and the first tracheal ring and extends caudally for a short distance between trachea and esophagus.

(3) Laryngeal vestibule and inlet. My observations and those of others (Lampert, 1926; Nèmai, 1926; Jordan, 1971; Kelemen, 1949; Kelemen and Sade, 1960; Starck and Schneider, 1960) indicate that the walls of the non-human primate laryngeal vestibule (Fig. 3) are considerably stiffened by the shape, size, and location of the epiglottis and the pair of arytenoid and cuneiform cartilages, and that all five cartilages are bound together by apparently firm fibrous connections and the interarytenoepiglottic muscular brace (Fig. 4).

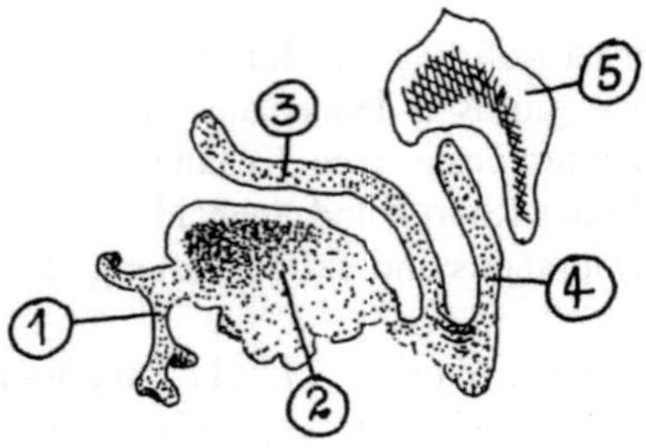

Figure 3. A right side view of the laryngeal cartilages; from left (dorsal) to right (ventral): 1, arytenoid; 2, cuneiform; 3, epiglottic cartilage; 4, thyroid cartilage; 5, hyoid bone.

The medially bulging cuneiforms constrict the middle of the vestibule creating an anterior and a posterior vestibular air channel. The anteroposterior diameter of the vestibule diminishes toward its inlet. The latter, T-shaped, narrow except in apparently gorillas, orangs, and siamangs (see Starck and Schneider, 1960), is made unyielding by the same fibro-cartilaginous scaffold. Given the disposition of the interarytenoepiglottic muscular brace around all five cartilages, and that of the inferior pharyngeal constrictor from one thyroid wing to the other (Fig. 4), approximation of the vestibular folds can be accomplished in non-human primates by either or both of the following maneuvers. In one, the inferior pharyngeal constrictor would adduct the thyroid wings and force the vestibular folds and lateral vestibular walls against each other; in another, the interarytenoepiglottic muscular brace would close the vestibule by approximating all five cartilages against each other.

(4) The thyrohyoid connections and the velo-pharyngeal morphology. With the only known exceptions of *Alouatta* (Schön, 1964), *Indri indri* (Starck and Schneider, 1960), and *Colobus polykomos vellerosus* (Hill and Booth, 1957), the hyoid bone appears to be either fully in front of, or rostrally very close to the thyroid cartilage (see Starck and Schneider, 1960). These two skeletal pieces are so placed and interconnected in other non-human primates that when either is moved, the other must be dragged along. This

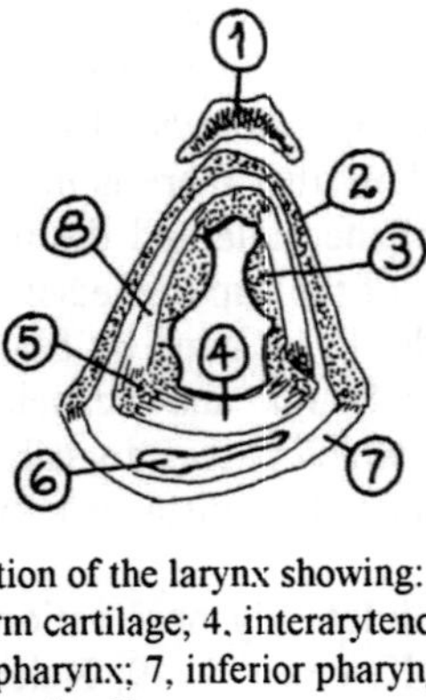

Figure 4. Cross-section of the larynx showing: 1, hyoid bone; 2, thyroid cartilage; 3, cuneiform cartilage; 4, interarytenoid muscle; 5, arytenoid cartilage; 6, laryngopharynx; 7, inferior pharyngeal constrictor; 8, arytenoepiglottic muscular fibers.

arrangement offsets distension of the vocal tract between the hyoid bone and the thyroid cartilage. In most cases the tips of the upper thyroid and greater hyoid horns are connected by the thickened dorsal free border of the thyrohyoid membrane (Starck and Schneider, 1960). Lemuriforms are said to lack upper thyroid horns (Hill, 1953), but the nature of their intercornual connection is not specified. In *Alouatta* the hyoid bulla is joined to the thyroid cartilage by a conspicuous and distensible membrane, and the tips of the greater hyoid and upper thyroid horns by a coiled fibrous band that differentiates in the free border of the membrane. The hyoid and larynx can be amply separated if either gyrates about a center that coincides with this band (see Schön, 1964).

According to Starck and Schneider (1960) and my own observations, the non-human primate epiglottis is in retrovelar, subvelar, or antevelar contact with the free margin of the velum, and the laryngeal inlet intranarial (Fig. 5). A pair of palatopharyngeal folds are almost colinear with the velum.

(5) The shape, muscular structure, and location of the tongue. According to my observations and those of Schneider (1958), the tongue of adult non-human primates is longer than wide, flat, and all within the also longer than wide oral cavity. The surface of its dorsum is congruent with the roof of the mouth in cadaveric specimens (Fig. 5). The root of the non-human primate tongue does not face the dorsal wall of the pharynx, as in modern adult humans; being coplanar with the surface of the lingual blade and tip, it faces the overlying velum and hard palate. The arrangement of the intrinsic tongue musculature in non-human primates (Schneider, 1958) does not appear to differ from that of adult man (Lowe, 1981).

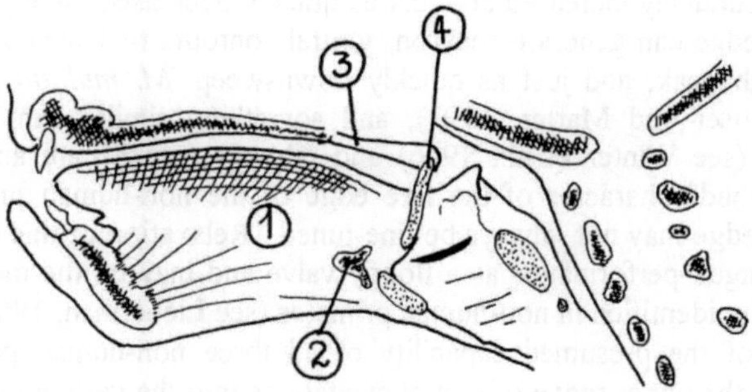

Figure 5. 1. tongue; 2. hyoid bone; 3. velum; 4. epiglottis.

(6) Length of glotto-buccal tract. My findings indicate that in 3 adult *M. mulatta* (average body weight 11 kg), the average glotto-labial length is 110.7 mm, S.D. 1.9; in 4 adult females (average body weight 9 kg), 86 mm, S.D. 6.9. In one adult, 3 kg male *Cebus albifrons*, the length is 26 mm; Lieberman (1969) reports a length of 178 mm for one adult male *Gorilla* and of 120 mm for one specimen of *Pan*. One adult, 6 kg male *Alouatta seniculus* of my series of howlers had a glotto-labial length of 170 mm. The length of the laryngeal vestibule (glottis vocalis to laryngeal inlet) is approximately 20% of the entire glotto-labial length in *M. mulatta*, and 45% in *A. seniculus*.

CONCLUSIONS

Given that the number of laboratory specimens and taxa I have studied was small, and that the information I found in the consulted references was at times restricted in details, at times in depth, my conclusions about the morphology of the vocal tract of non-human primates and its sound output capabilities must be taken as strictly preliminary.

(1) Vocal folds. The morphologies that I have identified appear capable of producing not only sounds such as are said to constitute the phonetic repertoire of non-human primates (see Lieberman, 1975, 1977) but also other potentially meaningful utterances. The sound-producing capabilities that I attribute to the peculiarly configured patterns of non-human primate vocal folds, are based on Titze's (1988) views that the lower the vocal fold tissue damping, and the greater the thickness of the fold, the lower is the subglottal pressure needed to make the fold vibrate (see Scherer, 1991). I ascribe low tissue damping to all three vocal fold patterns because of the presumed unleashed relationship of their respective free edge with respect to the underlying fibro-muscular parts of the fold; and because of the manifest localized enlargement of their lamina propria as it forms the framework of a vocal lip, or of an expanded Reinke's space filled with a large lump of amorphous connective tissue. Given these presumed structural properties of the non-human primate vocal fold's edge, if either kind of vocal fold pattern is much stretched, the oscillations of the cover layer at the edge of the fold can be expected to be very rapid, the glottal tone very high with separate and less energized harmonics, and the resulting vocalization perceived as high-pitched (e.g.,callithricids, Snowdon, 1989) and shrill (e.g., the shrill barks of rhesus macaques, Rowell and Hinde, 1962); if the fold is anteroposteriorly short, as it presumably is in small body sized individuals, and its stretching is extreme, an ultrasonic Fo can be generated (about *Callithrix jacchus*, see Epple, 1975; *Microcebus murinus*, Zimmermann and Lerch, 1993). If the external tension on the cover layer of a "pattern 1" or "pattern 2" vocal fold is first suddenly increased and just as quickly decreased, an equivalently strained and recoiled free-edge can generate chevron glottal contours that would start low, rapidly upsweep to a high peak, and just as quickly downsweep. *M. mulatta* accomplishes this acoustic feat (Hauser and Marler, 1993), and something similar can be recognized in squirrel monkeys (see Winter et al., 1966) and gibbons (see Mitani and Marler, 1989). Given the "unleashed" character of the free edge of the non-human primate vocal fold, vibrations of that edge may not always be fine-tuned. Relaxation of this edge facilitates its transient or prolonged performance as a floppy valve and may be the morphological basis for the Fo instability identified in non-human primates (see Lieberman, 1969).

As a result of the presumed capability of all three non-human primate vocal fold patterns to extend the upper range of glottal excitations into the very high frequencies, even ultrasonics if the membranous part of the fold is quite short, and the morphologically inherent constraints of all three patterns for fine-tuning glottal excitation sequences (see Pressman and Kelemen, 1955), the non-human primate larynx appears to have more phonatory range, but less phonatory precision than that of humans (for a contrary view, see Kirchner, 1993). Given this range-expanded capability for phonation, it is conceivable that non-human primates use their consequential Fo diversity to configure a vocal communication repertoire based more on Fo contrasts (see Green, 1975; May et al., 1988, 1989; Petersen and Jusczyk, 1984) than on formant patterning and transitions.

The different vibratory capabilities of the lip and non-lip parts of the cover layer enable diplophonia, a phonetic feature attributed at least to chimpanzees (see Kelemen, 1948) and worth investigating in other species. The "double" vocal fold of hylobatids may be a morphological elaboration for subserving diplophonia. The lip and thickening facilitate the production of an inhaled voice (Kelemen, 1948; Kelemen and Sade, 1960).

(2) Air sacs. The presence of these sacs does not unavoidably imply loud sound output. Gibbons produce loud vocalizations melodious to the human ear (see Mitani and Marler, 1989) but have no air sacs (Starck and Schneider, 1960). However, those monkeys and apes known to amplify low-pitched vocalizations (*Alouatta*, Kelemen and Sade, 1960; Schön Ybarra, 1986, 1988; colobines, Hill and Booth, 1957; great apes, Marler and Tenaza, 1977), do possess either a large superomedian sac (*Alouatta*, Kelemen and Sade, 1960; Schön Ybarra, 1986; colobines, Hill and Booth, 1957; cercopithecines, Gautier, 1971) or voluminous lateral air sacs (Brandes, 1931; Jordan, 1971; Kelemen, 1948; Starck and Schneider, 1960). I have observed in the field that the gular skin over the gap between the noticeable thyroid and thyroid bulla prominences of adult male *A. seniculus*, is elevated in concomitance with the emission of loud barks and roars (Schön Ybarra, 1986). These elevations I attributed to air filling of the laryngohyoid membranous tube that connects the laryngeal vestibule to the uncollapsible superomedian sac, and also of the lateral air sacs (see Schön, 1964, 1970, 1971). Gautier (1971) demonstrated in *Cercopithecus neglectus* that the intensity and sound quality of its vocalizations were noticeably reduced if the superomedian air sac could not be inflated after experimentally being punctured. Hill and Booth (1957) explained that, of the species of *Colobus*, those with a superomedian sac and an enlarged larynx can produce loud calls; von Böer and Wissdorf (1989) suggested the same for *Cercocebus torquatus torquatus*. The inferomedian sacs of callithricids, being rather small, may be tuned to intensify the high frequency fundamentals that these monkeys produce on account of their short vocal cords. Because of their subglottic location, posteromedian sacs are unlikely to have any role in resonating exhaled acoustic excitations of the tract. Their acoustic response capabilities remain obscure. The cheek pouches of Old World monkeys are sacculations of the mucosa of the oral vestibule; they are collapsed except when filled with food pieces, and not expected to play a role as resonators.

The use of the laryngeal air sacs as sound modifiers must be contingent on their being filled with air by means of vestibular maneuvers such as I suggested above. Activation of a pharyngeal pneumatic sphincter such as that which I described in *Alouatta* (Schön, 1964) would enhance the effect of the mentioned vestibular maneuvers. In howling monkeys, displacements forward of the hyo-epiglottic fatty pad against the ostium of the hyoid bone would impinge upon the volume capacity of the part of the superomedian sac that is contained inside the bulla, and, consequently, modulate the sac's resonance response.

(3) Laryngeal inlet and vestibule. The unyielding inlet and the contorted and strictured laryngeal vestibule of non-human primates configure what I estimate is another mechanism for pressurizing, therefore, for amplifying the glottal source, but one that adds turbulence to the laryngeal output as a consequence of fast air flow through the luminal strictures of the vestibule. This glottal source amplifying mechanism could subserve voice intensification in non-human primates that have no supraglottal air sacs; their loud vocalizations would consequentially tend to be noisy.

Sound amplifications such as are achieved via air sac inflations, or the pressurizing of intra-vestibular air, presumably subserve increase in the transmissibility of selected calls, hence localization of point sources in a tropical forest habitat where acoustic signals are degraded by environmental surfaces.

(4) Thyrohyoid and velo-pharyngeal morphology. The passive intranarial location of the laryngeal inlet in non-human primates, dictates that their laryngeal acoustic excitations be emitted via the nasal passages, unless that laryngo-narial disposition is interrupted, and, by a caudal retraction of the larynx, the latter made to open into the oral cavity through the oropharyngeal isthmus. Caudal retraction of the larynx is presumably achieved by a pull in that direction of the infrahyoid muscles on the thyro-hyoid skeletal complex, pull that must be concomitant to a relaxation of the geniohyoids and anterior digastric bellies given that these two pairs of muscles otherwise pull the hyoid bone forward. Non-nasalized sound emissions via the mouth require, in addition, occlusion of the nasopharyngeal port. This can be achieved by (1) velar elevation mediated by the levator veli palatini, (2) velar retraction allowed by a relaxation of the tensor veli palatini and a caudal pull of the palatopharyngeus, (3) the localized formation of a Passavant's ridge in the pharyngeal wall. Retrogression of all these maneuvers presumably re-establishes the intranarial disposition of the larynx. Maneuvers that support non-nasalized oral sound emissions, do not seem to configure a non-human primate oropharynx fronted by an arched root of the tongue. In non-human primates the hyoid bone can only be mobilized away and toward the oral aperture, not down from the skull base as it is needed to expand the height of the oropharyngeal isthmus and allow for significant arching of the root of the tongue.

(5) The tongue. Given the similarities in intrinsic tongue musculature between non-human primates and humans, I do not expect the intrinsic mobility of their tongues to differ significantly. However, because the tongue of non-human primates is flat, transversely narrow, and anteroposteriorly long, and their hyoid bone cannot be separated from the larynx (with the possible exception of *Alouatta* (Schön, 1964, 1970, 1971) and *Colobus polykomos vellerosus* (Hill and Booth, 1957), the overall gesticulatory capabilities of the non-human primate tongue appear to be much restricted in relation to those of modern adult man. This may be the morphological correlate of what so far appears to be the incapability of non-human primates to produce formant transitions and certain vowels. Cam down pivoting of the mandible on the temporomandibular articulation (Du Brul, 1967) can help maneuverability of its tip and blade.

(6) The circumlabial musculature. The gradual elaboration of these muscles, together with the unleashing of the upper lip and an increase in height of both lips, are features that, from prosimians to simians, indicate a progressively more elaborate capability to maneuver the lips and modify the length and cross-section of the bucco-labial part of the tract. This maneuverability of the simian lips presumably enhances the capability of simians over prosimians to lower basic formant patterns by lip protractions that enlarge that length, and alternatively, elevate basic formant patterns by labial commissure retractions that shorten the acoustically effective length of the oral chamber; and, by means of interlabial aperture constrictions, to lower F1 pronouncedly and Fo slightly (see Bickley and Stevens, 1987; Hauser et al., 1993; Hauser and Schön Ybarra, 1994). Wide openings of the oral aperture "flare" the oral cavity, and are achieved by mandibular drops. The maneuver has been shown to produce a concomitant symmetrical rise and then a fall of the dominant frequency (Hauser et al., 1993).

(7) Length of glotto-labial tract. My limited qualitative observations, and those of others (see Lieberman, 1969), suggest that, with the exception of howling monkeys, this parameter may be positively related to body size at, apparently, the inter-generic level. If that is the case, larger subjects can be expected to produce vocalizations the dominant frequency of which will be lower than that of vocalizations produced by lighter individuals.

Summarizing, available evidence indicates that the vocal tract of non-human primates has morphologically intrinsic faculties to produce, modulate and intensify a wide range of glottal excitations; that among the latter, some are likely to be unstable and some not finely tuned; that oral sound emissions require suspension of the passive intranarial position of the larynx; and that the tract of non-human primates can accomplish oral articulations not as yet recognized, but, so far as we know, none of those that accomplish some specific human vowels, nor, perhaps, formant transitions. Future investigations should be aimed at expanding basic data in terms of the number of the morphological parameters investigated, and the number of specimens and taxa studied. I suggest the following research topics. (1) Further investigations into the relations between vocal fold morphology and its intrinsic capabilities to produce different patterns of glottal excitations in different species: a line of research likely to provide for a better understanding of the morphological bases that subserve the phonatory versatility of non-human primates. (2) Given, (a) a presumed increase in tissue damping effect by the conus elasticus and vocal ligament on the vibratory capabilities of the cover layer, and (b), Sataloff's (1963) assertion that a vocal ligament exists only in living humans (see Hirano, 1991), it is important to determine the presence or absence of either in the vocal folds of different non-human primate species; the point may be significant for elucidating vocal fold operational capabilities related to the completion of Valsalva maneuvers during certain patterns of arboreal locomotion (see Negus, 1962). (3) The study of the volume capacity and natural resonance of the supraglottal air sacs in different species will help better comprehend how the sound quality and intensity of specific vocalizations can be shaped in non-human primates. (4) Further work on the disposition of (a) the interaryteno-epiglottic muscular sphincter in the vestibular walls, (b) the inferior pharyngeal constrictor with respect to the laryngeal and pharyngeal walls, (c) the intravestibular luminal strictures, should advance our understanding of the sound pressurizing mechanism used by forest-living non-human primates to locate point sources and avoid signal degradation. (5) Determination of the rotational capabilities of the arytenoid cartilage on the cricoid in different species in relation to their respective glottal shaping mechanisms for duly understanding glottal shaping mechanism in ourselves and our non-human primate relatives. (6) Study of body weight to specific morphometric parameter relations of the vocal tract, e.g., length of the membranous vocal folds (see Hauser, 1993), height of vocal lip, length of glotto-labial tract, in order to assess sound quality output capabilities in terms of body size. (8) Elucidation of the morphologically intrinsic movement faculties of the walls of the velo-pharyngeal and oropharyngeal isthmus regions, and, of the possible effect of the translatory capabilities of the hyoid bone on those movements, to reach a better comprehension of the mechanisms that subserve luminal configuration changes in this area that is so crucial for respiration, deglutition, and sound emissions. (9) The study of the disposition of extrinsic lingual muscles within the body of this organ in different species will add to our understanding of the non-human primate tongue maneuvering capabilities.

Availability of information such as detailed above, will help adjust and expand our views about the non-human primate vocal tract morphology and its acoustic capabilities as I have presented them in this chapter. In addition, it will supply morphological data for generating evolutionary considerations, and for the identification of non-human primate models to be used in experimental studies of vocal disturbances. Quantification of some of the morphological parameters to be investigated present difficulties that may be overcome with some of the current image capturing, digitizing and manipulating techniques. All morphological data will have to be integrated with recordings of vocalizations systematically obtained and analyzed.

ACKNOWLEDGEMENTS

I thank my colleague, Dr. R.H. Eaglen, for editorial help, Dr. E.L. Simons for the specimen of *Symphalangus* I studied, and the Caribbean Primate Research Center for those of *Cebus albifrons* and *Macaca mulatta*.

REFERENCES

Bauer, H.R., 1986, Frequency code: orofacial correlates of fundamental frequency, *Phonetica* 44: 173-191.

Bickley, C.A. and Stevens, K.N., 1987, Effect of a vocal tract constriction on the glottal source: data from voiced consonants, *in*: "Laryngeal Function in Phonation and Respiration", T. Baer, C. Sasaki, and K. Harris, eds., College Hall Publishers, Boston.

Bock, W.L. and von Wahlert, G., 1965, Adaptation and the form-function complex, *Evolution* 19:269-299.
Bowman, J.P., 1968, Muscle spindles in the intrinsicmuscles of the rhesus monkey's (*Macaca mulatta*) tongue, *Anat. Rec.* 161: 483-488.
Brandes, R., 1931, Über den Kehlkopf des Orang-Utan in verschiedenen Altersstadien mit besonderer Berücksichtigung der Kehlsackfrage, *Morph. Jahrb.*69:1-61.
Cave, A.J.E., 1967, Observations on the platyrrhine nasal fossa, *Amer. J. Phys. Anthrop.* 26: 277-288.
Doyle, W.J., Casselbrant, M.L., Swarts, J.D., and Bluestone, C.D., 1990, Observations on a role for the tensor veli palatini muscle in intrinsic palatal function, *Cleft Palate J.* 27: 317-318.
Du Brul, E.L., 1976, Biomechanics of speech sounds, *Ann. New York Acad. Sci.* 280: 631-642.
Epple, G., 1975, The behavior of marmoset monkeys (Callithricidae), *in*: "Primate Behavior: Developments in Field and Laboratory Research," vol. 4, L.A. Rosenblum, ed., Academic Press, New York.
Fiedler, W., 1956, Übersicht über das System der Primates, *in*: "Primatologia I", H. Hofer, A.H. Schultz, and D. Starck, eds., S. Karger, Basel.
Flügel, C. and Rohen, J.W., 1991, The craniofacial proportion and laryngeal position in monkeys and man of different ages. (A morphometric study cased on CT-scans and radiographs), *Mech. Aging Dev.* 61:65-83.
Frable, M.A., 1961, Computation of motion at the cricoarytenoid joint, *Arch. Otolaryngol.* 73:73-78.
Garret, J.D. and Larson, C.R., 1991, Neurology of the laryngeal system, *in*: "Phonosurgery: Assessment and Surgical Management of Voice Disorders", C.N. Ford and D.M. Bless, eds., Raven Press, New York.
Gautier, J.-P., 1971, Etude morphologique et fonctionelle des annexes extra-laryngées des cercopithecinae; liaison avec les cris d'espacement, *Biol. Gabonica* 7:230-267.
Geist, F.D., 1933, Nasal cavity, larynx, mouth, and pharynx, *in*: "Anatomy of the Rhesus Monkey", J. Hartman and W.L. Straus, Jr., eds., Williams & Wilkins, Baltimore.
Gray, S.D., Hirano, M., and Sato, K., 1993, Molecular and cellular structure of vocal fold tissue, *in*: "Vocal Fold Physiology", I.R. Titze, ed., Singular Publ. Group, San Diego.
Green, S., 1975, Variation of vocal pattern with social situation in the Japanese monkey (*Macaca fuscata*): a field study, *in*: "Primate Behavior", vol. 4, L.A. Rosenblum, ed., Academic Press, New York.
Hauser, M.D., 1989, Ontogenetic changes in the comprehension and production of vervet monkey (*Cercopithecus aethiops*) vocalizations, *J. Comp. Ethol.* 103: 149-158.
Hauser, M.D. 1993, The evolution of nonhuman primate vocalizations: effects of phylogeny, body weight, and social context. *Amer. Nat.* 142: 528-542.
Hauser, M.D. and Marler, P., 1993, Food associated calls in rhesus monkeys, I. Socioecological factors, *Behav. Ecol.* 4: 194-205.
Hauser, M.D. and Schön Ybarra, M.A., 1994, The role of lip configuration in monkey vocalizations: Experiments using xylocaine as a nerve block, *Brain Lang.* 46: 232-244.
Hauser, M., Evans, C.S., and Marler, P., 1993, The role of articulation in the production of rhesus monkey (*Macaca mulatta*) vocalizations, *Anim. Behav.* 45: 423-433.
Hershkovitz, P., 1949, Mammals of Northern Colombia, *Proc. U.S. Nat. Mus.* 98: 323-427.
Hill, W.C.O., 1952, The external and visceral anatomy of the olive colobus monkey (*Procolobus verus*), *Proc. Zool. Soc. Lond.* 122: 127-186.
Hill, W.C.O., 1953-70, "Primates. Comparative Anatomy and Taxonomy", vols. 1-5, University Press, Edinburgh; vols. 6-7, Wiley-Interscience, New York.
Hill, W.C.O. and Booth, A.H., 1957, Voice and larynx in African and Asiatic Colobidae, *J. Bombay Nat. Hist. Soc.* 54:309-321.
Hilloowala, R.A., 1975, Comparative anatomical study of the hyoid apparatus in selected primates, *Amer. J. Anat.* 142: 367-384.
Hirano, M., 1974, Morphological structure of the vocal cord as a vibrator and its variations, *Folia phoniat.* 26:89-94.
Hirano, M., 1991, Phonosurgical anatomy of the larynx, *in*: "Phonosurgery: Assessment and Surgical Management of Voice Disorders", C.N. Ford and D.M. Bless, eds, Raven Press, New York.
Huber, E., 1931, "Evolution of the Facial Musculature and Facial Expression", Johns Hopkins, Baltimore.
Jordan, J., 1971, Studies on the structure of the organ of voice and vocalizations in the chimpanzees, *Folia Morph.* (*Warz.*) 30 (1):97-126, (2):221-248, (3):69-122.
Jürgens, U., 1988, Central control of monkey calls, *in*: "Primate Vocal Communication", D.Todt, P. Goedeking, and D. Symmes, eds., Springer, Berlin.
Jürgens, U., 1990, Control of monkey calls, *in*: "Neurobiology of Comparative Cognition", R.P. Kesner and D.S. Olton, eds., Lawrence Erlbaum, Hillsdale, N.J.
Jürgens, U. and Lu, C.-L., 1993, Neuropharmacological studies on vocal behavior in the periaqueductal gray, *Primate Report* 36: 115-121.
Kelemen, G., 1948, The anatomical basis of phonation in the chimpanzee, *J. Morph.* 82: 229-256.

Kelemen, G., 1969, Anatomy of the larynx and the anatomical basis of vocal performance, *in*: "The Chimpanzee", vol. 1, J. Bourne, ed., S. Karger, Basel.
Kelemen, G. and Sade, J., 1960, The vocal organ of the howling monkey, *J. Morph.* 107:123-140.
Kirchner, J.A., 1993, The vertebrate larynx: Adaptations and aberrations, *Laryngoscope* 103: 1197-1201.
Laitman, J.T., Crelin, E.S., and Conlogue, G.J., 1977, The function of the epiglottis in monkey and man, *Yale J. Biol.* 50:43-48.
Lampert, H., 1926, Zur Kenntnis des Platyrrhinenkehlkopfes, *Gegenbaurs Morph. Jb.* 55:607-654.
Larson, C.R., 1985, The midbrain periaqueductal gray: a brain stem structure involved in vocalizations, *J. Speech Hear. Res.* 28:241-249.
Larson, C.R., Sutton, D., and Lindeman, R.C., 1974, Muscle spindles in nonhuman primate laryngeal muscles, *Folia Primatol.* 22:315-323.
Lieberman, P., 1969, Primate vocalizations and human linguistic ability, *J. Acoust. Soc. Amer.* 44:1574-1584.
Lieberman, P., 1975, "On the Origins of Language", Macmillan, New York.
Lieberman, P., 1977, The phylogeny of language, *in*: "How Animals Communicate", T.A. Sebeok, ed., Indiana University Press, Bloomington.
Lieberman, P., 1984, "The Biology and Evolution of Language", Harvard University Press, Cambridge.
Lieberman, P. and Blumstein, S.E., 1988, "Speech Physiology, Speech Perception, and Acoustic Phonetics", Cambridge University Press, New York.
Lieberman, P. and Laitman, J.T., 1993, The anatomy, physiology, acoustics and perception of speech: essential elements in an analysis of the evolution of human speech, *J. Hum. Evol.* 23:447-468.
Lieberman, P., Klatt, D.H., and Wilson, W.A., 1969, Vocal tract limitations on the vowel repertoires of rhesus monkey and other nonhuman primates, *Science* 164:1185-1187.
Lowe, A.A., 1981, The neural regulation of tongue movements, *Progr. Neurobiol.* 15: 295-344.
Marler, P., 1969, Vocalizations of chimpanzees, *Rec. Advan. Primatol.* 1: 94-100.
Marler, P. and Tenaza, R., 1977, Signaling behavior of apes with special reference to vocalization, *in*: "How Animals Communicate", T.A. Sebeok, ed., Indiana University Press, Bloomington.
Masters, J.C., 1991, Loud calls of *Galago crassicaudatus* and *G. garnettii* and their relation to habitat structure, *Primates* 32: 153-167.
Maue, W.M. and Dickson, D.R., 1971, Cartilages and ligaments of the adult human larynx, *Arch. Otolaryngol.* 94: 432-439.
May, B., Moody, D., and Stebbins, W.C., 1988, Significant features of Japanese macaque communication sounds: a psychophysical study, *Anim. Beh.* 36: 1432-1444.
May, B., Moody, D., and Stebbins, W.C., 1989, Categorical perception of conspecific communication sounds by Japanese macaques, *Macaca fuscata, J. Acoust. Soc, Amer.* 85: 837-847.
Mitani, J. and Marler, P., 1989, A phonological analysis of male gibbon singing behavior, *Behaviour* 109:20-45.
Napier, J.R. and Napier, P.H., 1967, "A Handbook of Living Primates", Academic Press, London.
Negus, V., 1962, "The Comparative Anatomy and Physiology of the Larynx." Hafner Publ., New York.
Némai, J., 1926, Das Stimmorgan der Primaten, *Z. Anat. Entwicklungsgesch.* 81:657-672.
Newman, J.D., 1988, "The Physiological Control of Mammalian Vocalization", Plenum Press, New York.
Petersen, M.R. and Jusczyk, P., 1984, On perceptual predisposition for human speech and monkey vocalizations, *in*: "The Biology of Learning", P. Marler and D.B. Moody, eds., Springer-Verlag, Berlin.
Pickett, J.M., 1980, "The Sounds of Speech Communication", Allyn and Bacon, Boston.
Ploog, D., 1988, Neurobiology and pathology of subhuman vocal communication and human speech, *in*: "Primate Vocal Communication", D. Todt, P. Goedeking, and D. Symmes, eds., Springer, Berlin.
Pressman, J. and Kelemen, G., 1955, Physiology of the larynx, *Physiol. Rev.* 35:506-554.
Raman, R. and Darmanadan, M.S., 1989, Muscle receptors: content of the extrinsic and intrinsic muscles of the larynx in the bonnet monkey (*Macaca radiata*), *Anat. Rec.* 223: 433-436.
Rowell, T.E. and Hinde, R.A., 1962, Vocal communication by the rhesus monkey, *Proc. Zool. Soc., Lond.* 138: 279-294.
Sapir, S., Campbell, C., and Larson, C., 1981, Effect of geniohyoid, cricothyroid and sternothyroid muscle stimulation on voice fundamental frequency of electrically elicited phonation in rhesus macaque, *Laryngoscope* 91:457-468.
Sataloff, R.T., 1993, Molecular and cellular structure of vocal fold tissues: Response, *in*: "Vocal Fold Physiology", I.R. Titze, ed., Singular Publ. Group, San Diego.
Satoda, T., Uemura-Sumi, M., Tashiro, T., Takahashi, O., Matsushima, R., and Mizuno, N., 1990, Localization of motoneurons innervating the stylohyoid muscle in the monkey, cat, rabbit, rat and shrew, *J. Hirnforsch.* 31: 731-737.

Scherer, R.C., 1991, Physiology of phonation: a review of basic mechanics, *in*: "Phonosurgery: Assessment and Surgical Management of Voice Disorders", C.N. Ford and D.M. Bless, eds., Raven Press, New York.
Schneider, R., 1958, Zunge und weicher Gaumen, *in*: "Primatologia III/1", H. Hofer, A.H. Schultz, and D. Starck, eds., S. Karger, Basel.
Schön, M.A., 1964, Possible function of some pharyngeal and lingual muscles of the howling monkey, *Acta Anat.* 58:271-283.
Schön, M.A., 1970, On the mechanism of modulating the volume of the voice in the howling monkey, *Acta Otolaryngol.* 70:443-447.
Schön, M.A., 1971, The anatomy of the resonating mechanism in the howling monkey, Folia Primatol. 15:117-132.
Schön Ybarra, M.A., 1986, Loud calls of adult male red howling monkeys (*Alouatta seniculus*), *Folia Primatol.* 47:204-216.
Schön Ybarra, M.A., 1988, Morphological adaptations for loud phonations in the vocal organ of howling monkeys, *Primate Report* 22:19-24.
Seyfarth, R.M. and Cheney, D.L., 1986, Vocal development in vervet monkeys, *Anim. Behav.* 34:1640-1665.
Snowdon, C.T., 1989, Vocal communication in New World monkeys, *J. Hum. Evol.* 18: 611-633.
Snowdon, C.T., Brown, C.H., and Petersen, M.R., 1982, "Primate Communication", Cambridge University Press, Cambridge.
Sonesson, B., 1959, Die funktionelle Anatomie des Cricoarytaenoidgelenkes, *Z. Anat. Entwicklungsgesch.* 121:292-303.
Sonninen, A., 1956, The role of the external laryngeal muscles in length adjustment of the vocal cords, *Acta Otolaryngol.* Suppl. 130:9-102.
Starck, D. and Schneider, R., 1960, Respirationsorgane, *in*: "Primatologia III/2", H. Hofer, A.H. Schultz, and D. Starck, eds., S. Karger, Basel.
Strutz, J., Hammerich, T., and Amedee, R., 1988, The motor innervation of the soft palate. An anatomical study in the guinea pigs and monkeys, *Arch. Otorhinolaryngol.* 245:180-184.
Sutton, D., Larson, C.R., Lovell, M.E., and Lindeman, R.C., 1977, Facial muscle innervation in cebid (New World) and macaque (Old World) monkeys via a ramus communicating with the trigeminal nerve, *Arch. Oral Biol.* 22: 59-64.
Taylor, E.M., Sutton, D., and Lindeman, R.C., 1976, Dimensions of the infant monkey upper airway, *Growth* 40:69-74.
Titze, I.R., 1988, The physics of small-amplitude oscillation of the vocal folds, *J. Acoust. Soc. Amer.* 83:1536-1552.
Todt, D., Goedeking, P., and Symmes, D., 1988, "Primate Vocal Communication", Springer, Berlin.
von Leden, H. and Moore, P., 1961, The mechanics of the cricoarytenoid joint, *Arch. Otolaryngol.* 73:63-72.
von Böer, M. and Wissdorf, H., 1989, Beitrag zur Anatomie und Funktion des aus dem Recessus laryngis lateralis hervorgehenden Kehlsackes bei der Rotkopfmangabe (*Cercopithecus torquatus torquatus*). *Verhandlungsber. 31st Internat. Symp. Erkrank. Zoo Wildt.*, Dortmund.
West, R. and Larson, C.R, 1993, Laryngeal and respiratory activity during vocalization in macaque monkeys, *J. Voice* 7:54-68.
Winter, P., Ploog, D., and Latta, J., 1966, Vocal repertoire of the squirrel monkey (*Saimiri sciureus*), its analysis and significance, *Exp. Brain Res.* 1: 359-384.
Woodson, G.B., 1992, Voice, *Otolaryngol. Head Neck Surg.*106:20-21.
Yoshida, Y., Mitsumasu, T., Hirano, M., and Kanaseki, T., 1985, Somatotopic representation of the laryngeal motoneurons in the medulla of monkeys, *Acta Otolaryngol.* 100: 299-303.
Zimmermann, E., 1981, First record of ultrasound in two prosimian species, *Naturwissenschaften* 68: 531.
Zimmermann, E. and Lerch, C., 1993, The complex acoustic design of an advertisement call in male mouse lemurs (*Microcebus murinus*, Prosimii, Primates) and sources of its variation, *Ethology* 93: 211-224.

NEURONAL CONTROL OF VOCAL PRODUCTION IN NON-HUMAN AND HUMAN PRIMATES

Uwe Jürgens

German Primate Center
Kellnerweg 4
3400 Göttingen, Germany

INTRODUCTION

Vocal production is organized on different levels of complexity. The lowest level is represented by vocal reactions that are genetically determined in their acoustic structure and are elicited in a reflex-like manner by external or internal stimuli. An example is pain shrieking. A heavy blow against the body, for instance, will elicit shrieking from birth on. A monkey or human infant does not need to hear shrieking from other conspecifics in order to be able to produce it: even deaf infants shriek (Eibl-Eibesfeldt, 1973; Winter et al., 1973). There is also no need for prior experience with such a stimulus in the form of a pairing with another, unconditioned stimulus in the Pavlovian sense.

A somewhat higher level of vocal behavior is represented by calls the acoustic structure of which is genetically determined, the release of which, however, is under voluntary control. Shrieking, for instance, can be produced by humans also in the absence of pain - e.g., on the stage by an actor mimicking pain. On the other hand, shrieking can be suppressed - even if pain is severe. Similar holds for non-human primates. Rhesus monkeys, for instance, can be trained in a vocal operant conditioning task to increase their vocalization rate, if vocalization is rewarded with food (Sutton et al., 1973). They even learn to utter one specific call type ("coo") during the presentation of a coloured stimulus light, utter a second call type ("bark") during the presentation of a differently coloured light, and refrain from vocalizing during presentation of a third stimulus light (Sutton, 1979).

A still higher level of vocal motor control, is characterized by vocalizations the voluntary control of which is not limited to their initiation and inhibition, but also includes their acoustic structure. This level is represented by vocal patterns learned by imitation as well as patterns created anew by invention. Both types of vocal behavior seem to exist in a well-developed form only in human, not in subhuman primates. In the following, it will be shown that the number of brain structures involved in vocal production depends upon the level at which communication takes place.

INNATE VOCAL REACTIONS

The brain structures responsible for the motor control of innate vocal reactions seem to be limited essentially to the brainstem. Support for this conclusion comes mainly from two observations. One derives from anencephalic human infants, that is, infants born without forebrain and cerebellum, having only a more or less intact brainstem. Such infants, despite the fact that they lack the whole cortex as well as the basal ganglia, thalamus and cerebellum, nevertheless react to painful stimuli with crying (Monnier and Willis, 1953). The other observation comes from systematic brain transection experiments in cats. In these experiments, it has been found that as long as the transection is made rostral to the mesencephalon, the cats still react to tail pinching with meowing, growling and hissing. If the transection is made caudal to the midbrain, vocalization can no longer be obtained (Bazett and Penfield, 1922). Both observations suggest (1) that in man as well as subhuman mammals, the forebrain and cerebellum are dispensable for the production of at least some innate vocal reactions; (2) that the mesencephalon represents the highest level of the neuraxis indispensable for vocalization.

Systematic lesioning experiments in the squirrel monkey have revealed that the critical structure within the mesencephalon is the periaqueductal gray and laterally bordering tegmentum. The destruction of this area alone is sufficient to produce mutism (Jürgens and Pratt, 1979a). Similar holds for man. There is a case report in the neurological literature describing a patient who, after a small lesion restricted to the periaqueductal gray and bordering tegmentum, became completely mute until his death two months later (Botez and Carp, 1968). This patient did not show a general akinesia, but was able to walk around and take his meals without help.

The periaqueductal region is necessary but, of course, not sufficient for the production of vocalization. It requires an input from sensory brain structures capable of initiating vocalization and an output to the motor nuclei innervating the laryngeal, oral and respiratory muscles involved in phonation. Neuroanatomical studies have shown that the periaqueductal gray receives sensory input of different modalities (Fig. 1). Somatosensory information from the trunk and extremities reach the periaqueductal gray via collaterals of the spinothalamic tract originating in the dorsal horn of the spinal cord (Harman et al., 1988; Yezierski et al., 1987). Gustatory and somatosensory stimuli from the viscera reach it via the solitary tract nucleus (Beckstead et al., 1980). Somatosensory stimuli from the face region enter the periaqueductal gray via the sensory trigeminal nuclei (Mantyh, 1982). Auditory information comes from the inferior colliculus, visual information from the adjacent superior colliculus (Beitz, 1982; Meller and Dennis, 1986).

On the output side, neuroanatomical studies have shown that not all motoneuron pools involved in phonation receive their information directly from the periaqueductal region (Holstege et al., 1977; Travers and Norgren, 1983). All phonatory motor nuclei, however, get an input from the reticular formation around the nucleus ambiguus and retroambiguus which, in turn, is directly connected with the periaqueductal region (Fig. 1; Holstege, 1989; Thoms and Jürgens, 1987). Small regions within the peri-ambigual reticular formation, unlike those in the periaqueductal region, do not abolish vocalization but cause a deterioration of the acoustic structure (Kirzinger and Jürgens, 1985). Single-unit recording within this structure reveals a number of neurons which fire specifically during and immediately before vocalization (Kirzinger and Jürgens, 1991). Electrical stimulation of the peri-ambigual reticular formation during phonation causes a change of phonatory pattern (Dressnandt and Jürgens, 1992). These findings suggest that the periambigual reticular formation represents a vocal motor coordination center at which the activity of the laryngeal, articulatory and respiratory muscles involved in phonation is coordinated.

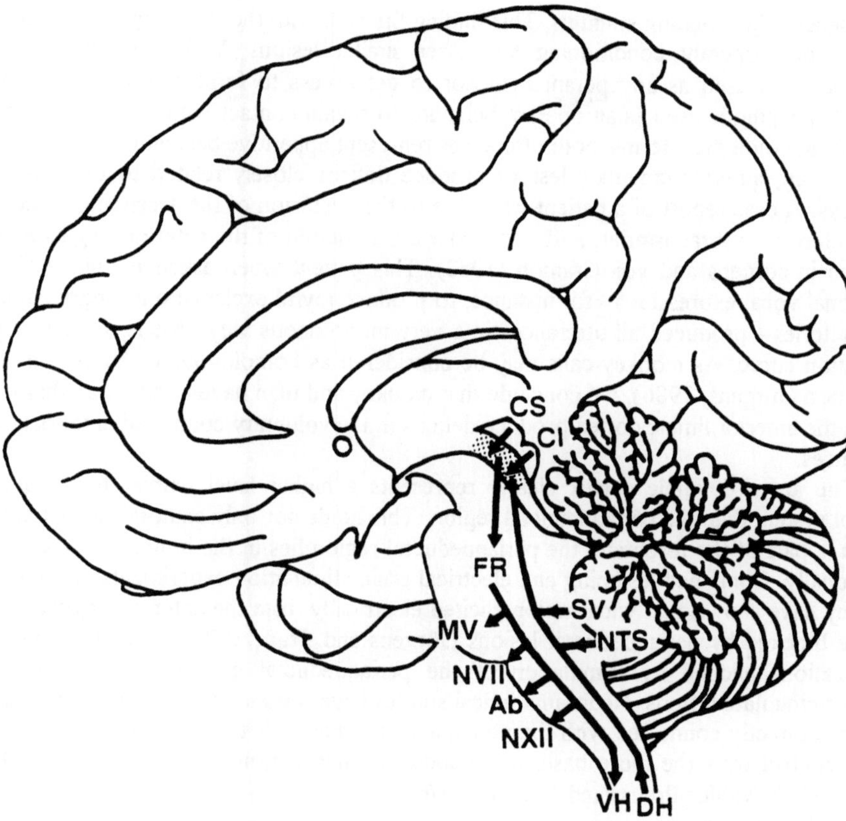

Figure 1. Sagittal section of the human brain showing some of the sensory input and motor output pathways of the periaqueductal gray. The periaqueductal gray is marked by a dot raster. Abbreviations: Ab nucl. ambiguus; CI colliculus inferior; CS colliculus superior; DH dorsal horn of spinal cord; FR formatio reticularis; MV trigeminal motor nucleus; NTS nucl. solitarius; N VII nucl. facialis; N XII nucl. hypoglossus; S V sensory trigeminal nuclei; VH ventral horn of spinal cord.

VOLUNTARY CONTROL OF VOCAL REACTIVITY

Rhesus monkeys trained to master a vocal operant conditioning task loose their capability to produce vocalization voluntarily after anterior cingulate cortex lesions (Aitken, 1981; Sutton et al., 1974). These animals are not mute. They still react to threatening stimuli with agonistic calls. But they are unable to produce calls as an operant behavior to get access to food or to postpone electric shocks in a Sidman avoidance schedule. On the other hand, these animals are able to master operant conditioning tasks if the operant behavior is non-vocal, that is, lever-pressing or changing from one compartment into another in a shuttle box.

In the squirrel monkey, MacLean and Newman (1988) found that ablation of the anterior cingulate cortex and subjacent subcallosal cortex abolishes spontaneous isolation peeping, a long-distance contact call uttered by isolated animals. If such animals are exposed to play-backs of isolation peeps, they respond to these calls with isolation peeps. Destruction of the anterior limbic cortex thus does not prevent the production of isolation peeps in general. It only abolishes self-initiated isolation peeping, that is, calls produced in the absence of an eliciting stimulus. This finding fits well with the rhesus monkey's failure to master vocal operant conditioning task after similar lesions. While in the latter case, vocalization is used as an operant behavior to get access to food, the squirrel monkey's isolation peeping is used as an operant behavior to regain contact with group members out of sight. In ethological terms, both utterances represent appetitive behavior.

In man, anterior cingulate lesions produce deficits closely related to those found in monkeys. A case report of a patient who, due to the occlusion of the ascending branches of both anterior cerebral arteries, suffered bilateral destruction of the anterior cingulate cortex is given in Jürgens and von Cramon (1982). This patient when asked to mimic different emotional vocal expressions - for instance, to produce joyful exclamations, angry curses, or pain outcries - produced all utterances in a very monotonous way, that is, with a very flat intonation curve. As monkey calls may be considered as homologous to human emotional intonation (Jürgens, 1986), we conclude that monkey and man have in common that lesions within the anterior limbic cortex produce deficits in the voluntary control of emotional vocal utterances.

The anterior limbic cortex clearly represents a higher level within the vocalization control system than the periaqueductal region. This holds not only anatomically - the limbic cortex is part of the forebrain, the periaqueductal region lies in the brainstem; it also holds functionally. Combined lesioning and electrical brain stimulation experiments in the squirrel monkey have shown that vocalization elicited electrically from the anterior cingulate cortex can be blocked by periaqueductal lesions (Jürgens and Pratt, 1979b). On the other hand, vocalizations elicited by stimulation of the periaqueductal gray remain unaffected by anterior cingulate lesions. Neuroanatomical studies have revealed that the anterior cingulate cortex is directly connected with the periaqueductal region; it thus is in a position to exert direct control over the more basic periaqueductal vocalization area (Jürgens and Müller-Preuss, 1977; Müller-Preuss and Jürgens, 1976).

LEARNED VOCAL PATTERNS

Subhuman primates seem to lack the capacity for vocal imitation - and thus vocal learning - more or less completely (Hayes and Hayes, 1951; Kellogg, 1968; Owren et al., 1992). From this the question arises, what differences exist in the brain anatomy of human and non-human primates which could explain the differences in vocal learning capacity? Our own neuroanatomical studies have shown that monkeys (and the same seems to hold for other non-human mammals as well; Newman et al., 1989; Szentagothai and Rajkovits, 1958) lack a direct connection between the primary motor cortex and the nucleus ambiguus, that is, the site of the laryngeal motoneurons (Jürgens, 1976). Such a direct connection does exist in humans (Kuypers, 1958a). As the primary motor cortex plays a crucial role in voluntary motor control in general, the lack of a direct connection in the monkey fits with the monkey's poor capacity for voluntary vocal control. This interpretation gets further support from another observation. While monkeys lack a direct connection between motor cortex and laryngeal motoneurons, they do have a direct connection between motor cortex and jaw muscle motoneurons, which lie in the trigeminal motor nucleus (Kuypers, 1958b; Sirisko and Sessle, 1983). Conditioning experiments show that monkeys are able to learn a controlled sensible jaw movement, namely, to exert a predetermined constant biting force for a fixed period (Luschei and Goodwin, 1975). Corresponding holds for finger movements. Monkeys are able to perform delicate learned finger movements (Hepp-Reymond and Wiesendanger, 1972). Again, there is a direct connection between the motor cortex and those motoneurons in the cervical spinal cord innervating the finger muscles - a connection lacking, for instance, in the cat, which also lacks delicate finger movements (Kuypers, 1981; Wiesendanger, 1981). These observations suggest that an improved capability for voluntary fine motor control is paralleled by the formation of direct cortico-motoneuronal connections.

The crucial role of the primary motor cortex for voluntary motor control is also exemplified by the effect of lesions in this area. Bilateral destruction of the motor cortex in man abolishes the capacity to produce learned vocal utterances, such as speech or humming of melodies (Foerster, 1936; Groswasser et al., 1988; Leicester, 1980). Bilateral destruction of the same area in the monkey is without effect on vocalization (Jürgens et al., 1982). The same holds for those brain structures which serve as the main input and output structures of the primary motor cortex. The primary motor cortex receives its information mainly from the cerebellum via the ventrolateral thalamus and sends its information predominantly to the putamen, a part of the basal ganglia (Fig. 2). Lesions in all three structures cause severe disturbances in speech production but do not affect the acoustic structure of monkey calls (Andrew et al., 1983; Bell, 1968; Gurd et al., 1988; Kent et al., 1979; Kirzinger, 1985; Kirzinger and Jürgens, 1985; Krayenbühl et al., 1963; Lechtenberg and Gilman, 1978; Metter et al., 1986; Samra et al., 1969). Apparently, a number of brain structures are involved in the production of learned vocal patterns that are dispensable for the production of innate vocal patterns.

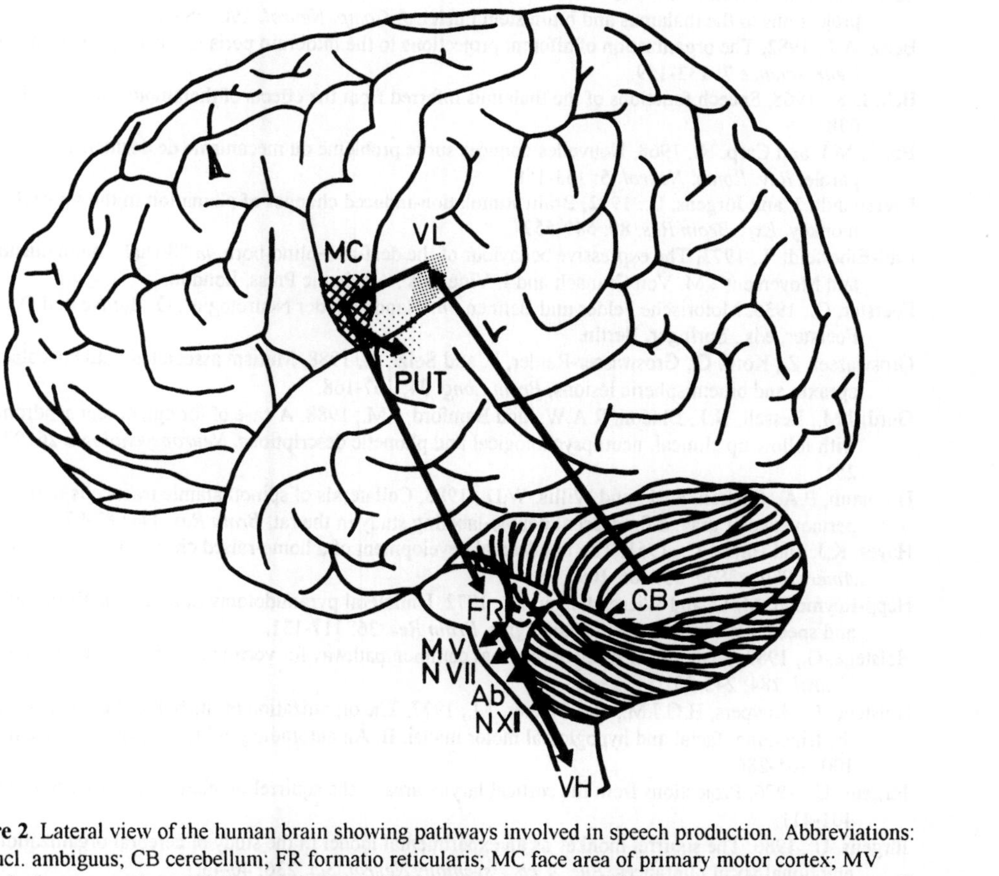

Figure 2. Lateral view of the human brain showing pathways involved in speech production. Abbreviations: Ab nucl. ambiguus; CB cerebellum; FR formatio reticularis; MC face area of primary motor cortex; MV trigeminal motor nucleus; N VII nucl. facialis; N XII nucl. hypoglossus; PUT putamen; VH ventral horn of spinal cord; VL nucl. ventralis lateralis thalami.

CONCLUSION

The motor coordination of the different laryngeal, respiratory and articulatory muscles involved in phonation seems to take place in the reticular formation of the ventrolateral medulla. In man, this area receives an input from the motor cortex which, in collaboration with the cerebellum, ventrolateral thalamus and basal ganglia, enables the production of learned vocal patterns. In monkey as well as man, the reticular vocalization area receives a second input from the periaqueductal gray of the midbrain which, in turn, receives afferents from the anterior cingulate cortex. This input seems to be responsible for the sensorily and voluntarily driven initiation of vocalization rather than its motor coordination.

REFERENCES

Aitken, P.G., 1981, Cortical control of conditioned and spontaneous vocal behavior in rhesus monkeys, *Brain Lang.* 13: 171-184.

Andrew, J., Fowler, C. J., and Harrison, M. J. G., 1983, Stereotaxic thalamotomy in 55 cases of dystonia, *Brain* 106: 981-1000.

Bazett, H.C. and Penfield, W.G., 1922, A study of the Sherrington decerebrate animal in the chronic as well as the acute condition, *Brain* 45: 185-265.

Beckstead, R.M., Morse, J.R., and Norgren, R., 1980, The nucleus of the solitary tract in the monkey: projections to the thalamus and brain stem nuclei, *J. Comp. Neurol.* 190: 259-282.

Beitz, A.J., 1982, The organization of afferent projections to the midbrain periaqueductal gray of the rat, *Neuroscience* 7: 133-159.

Bell, D.S., 1968, Speech functions of the thalamus inferred from the effects of thalamotomy, *Brain* 91: 619-638.

Botez, M.I. and Carp, N., 1968, Nouvelles données sur le problème du mécanisme de déchlenchement de la parole, *Rev. Roum. Neurol.* 5: 153-158.

Dressnandt, J. and Jürgens, U., 1992, Brain stimulation-induced changes of phonation in the squirrel monkey, *Exp. Brain Res.* 89: 549-559.

Eibl-Eibesfeldt, I., 1973, The expressive behaviour of the deaf-and-blind-born, *in*:"Social Communication and Movement", M. Von Cranach and I. Vine, eds., Academic Press, London.

Foerster, O., 1936, Motorische Felder und Bahnen, *in*:"Handbuch der Neurologie", O. Bumke und O. Foerster, eds., Springer, Berlin.

Groswasser, Z., Korn, C., Groswasser-Reider, I., and Solzi, P., 1988, Mutism associated with buccofacial apraxia and bihemispheric lesions, *Brain Lang.* 34: 157-168.

Gurd, J.M., Bessell, N.J., Bladon, R.A.W., and Bamford, J.M., 1988, A case of foreign accent syndrome, with follow-up clinical, neuropsychological and phonetic descriptions. *Neuropsychologia* 26: 237-251.

Harmann, P.A., Carlton, S.M., and Willis, W.D., 1988, Collaterals of spinothalamic tract cells to the periaqueductal gray: a fluorescent double-labeling study in the rat, *Brain Res.* 441: 87-97.

Hayes, K.J, and Hayes, C., 1951, The intellectual development of a home-raised chimpanzee, *Proceed. Amer. Philos. Soc.* 95: 105-109.

Hepp-Reymond, M.-C. and Wiesendanger, M., 1972, Unilateral pyramidotomy in monkeys. Effect on force and speed of a conditioned precision grip, *Brain Res.* 36: 117-131.

Holstege, G., 1989, Anatomical study of the final common pathway for vocalization in the cat, *J. Comp. Neurol.* 284: 242-252.

Holstege, G., Kuypers, H.G.J.M., and Dekker, J.J., 1977, The organization of the bulbar fibre connections to the trigeminal facial and hypoglossal motor nuclei. II. An autoradiographic tracing study in cat, *Brain* 100: 265-286.

Jürgens, U., 1976, Projections from the cortical larynx area in the squirrel monkey, *Exp. Brain Res.* 25: 401-411.

Jürgens, U., 1986, The squirrel monkey as an experimental model in the study of cerebral organization of emotional vocal utterances, *Eur. Arch. Psychiatr. Neurol. Sci.* 236: 40-43.

Jürgens, U., Kirzinger, A., and von Cramon, D., 1982, The effects of deep-reaching lesions in the cortical face area on phonation. A combined case report and experimental monkey study, *Cortex* 18: 125-140.

Jürgens, U. and Müller-Preuß, P., 1977, Convergent projections of different limbic vocalization areas in the squirrel monkey, *Exp. Brain Res.* 29: 75-83.

Jürgens, U., and Pratt, R., 1979a, Role of the periaqueductal grey in vocal expression of emotion, *Brain Res.* 167: 367-378.
Jürgens, U. and Pratt, R., 1979b, The cingular vocalization pathway in the squirrel monkey, *Exp. Brain Res.* 34: 499-510.
Jürgens, U. and von Cramon, D., 1982, On the role of the anterior cingulate cortex in phonation: a case report, *Brain Lang.* 15: 234-248.
Kellogg, W.N., 1968, Communication and language in the home-raised chimpanzee, *Science* 162: 423-427.
Kent, R.D., Netsell, R., and Abbs, J.H., 1979, Acoustic characteristics of dysarthria associated with cerebellar disease, *J. Speech Hear. Res.* 22: 627-648.
Kirzinger, A., 1985, Cerebellar lesion effects on vocalization of the squirrel monkey, *Behav. Brain Res.* 16: 177-181.
Kirzinger, A. and Jürgens, U., 1985, The effects of brain stem lesions on vocalization in the squirrel monkey, *Brain Res.* 358: 150-162.
Kirzinger, A. and Jürgens, U., 1991, Vocalization-correlated single-unit activity in the brain stem of the squirrel monkey, *Exp. Brain Res.* 84: 545-560.
Krayenbühl, H., Siegfried, J., and Yasargil, M. G., 1963, Résultats tardifs des opérations stéréotaxiques dans le traitement de la maladie de Parkinson, *Rev. Neurologique* 108: 485-494.
Kuypers, H.G.J.M., 1958a, Corticobulbar connexions to the pons and lower brain-stem in man, *Brain* 81: 364-388.
Kuypers, H.G.J.M., 1958b, Some projections from the peri-central cortex to the pons and lower brain stem in monkey and chimpanzee, *J. Comp. Neurol.* 110: 221-255.
Kuypers, H.G.J.M., 1981, Anatomy of the descending pathways, *in*: "Handbook of Physiology. The Nervous System. Vol. II. Motor Control, Part I.", J.M. Brookhart, V.B. Mountcastle, V.B. Brooks, and S.R. Geiger, eds., American Physiological Society, Bethesda, Md.
Lechtenberg, R. and Gilman, S., 1978, Speech disorders in cerebellar disease, *Ann. Neurol.* 3: 285-290.
Leicester, J., 1980, Central deafness and subcortical motor aphasia, *Brain Lang.* 10: 224-242.
Luschei, E. S. and Goodwin, G. M., 1975, Role of monkey precentral cortex in control of voluntary jaw movements, *J. Neurophysiol.* 38: 146-157.
MacLean, P.D. and Newman, J.D., 1988, Role of midline frontolimbic cortex in production of the isolation call of squirrel monkeys, *Brain Res.* 450: 111-123.
Mantyh, P.W., 1982, The ascending input to the midbrain periaqueductal gray of the primate, *J. comp. Neurol.* 211: 50-64.
Meller, S.T. and Dennis, B.J., 1986, Afferent projections to the periaqueductal gray in the rabbit, *Neuroscience* 19: 927-964.
Metter, E.J., Jackson, C., Kempler, D., Riege, W. H., Hanson, W.R., Mazziotta, J.C., and Phelps, M.E., 1986, Left hemisphere intracerebral hemorrhages studied by (F-18)-fluordeoxyglucose PET, *Neurology* 36: 1155-1162.
Monnier, M. and Willis, H., 1953, Die integrative Tätigkeit des Nervensystems beim meso-rhombo-spinalen Anencephalus (Mittelhirnwesen), *Monatsschr. Psychiat. Neurol.* 126: 239-273.
Müller-Preuß, P., and Jürgens, U., 1976, Projections from the "cingular" vocalization area in the squirrel monkey, *Brain Res.* 103: 29-43.
Newman, D.B., Hilleary, S.K., and Ginsberg, C. Y., 1989, Nuclear terminations of corticoreticular fiber systems in rats, *Brain Behav. Evol.* 34: 223-264.
Owren, M.J.; Dieter, J.A., Seyfarth, R.M., and Cheney, D.L., 1992, "Food calls" produced by adult female rhesus (Macaca mulatta) and Japanese (M. fuscata) macaques, their normally-raised offspring, and offspring cross-fostered between species, *Behaviour* 120: 218-231.
Samra, K., Riklan, M., Levita, E., Zimmerman, J., Waltz, J.M., Bergmann, L., and Cooper, I.S., 1969, Language and speech correlates of anatomically verified lesions in thalamic surgery for parkinsonism, *J. Speech Hearing Res.* 12: 510-540.
Sirisko, M.A. and Sessle, B.J., 1983, Corticobulbar projections and orofacial and muscle afferent inputs of neurons in primate sensorimotor cerebral cortex, *Exp. Neurol.* 82: 716-720.
Sutton, D., 1979, Mechanisms underlying learned vocal control in primates, *in*: "Neurobiology of Social Communication in Primates: An Evolutionary Perspective", H.D. Steklis and M.J. Raleigh, eds., Academic Press, New York.
Sutton, D., Larson, C., and Lindeman, R.C., 1974, Neocortical and limbic lesion effects on primate phonation, *Brain Res.* 71: 61-75.
Sutton, D., Larson, C., Taylor, E. M., and Lindeman, R. C., 1973, Vocalization in rhesus monkeys: conditionability, *Brain Res.* 52: 225-231.
Szentâgothai, J. and Rajkovits, K., 1958, Der Hirnnervenanteil der Pyramidenbahn und der prämotorische Apparat motorischer Hirnnervenkerne, *Arch. Psychiat. Nervenkr.* 197: 335-354.

Thoms, G. and Jürgens, U., 1987, Common input of the cranial motor nuclei involved in phonation in squirrel monkey, *Exp. Neurol.*95: 85-99.

Travers, J.B. and Norgren, R., 1983, Afferent projections to the oral motor nuclei in the rat, *J. Comp. Neurol.* 220: 280-298.

Wiesendanger, M., 1981, The pyramidal tract. Its structure and function. *in*: "Handbook of Behavioral Neurobiology. Vol. 5: Motor Coordination", A.L. Towe, and E.S.C. Luschei, eds., Plenum, New York.

Winter, P., Handley, P., Ploog, D., and Schott, D., 1973, Ontogeny of squirrel monkey calls under normal conditions and under acoustic isolation, *Behaviour* 47: 230-239.

Yezierski, R.P., Sorkin, L.S., and Willis, W. D., 1987, Response properties of spinal neurons projecting to midbrain or midbrain-thalamus in the monkey, *Brain Res.* 437: 165-170.

ROLE OF AUDITORY CORTEX IN THE PERCEPTION OF VOCALIZATIONS BY JAPANESE MACAQUES

Henry E. Heffner and Rickye S. Heffner

Laboratory of Comparative Hearing
and the Department of Psychology
University of Toledo
Toledo, OH 43606, USA

INTRODUCTION

Ever since the latter half of the 19th century when it was discovered that sensory and motor functions could be localized in the cerebral cortex, the role of the cortex in auditory processing has been studied by observing the effects of brain damage in both humans and animals. Initially, it was envisioned that the study of experimental lesions in animals would provide a basis for inferring cortical functioning in humans. This was based on the fact that lesions in animals could be precisely controlled thus avoiding the complications which arise when analyzing the effects of lesions which are either too small to result in distinct deficits or else are too large and result in multiple deficits which may be difficult to differentiate (e.g., Ferrier, 1876b). Indeed, the utility of animal research was soon demonstrated when ablation studies narrowed the search for auditory cortex in monkeys to the posterior two thirds of the superior temporal gyrus (Ferrier 1876a).

Yet despite early successes, studies of auditory processing in humans and animals have for the most part been pursued independently with little interaction between the two fields (see Neff et al., 1975). Two factors have contributed to this state of affairs. First, the study of the effects of auditory lesions in humans and animals has occasionally resulted in contradictory findings. The primary example of this is the question of whether cortical lesions result in a hearing loss. Although a number of cases of "cortical deafness" following bilateral temporal lobe lesions have appeared in the clinical literature, the failure (until recently) to find significant hearing losses in animals cast doubt on the validity of the human results and led animal researchers to believe that the cortex did not play a role in the detection of sound (e.g., Masterton and Berkley, 1974; for reviews, see Heffner and Heffner, 1986b, 1990c).

A second factor contributing to the separation of human and animal studies has been the difference in the focus of their investigations. While animal research has concentrated on

Current Topics in Primate Vocal Communication
Edited by E. Zimmermann *et al.*, Plenum Press, New York, 1995

the role of cortex in the discrimination of the physical aspects of sounds (e.g., frequency, intensity and temporal discriminations), much of the human research has been directed to

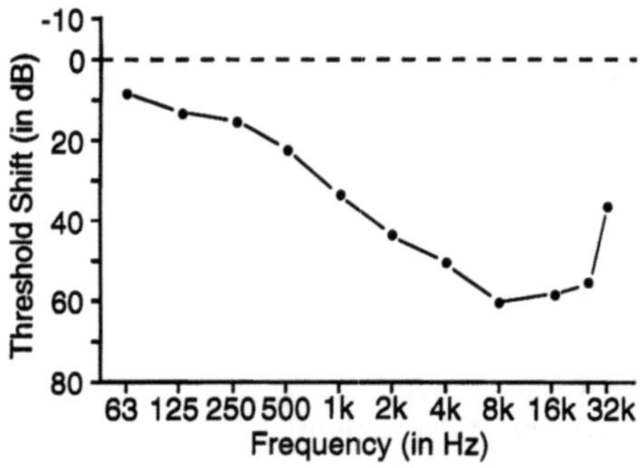

Figure 1. Average increase in pure-tone thresholds following complete bilateral auditory cortex lesions in 4 Japanese macaques. The dashed line at zero is the preoperative hearing level of the animals. Note that the hearing loss is small at low frequencies, gradually increases in the mid-frequency range, and then declines at the high frequencies, a characteristic which presumably reflects the proportion of cortical neurons responding to the different frequencies (modified from Heffner and Heffner, 1990c).

the study of the role of the cortex in the perception of speech. Although there has long been great interest in the possibility that the neural mechanisms underlying the perception of vocalizations in non-human primates might bear some similarity to that underlying human speech perception, the actual amount of research on this topic has been small (cf. Steklis and Raleigh, 1979).

Functional Similarities Between Macaque and Human Auditory Cortex

In the last few years, however, two findings have appeared which demonstrate that there are close similarities between humans and other primates, particularly macaques, with regard to cortical auditory processing. First, studies of cortical lesions in Japanese macaques have demonstrated that bilateral lesions of auditory cortex result in a hearing loss similar to the cortical deafness observed in humans (e.g., Jerger et al., 1969). Specifically, bilateral lesions of the superior temporal gyrus in macaques result in a substantial and permanent hearing loss as demonstrated by pure-tone audiometry (Fig. 1; Heffner and Heffner, 1986b, 1990c). Furthermore, the pattern of the hearing loss is similar to that often observed in humans in that it is greater at middle frequencies than at low frequencies (the degree of loss at high frequencies in humans has not been determined) (Leicester, 1980; Michel and Peronnet, 1980). This finding may reflect the fact that there are more cortical neurons tuned to middle than to low and high frequencies (e.g., Merzenich et al., 1976) with the result that removal of the cortical influence on the lower auditory system, which mediates the detection of sound, is in proportion to the cortical frequency representation. Moreover, research on macaques has extended what is known about cortical deafness by revealing that unilateral lesions result in a small, but reliable hearing loss in the ear contralateral to the lesion (Heffner and Heffner, 1989c).

Previous animal studies, which used rats, cats, ferrets, and dogs, did not find a hearing loss following cortical ablation, apparently because cortical deafness does not occur in these species. The fact that cortical deafness has so far been observed only in humans and macaques suggests that it may be peculiar to primates (for reviews, see Heffner and Heffner 1986b, 1990c). Thus, by using macaques, it has been possible not only to verify the clinical observations of cortical deafness following bilateral temporal lobe lesions, but to extend these findings by demonstrating that unilateral lesions result in a hearing loss in the contralateral ear - a result which clinical studies had been unable to demonstrate conclusively (Heffner and Heffner, 1990b).

The second finding indicating a similarity between humans and macaques is that macaques possess a "cortical mechanism" for the perception of their species-specific vocalizations that is lateralized to the left hemisphere. That is, Japanese macaques have been shown to have a right ear advantage for the perception of two forms of their "coo" calls (e.g., Petersen et al., 1978). Furthermore, left, but not right, auditory cortex lesions have a marked, although transient, effect on their ability to discriminate these calls (e.g., Heffner and Heffner, 1986a). These results suggest that Japanese macaques have a specialization of the left temporal lobe analogous to that in humans for the perception of speech (i.e., Wernicke's area). Thus, it is possible to use macaques to study the role of the cortex in the perception of species-specific vocalizations and, perhaps, to generalize the results to the study of speech perception in humans.

In short, there appear to be important similarities in the role of the cortex in humans and Japanese macaques. Not only does the cortex of both species play a similar role in the detection of sound, but both possess a left hemisphere specialization for the perception of species-specific vocalizations. It is this latter aspect of cortical functioning which is the topic of this chapter.

PERCEPTION OF VOCAL COMMUNICATIONS BY JAPANESE MACAQUES

For over a hundred years it has been known that lesions of the posterior portion of the superior temporal gyrus of the left hemisphere in humans almost invariably result in the inability to understand language (for reviews, see Geschwind, 1965; Passingham, 1979). This fact has been used to support the idea that there is a cortical area specialized for the perception of language that, in most people, is located in the left temporal lobe (e.g., Wernicke, 1874; Geschwind, 1965). Language, however, can be viewed as an example of species-specific communication (e.g., Steklis and Raleigh, 1979) and its mechanism is unlikely to have arisen de novo with humans. Thus, the question arises as to whether other primates possess a similar cortical specialization for the perception of their vocal communications and whether that mechanism might also be lateralized to the left hemisphere.

Right-Ear Advantage in Macaques

Some years ago, there appeared evidence that the left hemisphere of Japanese macaques (*Macaca fuscata*) is specialized for the perception of species-specific communication sounds. This evidence came from the work of Petersen and his colleagues who studied the ability of Japanese macaques to discriminate two forms of their "coo" vocalizations (e.g., Petersen, 1981; Petersen et al., 1978). Briefly, they trained monkeys to discriminate seven "smooth early high" (SE) coos from eight "smooth late high" (SL) coos using a go/no-go procedure. These coos are distinguished by the temporal position of the peak fundamental frequency in the frequency-modulation portion of the call. The two types of coos are acoustically similar, but appear to have different meanings the SE coo is emitted by both sexes and is a "contact-seeking" call, whereas the SL coo is produced primarily by estrous females seeking male consorts (Green, 1975). When the monkeys were trained to discriminate the coos, which were presented through headphones, Petersen found that the Japanese macaques showed a consistent right-ear advantage. Because other species of macaques failed to show a right-ear advantage when trained to discriminate the Japanese macaque coos, it was concluded that this effect was specific to the species whose vocalizations were being discriminated (Petersen, 1981; Petersen et al., 1978). Indeed, recent evidence has suggested that rhesus macaques may show a right-ear advantage when tested with their own species-specific vocalizations (Hauser and Andersson, in press).

The right-ear advantage of Japanese macaques for the perception of their coos may be analogous to that shown by humans for the perception of speech sounds (e.g., Kimura, 1961, 1967). The human right-ear advantage is explained by the fact that input from each ear is dominant in the contralateral hemisphere. As a result, input from the right ear will predominate in the left hemisphere where, in most people, Wernicke's area is located. Thus, speech sounds presented to the right ear have an advantage over those presented to the left ear when competing for analysis in Wernicke's area. Accordingly, the presence of a right-ear advantage in Japanese macaques suggested that they might also have a left hemisphere specialization for the perception of their vocal communications.

Once it had been established that Japanese macaques have a right-ear advantage, the next step was to examine the effect of unilateral temporal lobe lesions on the perception of their vocalizations. If the monkeys possessed a left hemisphere specialization then one would expect left hemisphere lesions to have a greater effect than right hemisphere lesions on their perception of the coos.

Effect of Cortical Lesions on the Perception of Coos

Some years ago, we began to investigate the effect of superior temporal gyrus lesions on the ability of Japanese macaques to categorize their vocalizations using the same stimuli that Petersen and his colleagues used to demonstrate the right-ear advantage (Petersen, 1981; Petersen et al., 1978). In our study, the monkeys were preoperatively trained to classify examples of the two types of coos presented from a loudspeaker placed in front of them using a conditioned-avoidance procedure. Briefly, a thirsty monkey was seated in a standard primate chair and accustomed to drinking out of a water spout mounted on the chair. Initial auditory training consisted of presenting one of the SE coos at random intervals from 7 to 49 sec apart. Each presentation consisted of playing the coo three times, following which a mild electric shock was delivered through the water spout. After several stimulus presentations, the monkeys learned to avoid the shock by breaking contact with the spout whenever the coo was presented and to return to the spout after the shock had been presented.

The animals were then trained to distinguish one SE coo from one SL coo by presenting one of the two coos every 7 sec. Because only the SE coo was followed by shock, the animals learned to break contact whenever they heard the SE coo, but to maintain contact whenever the SL coo was presented. Note that in this initial training, only one example of each type of coo was used and, at this point, the monkeys could distinguish between them on the basis of any number of physical differences such as frequency, intensity, or duration.

Once an animal had learned to break contact only when the SE coo was presented, additional examples of both types of coos were presented until the animal had learned to discriminate the entire set of 15 coos. Thus, the animals demonstrated the ability to distinguish the two types of coos by breaking contact with the spout following presentation of an SE coo and maintaining contact after presentation of an SL coo. This task could not easily be accomplished by discriminating on the basis of some physical aspect of the sounds because each of the two sets of coos contained individual coos which varied noticeably in duration, intensity, spectrum, etc. Although the animals could conceivably have memorized each of the 15 coos, this strategy would be expected to increase the difficulty of the task as

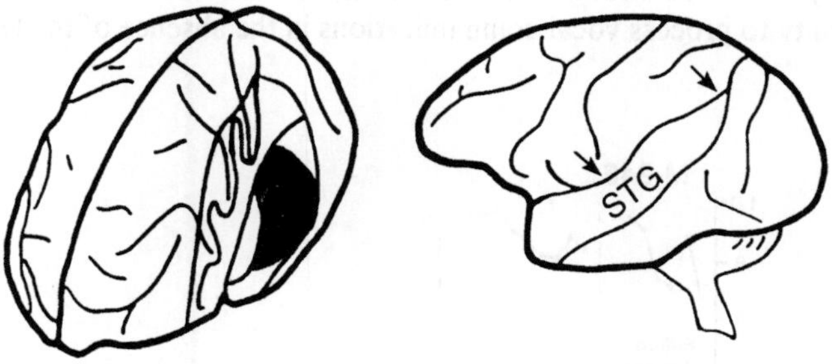

Figure 2. Drawing of a Japanese macaque brain indicating the location of primary auditory cortex and the surrounding auditory fields (also referred to as secondary auditory cortex). Left: view with the parietal operculum removed to show the location of primary (blackened) and the surrounding auditory fields (stippled). Right: lateral view; arrows indicate the extent of auditory cortex which is buried within the depths of the Sylvian fissure. STG, superior temporal gyrus (drawings based on the electrophysiological study of Merzenich and Brugge, 1973).

more examples were added. However, just the opposite occurred the discrimination appeared to become easier as more examples were added. That is, while learning to discriminate the first pair of coos took an average of 6.8 sessions, learning the next 6 pairs (plus one additional SL coo) took only an additional 25.9 sessions, with one animal learning the additional coos in only 10 sessions. Moreover, the monkeys did not forget the discrimination and could successfully perform it even after a hiatus of over 3 months. Thus, the monkeys' performances suggest that they were performing a biologically simple discrimination in spite of the fact that the stimuli were acoustically complex signals.

Following training, five monkeys received lesions of the left superior temporal gyrus, which included primary auditory cortex and the surrounding auditory fields (Fig. 2), while 5 other monkeys received lesions of the right superior temporal gyrus. Analysis of the lesions indicated that primary and secondary auditory cortex was removed in each case as demonstrated by cortical reconstruction and analysis of the resulting thalamic degeneration (Heffner and Heffner, 1986a).

The behavioral results showed that right auditory cortex lesions had no noticeable effect on the discrimination (Fig. 3). That is, the animals with right hemisphere lesions were able to perform the discrimination at normal levels when tested within 3 to 8 days following surgery. On the other hand, left auditory cortex lesions consistently resulted in a noticeable, although transient, deficit in the ability of the animals to perform the discrimination. That is, when tested within 4 to 11 days after surgery, the performance of the animals with left hemisphere lesions was below preoperative levels for the first 5 to 15 sessions. This difference between the effects of left and right auditory cortex lesions suggests that the left hemisphere in Japanese macaques is dominant for the perception of species-specific vocalizations.

The difference between the performances of the two groups suggests that the initial impairment shown by the left unilateral cases was not due to any general postoperative malaise. Nor did the difference appear to be due to such factors as amount of recovery time before testing, overall time between pre- and postoperative testing, or to differences in the size of the lesions (Heffner and Heffner, 1984, 1986a, 1989b, 1990a).

However, as noted, the left-hemisphere deficit for this task proved to be impermanent and these animals were able to regain preoperative performance levels within a few weeks. The discovery that the deficit is transient suggested two possibilities. The first is that the lateralization may not be as strong as that found in humans and that the right hemisphere possesses the ability to process vocal communications in the absence of the left hemisphere.

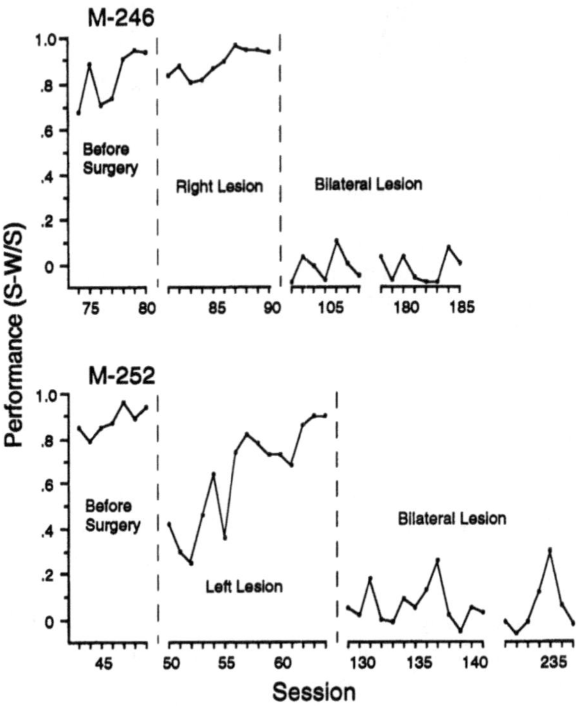

Figure 3. Effect of cortical ablation in two monkeys (M-246 and M-252) on the ability to discriminate the coo vocalizations. Performance on this task can range from 1.0 (perfect discrimination of the two sets of coos) to around 0 (failure to discriminate). Note that left, but not right, auditory cortex lesions result in a transient deficit. Bilateral lesions result in a permanent deficit which is still evident months after the surgery. S represents the average of the SL coo (safe) scores and W is the average of the SE coo (warning) scores (modified from Heffner and Heffner, 1986a).

The second possibility is that there is in fact a permanent deficit following left unilateral lesions, but that the specific task used here is relatively undemanding and that a more stringent test is needed to demonstrate it.

Nevertheless, it is clear that the perception of vocal communications by Japanese macaques is a cortically-dependent task even if it is not as strongly lateralized as speech perception is in humans. This is demonstrated by the effect of bilateral ablation on the discrimination of the coos. As the examples in Fig. 3 indicate, when a second auditory cortex lesion is made in the other hemisphere, the animals are rendered totally and permanently unable to discriminate the coos (Heffner and Heffner, 1986a). This result supports the interpretation that the recovery of the monkeys with left auditory cortex lesions occurs because the right hemisphere is able to mediate the discrimination. However, once both auditory cortices have been removed, not only are the animals unable to discriminate the two sets of coos (i.e., all 15 coos), but they cannot even perform the task when it is made easier by reducing the discrimination to a single pair of SE and SL coos. Thus, removal of auditory cortex changes the discrimination from an acoustically complex yet biologically simple task to one that is both biologically and acoustically complex.

Although unable to discriminate the coos, the monkeys with bilateral lesions were still capable of making other auditory discriminations (Heffner and Heffner, 1986a). Not only were they easily able to detect the coos (their hearing loss was not total), but they could discriminate them from noise bursts. However, when tested on their ability to discriminate the coos from tones, they could easily discriminate the coos from 2 and 4 kHz tones, but had difficulty discriminating them from a 1 kHz tone and were never able to discriminate them from a 500 Hz tone. Because 500 and 1000 Hz are close to the fundamental frequencies of the coos which had beginning frequencies from about 500 to 750 Hz, this result suggests that the monkeys had a deficit in frequency discrimination.

Because numerous studies have demonstrated that monkeys with auditory cortex lesions retain the ability to discriminate frequency (Heffner and Masterton, 1978; Massopust et al., 1971), it is difficult to account for their failure to discriminate the coos from a 500-Hz tone in terms of a general inability to discriminate frequency. However, a characteristic of the coos is that they are frequency-modulated signals - that is, the fundamental frequency of a coo rises and then falls with the peak frequency occurring earlier in the SE coos than in the SL coos. This leads to the possibility that the deficit in discriminating the coos may be the result of an inability to discriminate rapid changes in frequency, a deficit that a typical frequency-discrimination task using comparatively long-duration tone pips might miss (cf. Heffner and Masterton, 1978). If so, then this represents another possible similarity between monkeys and humans as it has been suggested that the hemispheric specialization for speech perception in humans reflects left-hemispheric dominance in the processing of rapidly changing signals (Schwartz and Tallal, 1980).

It should be noted that the permanent deficit seen following a bilateral lesion does not appear to be a secondary result of a cortical hearing loss. This was demonstrated by a study in which monkeys with bilateral lesions were tested to determine if amplifying the coos would improve the animals' performances (Heffner and Heffner, 1989a). The results of this study showed that even if the different frequency bands of the coos were selectively amplified to compensate, as closely as possible, for each animal's individual hearing loss, the monkeys still could not discriminate the coos. Moreover, when normal monkeys were presented with coos which had been attenuated and filtered to simulate a cortical hearing loss, they were still able to discriminate the coos at normal levels. These two findings indicate that the loss in the ability to discriminate the coos which follows bilateral lesions cannot be explained in terms of a simple hearing loss.

Nor does this deficit appear to be due to any general impairment in auditory learning or memory. This is demonstrated by the fact that bilateral lesions, which abolish the ability to discriminate the coos, do not abolish the ability to discriminate other auditory stimuli, such as frequency and intensity. Therefore, it would appear that bilateral auditory cortex lesions do not have any general effect on auditory learning or memory. Moreover, lesions which produce a specific short-term memory deficit in monkeys (i.e., superior temporal gyrus lesions which spare auditory cortex) do not abolish the ability to categorize the coos (cf. Colombo et al., 1990; Heffner and Heffner, 1989b). Finally, although the monkeys in our studies were tested with a conditioned-avoidance procedure, the results cannot be explained in terms of an inability of the animals to establish fear conditioning (LeDoux et al., 1984) since, as previously noted, the monkeys were able to condition to frequency differences using the same avoidance task. Thus, while the deficit could conceivably be the result of some sort of learning or memory deficit which affects some, but not all, auditory discriminations, there is currently little evidence for such a possibility.

Location of the Neural Mechanism

Given the discovery that bilateral removal of the superior temporal gyrus results in a total inability to discriminate the coos, the question arises as to whether smaller lesions might have the same effect. In order to address this question, it is first necessary to review what is known concerning the location of auditory cortex.

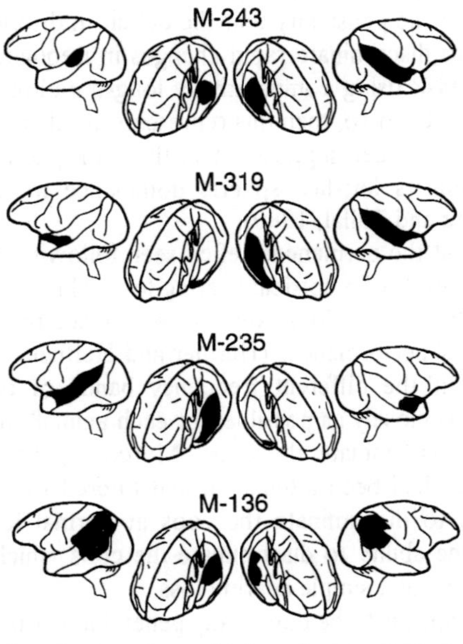

Figure 4. Cortical reconstructions for 4 monkeys which received partial lesions of the superior temporal gyrus. Shaded areas indicate the extent of the lesions. The lesions are arranged in descending order of their effect on the ability to discriminate the coos (cf. Fig. 5). Note that only M-243 showed a total and permanent deficit in the ability to discriminate the coos (modified from Heffner and Heffner, 1989b).

Although the exact boundaries of auditory cortex can be difficult to define, both cytoarchitectural and electrophysiological studies have indicated that there is a central core area (primary auditory cortex) and a surrounding belt of secondary auditory fields. Primary auditory cortex lies in the depth of the Sylvian fissure on the middle third of the superior temporal plane and is surrounded by four secondary auditory fields (Fig. 1; Merzenich and Brugge, 1973; for a review see Heffner and Heffner, 1986b). Other parts of the superior temporal gyrus also appear to be auditory. Auditory responses have been recorded on the

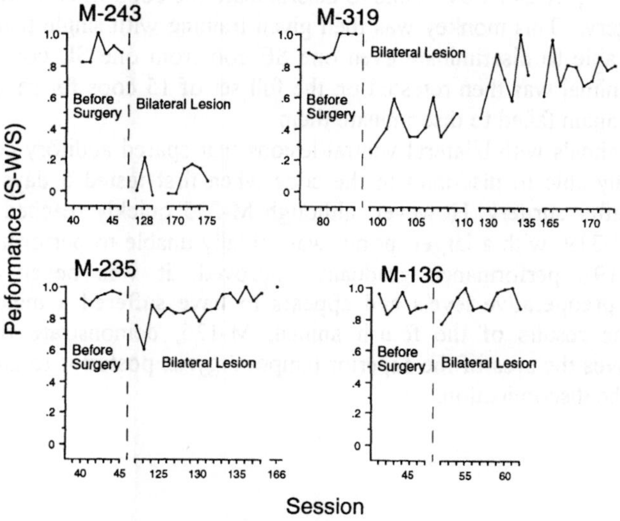

Figure 5. Effect of restricted superior temporal gyrus lesions on the ability to discriminate the coos. See Fig. 4 for drawings of the lesions. Note that only M-243 was completely unable to perform the discrimination (modified from Heffner and Heffner, 1989b).

superior temporal plane rostral to the secondary fields and on the lateral surface of the superior temporal gyrus indicating that there are at least two other auditory fields (Merzenich and Brugge, 1973). In addition, data on the cytoarchitectonics and cortical connections of the auditory fields suggests that all of the superior temporal gyrus with the exception of the caudal (posterior) end, which receives no significant projections from the medial geniculate, may have auditory functions (Galaburda and Pandya, 1983; Mesulam and Pandya, 1973).

The effect of restricted bilateral lesions on the ability to discriminate the coos was determined using 4 animals (Heffner and Heffner, 1989b). In this study, 3 of the monkeys received a large lesion of the superior temporal gyrus in one hemisphere followed by a restricted lesion of the superior temporal gyrus in the other hemisphere (Fig. 4). Because unilateral lesions do not result in a permanent deficit, any effect would be the result of removing the same region on both sides, i.e., the functional lesion would be that area which was removed in both hemispheres. The functional lesions received by the 4 monkeys were: primary and secondary auditory cortex was removed in M-243, a portion of the superior temporal gyrus rostral to the primary and secondary auditory fields was removed in M-319, the rostral portion of the superior temporal gyrus outside the primary and secondary auditory fields was removed in M-235 (resulting in a similar, though slightly smaller functional lesion than in M-319), and the superior temporal gyrus caudal to primary and secondary auditory cortex and the immediately adjacent parietal, occipital, and temporal area were removed in M-136.

The results of this study showed that only the monkey in which primary and secondary auditory cortex were removed bilaterally (M-243) had a complete and permanent deficit. As shown in Fig. 5, M-243 was unable to discriminate the coos when tested during the first week after surgery. This monkey was then given training with single pairs of coos, but at no time was it able to discriminate even one SE coo from one SL coo despite weeks of training. The animal was then retested on the full set of 15 coos for seven more sessions, during which it again failed to discriminate them.

The two animals with bilateral ventral lesions that spared auditory cortex on one side were successfully able to discriminate the coos when first tested 5 days (M-235) and 10 days (M-319) after surgery. However, although M-235 quickly reached normal levels of performance, M-319, with a larger lesion, was initially unable to perform at normal levels. Although M-319's performance gradually improved, it was never able to perform consistently at preoperative levels and appears to have suffered a minor but permanent impairment. The results of the fourth animal, M-136, demonstrate that a lesion that bilaterally removes the area of the superior temporal gyrus posterior to auditory cortex has little effect on the discrimination.

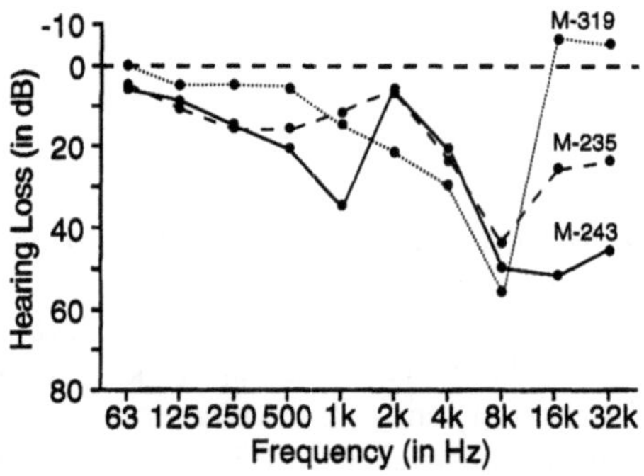

Figure 6. Effect of restricted superior temporal gyrus lesions on pure-tone thresholds. Dashed line at zero is the average hearing level for normal macaques. Note that although all three of the monkeys had a hearing loss, only M-243 was unable to discriminate the coos (cf. Fig. 5) (Data from Heffner and Heffner, 1989b).

Interestingly, 3 of the monkeys showed a hearing loss following bilateral lesions. As shown in Fig. 6, M-235, M-243, and M-319 each showed a hearing loss that followed the same general pattern as that shown in Fig. 2 for monkeys with large superior temporal gyrus lesions. However, their hearing losses were not as severe evidently due to the fact that their lesions were smaller. Moreover, the hearing losses of these 3 animals are similar to each other, with the exception of frequencies above 8 kHz, yet only one of them, M-243, was unable to discriminate the coos. The fact that two of the monkeys showed a hearing loss, but were still able to discriminate the coos supports the conclusion that the deficit in the discrimination of the coos cannot be accounted for by the hearing loss - because the coos did not contain any energy above 8 kHz, the difference in the audiograms above this point was not a factor. This conclusion is further supported by the fact that increasing the intensity of the coos to compensate for M-243's hearing loss did not improve its performance. The fact that a superior temporal gyrus lesion can result in a hearing loss without abolishing the ability to discriminate the coos demonstrates that these two deficits can be partially dissociated.

In summary, removal of primary auditory cortex and the surrounding auditory fields in both hemispheres abolishes the ability to discriminate the coos. In contrast, lesions of the superior temporal gyrus that include bilateral removal of areas rostral or caudal to auditory cortex, but spare auditory cortex on one side, have little or no effect on the discrimination. Although we cannot rule out the possibility that the neural mechanism for the perception of species-specific vocal communications may be located outside the superior temporal gyrus and that the deficit observed in M-243 may be caused by the disruption of auditory input from auditory cortex to that area, this study demonstrates that the deficit may be produced by small lesions restricted to the middle third of the superior temporal plane.

CONCLUSION

The perception of vocal communications by Japanese macaques is a cortically-dependent task which is lateralized to some extent to the left hemisphere. This fact suggests that macaques possess a cortical mechanism analogous to the cortical speech perception mechanism in humans which, although predominately an unilateral mechanism, does appear in bilateral form in some individuals (e.g., Courtois et al., 1983). Given the current state of knowledge concerning the cortical mechanism in macaques, there are at least three questions which need to be addressed.

The first question concerns the degree to which the mechanism is lateralized to the left hemisphere. As previously noted, although left hemisphere lesions result in an immediate deficit in categorizing the coos, the right hemisphere is eventually able to mediate the task at normal levels. While this result may indicate that the mechanism is only weakly lateralized to the left hemisphere, it may also be a reflection of the simplicity of the task itself. That is, the coo classification task used so far employs only 15 sounds and the animals typically receive several months of preoperative training using these sounds. Thus, a more demanding task might well reveal a permanent deficit following left hemisphere lesions. For example, the task could be made more difficult by presenting competing stimuli to the two ears using headphones as well as by including additional examples of coos to which the monkeys have not been exposed preoperatively.

The second question concerns the nature of the deficit, that is, whether the permanent deficit seen following bilateral lesions is specific to vocal communications or whether it represents a general deficit in the discrimination of the physical parameters of sound.

Although it is well established that auditory cortex ablation does not abolish the ability to discriminate frequency, intensity, or the temporal aspects of sound, few studies have examined these abilities using stimuli which mimic natural sounds such as vocalizations. Thus, although the ability to discriminate the frequency of long duration sounds may be normal, cortical ablation may affect the ability to perform such a discrimination on a short or rapidly changing sound (e.g., Symmes, 1966). Furthermore, a number of these studies were performed on carnivores and it is now apparent that the effects of cortical lesions in primates and carnivores may differ a significant finding for which we as yet have no explanation (e.g., Heffner and Heffner, 1986b).

The third question concerns the issue of whether the cortical deficit is "species-specific". As demonstrated by Petersen and his colleagues, although other species of macaques, such as *Macaca nemestrina* and *Macaca radiata*, can learn to categorize Japanese macaque coos, they do not show a right-ear advantage (e.g., Petersen, 1981). This result suggests that macaques process the vocalizations of other species differently from those of their own species. If so, then it might be expected that non-Japanese macaques would not show a deficit following left hemisphere lesions and, perhaps would not show a deficit following a bilateral lesion.

ACKNOWLEDGEMENTS

This work was supported by NIH grant NS30539. The authors thank G. Koay, L. Boester, and D. Koay for their help in preparing this chapter.

REFERENCES

Colombo, M., D'Amato, M.R., Rodman, H.R., and Gross, C.G., 1990, Auditory association cortex lesions impair auditory short-term memory in monkeys, *Science* 247:336-338.

Courtois, G.A., Lecours, A.R., and Lhermitte, F., 1983, Cerebral dominance and language, *in*: "Aphasiology", A.R. Lecours, F. Lhermitte and B. Bryans, eds., Bailliere Tindal, London.

Ferrier, D., 1876a, The Croonian lecture: Experiments on the brain of monkeys: Second series, *Philos. Trans. R. Soc. Lond. Biol.* 165:433-488.

Ferrier, D., 1876b, "The Functions of the Brain", Smith Elder, London.

Galaburda, A.M. and Pandya, D.N., 1983, The intrinsic architectonic and connectional organization of the superior region of the rhesus monkey, *J. Comp. Neurol.* 221:169-184.

Geschwind, N., 1965, Disconnection syndromes in animals and man, *Brain* 88:237-294.

Green, S., 1975, Variation of vocal pattern with social situation in the Japanese monkey (*Macaca fuscata*): A field study, *in*: "Primate Behavior: Vol. 4", L.A. Rosenblum, ed., Academic Press, New York.

Hauser, M.D. and Andersson, K., in press, Left hemisphere dominance for processing vocalizations in adult, but not infant rhesus monkeys: Field experiments. *Proc. Nat. Acad. Sci.*

Heffner, H.E. and Heffner, R.S., 1984, Temporal lobe lesions and perception of species specific vocalizations by Japanese macaques, *Science* 226:75-76.

Heffner, H.E. and Heffner, R.S., 1986a, Effect of unilateral and bilateral auditory cortex lesions on the discrimination of vocalizations by Japanese macaques, *J. Neurophysiol.* 56:683-701.

Heffner, H.E. and Heffner, R.S., 1986b, Hearing loss in Japanese macaques following bilateral auditory cortex lesions, *J. Neurophysiol.* 55:256-271.

Heffner, H.E. and Heffner, R.S., 1989a, Cortical deafness cannot account for the inability of Japanese macaques to discriminate species-specific vocalizations, *Brain Lang.* 36:275-285.

Heffner, H.E. and Heffner, R.S., 1989b, Effect of restricted cortical lesions on absolute thresholds and aphasia-like deficits in Japanese macaques, *Behav. Neurosci.* 103:158-169.

Heffner, H.E. and Heffner, R.S., 1989c, Unilateral auditory cortex ablation in macaques results in a contralateral hearing loss, *J. Neurophysiol.* 62:789.

Heffner, H.E. and Heffner, R.S., 1990a, Role of primate auditory cortex in hearing, *in*: "Comparative Perception, Vol. 2: Complex Signals", W.C. Stebbins and M.A. Berkley, eds., Wiley, New York.

Heffner, H.E. and Heffner, R.S., 1990b, Unilateral auditory cortex lesions in macaques result in a contralateral hearing loss, *J. Neurophysiol.* 62, 789-801.

Heffner, H.E. and Heffner, R.S., 1990c, Effect of bilateral auditory cortex lesions on absolute thresholds in Japanese macaques, *J. Neurophysiol.* 64, 191-205.

Heffner, H. and Masterton, B., 1978, Contribution of auditory cortex to hearing in the monkey (*Macaca mulatta*), *in*: "Recent Advances in Primatology", D.J. Chivers and J. Herbert, eds., Academic Press, New York.

Jerger, J., Weikers, N.J., Sharbrough, F.W. III, and Jerger, S., 1969, Bilateral lesions of the temporal lobe: A case study, *Acta Oto-Laryngol. Suppl.* 258:1-51.

Kimura, D., 1961, Cerebral dominance and the perception of verbal stimuli, *Canadian J. Psychol.* 15: 166-171.

Kimura, D., 1967, Functional asymmetry of the brain in dichotic listening, *Cortex*, 3, 163-178.

LeDoux, J.E., Sakaguchi, A., and Reis, D.J., 1984, Subcortical efferent projections of the medial geniculate nucleus mediate emotional responses conditioned to acoustic stimuli, *J. Neurosci.* 4:683-698.

Leicester, J., 1980, Central deafness and subcortical motor aphasia, *Brain Lang.* 10:224-242.

Massopust, L.C., Jr., Wolin, L.R., and Frost, V., 1971, Frequency discrimination thresholds following auditory cortex ablations in the monkey, *J. Aud. Res.* 11:227-233.

Masterton, R.B. and Berkley, M.A., 1974, Brain function: Changing ideas on the role of sensory, motor, and association cortex in behavior, *Ann. Rev. Psychol.* 25:277-312.

Merzenich, M.M. and Brugge, J.F., 1973, Representation of the cochlear partition on the superior temporal plane of the macaque monkey, *Brain Res.* 50:275-296.

Merzenich, M.M., Kaas, J.H., and Roth, G.L., 1976, Auditory cortex in the grey squirrel: Tonotopic organization and architectonic fields, *J. Comp. Neurol.* 166:387-401.

Mesulam, M.-M. and Pandya, D.N., 1973, The projections of the medial geniculate complex within the Sylvian fissure of the rhesus monkey, *Brain Res.* 60:315-333.

Michel, F. and Peronnet, F., 1980, A case of cortical deafness: Clinical and electrophysiological data, *Brain Lang.* 20:367-377.

Neff, W.D., Diamond, I.T., and Casseday, J.H., 1975, Behavioral studies of auditory discrimination: Central nervous system, *in*: "Handbook of Sensory Physiology. Auditory System, vol. V, pt. 2", W.D. Keidel and W.D. Neff, eds., Springer-Verlag, New York.

Passingham, R.E., 1979, Specialization and the language areas, *in*: "Neurology of Social Communication in Primates", H.D. Steklis and M.J. Raleigh, eds., Academic Press, New York.

Petersen, M.R., 1981, Perception of acoustic communication signals by animals: Developmental perspectives and implications, *in*: "Development of perception: Vol. 1", R.N. Aslin, J.R. Alberts and M.R. Petersen, eds., Academic Press, New York.

Petersen, M.R., Beecher, M.D., Zoloth, S.R., Moody, D.B., and Stebbins, W.C., 1978, Neural lateralization of species-specific vocalizations by Japanese macaques (*Macaca fuscata*), *Science* 202:324-327.

Schwartz, J. and Tallal, P., 1980, Rate of acoustic change may underlie hemispheric specialization for speech perception, *Science* 207:1380-1381.

Steklis, H.D. and Raleigh, M.J., 1979, "Neurobiology of Social Communication in Primates", Academic Press, New York.

Symmes, D., 1966, Discrimination of intermittent noise by macaques following lesions of the temporal lobe, *Exp. Neurol.* 16:201-214.

Wernicke, C., 1874, "Der Aphasische Symptomencomplex. Eine psychologische Studie auf anatomischer Basis", Cohn & Weigert, Breslau.

AMPLITUDE ENVELOPE ENCODING AS A FEATURE FOR TEMPORAL INFORMATION PROCESSING IN THE AUDITORY CORTEX OF SQUIRREL MONKEYS

Armin Bieser

German Primate Center
Division of Neurobiology
37077 Göttingen, Germany

INTRODUCTION

Auditory signals represent complex patterns of information in the frequency and intensity domain over time. How do neurons encode the complex frequency-time varying features of a natural vocalization? Where and how are the complex acoustic signals transferred and coded in the animal's brain? Are there specialized areas for the extraction of communicatively meaningful sound parts? For the neurophysiological experimentalist the situation seems at first sight frustrating. The answers obtained with single unit recordings are only simple yes/no codes (i.e, action potentials). First, the interpretation of the action-potential frequencies with respect to different complex stimuli seemed to be impossible at the single neuron level. Therefore, when the first monkey studies started, the only promising concept seemed the search for highly specialized cortical neurons which may act as "vocalization detectors". Such neurons should answer only to a very small set of vocalizations but not - or not in the same way - to noise or tone bursts. Neurons with such feature specificity have never been clearly identified. Even complex vocalizations appeared to evoke neural responses only by producing spike trains following the time structure of the transient elements in the stimulus (Newman, 1988). Consequently, it is not surprising that previous electrophysiological research with species-specific sounds gave rise to more questions than answers. Moreover, the theoretical background for these experiments was poorly described. Highly specialized "vocalization detector" neurons are only to be expected if one assumes one of the following mechanisms: an innate neural network specialized for filtering the species' own vocalizations, or, alternatively, a sensory serial information processor leading to the extraction of information-bearing sounds. Both mechanisms could produce higher-order neurons exclusively restricted in their activity to vocalizations as adequate stimuli. An argument against a simple innate neural network comes from behavioral studies. The meaning of an animal's vocalization can vary with context, even though the acoustic morphology of the sound remains unchanged (Symmes, 1981; Cheney and Seyfarth, 1990). An argument against the second alternative is the anatomically established

Current Topics in Primate Vocal Communication
Edited by E. Zimmermann *et al.*, Plenum Press, New York, 1995

centrifugal pathways which parallel the centripetal auditory pathways (Pickles, 1982). On the other hand, electrophysiologists have documented several neurons that are, for example, specialized to small intensity or frequency ranges. This can be interpreted as support for the view that neurons are able to extract features, which may be information-bearing parameters in a complex stimulus. Such feature encoding seems an essential framework for the brain in complex sound recognition. To offer an example, it is an every day experience that in speech perception the meaning of a word is often known despite an unclear pronunciation or against a noisy environment. We complete a familiar word if we detect some features or meaningful parts of the word. Therefore, many studies regarding the auditory perception of sounds focus on experiments with relatively simple artificial sounds, which may act as information-bearing elements in a complex sound.

A COMMON FEATURE OF COMPLEX SOUNDS: TEMPORALLY CHANGING INTENSITIES

What features should be used in an experimental set-up? Pure sinusoidal tones are very unusual in a natural environment, and the other short stimuli like clicks and noise bursts lack the complexity of an animal sound. However, there is one feature found in all complex sounds: the temporal patterning of the signal's amplitude modulation. In humans, for example, the audible sensation of rhythm depends primarily on variations in the temporal envelope of a sound (Zwicker and Fastl, 1990).

The ability of the cortical neural circuits to code such changing features within a sound seems a necessary aspect of acoustic information processing. The only available technique to measure how neurons respond to brief changes in intensity is the single unit recording technique, using small microelectrodes. The spatial resolution of this technique allows one to characterize the response properties of single neurons or neuron clusters. If the amplitude envelope is, as assumed, an important feature for sound perception, a high number of cortical neurons should code this feature in a reproducible spike pattern. The spatial resolution of the electrophysiological technique should in practice allow one to separate areas where this information processing takes place.

Figure 1 illustrates two squirrel monkey calls. The upper part shows the time-amplitude waveform of a "trill" and below it the amplitude envelope of this vocalization. Modulation of sound intensity can be studied with relatively simple sounds, such as sinusoidally modulated tones or noise. Periodic modulations of the envelope arise in sounds when two sinusoidal waves with different frequencies (f1 and f2) are added. The resulting signal has a sinusoidal envelope fluctuation when $f2>f1$. The envelope fluctuation frequency is $f_{mod} = f2-f1$. The spectral properties of the signal show one or, when adding a third signal f3, two sidebands and f2 as carrier frequency. The modulation frequency f_{mod} shows no sideband.

Psychophysical data evaluated in humans have shown that such amplitude-modulated sounds (**AM-sound**) produce two different hearing sensations: sounds with frequencies below 20 Hz produce the hearing sensation "fluctuation strength" and AM-sounds in the range between about 15 to 150 Hz elicit the hearing sensation "roughness" (Terhardt, 1974; Fastl, 1977). Thus, the temporal patterning of signal amplitude is a critical aspect of auditory information processing.

To find out where and how these temporal parameters are extracted is a major goal of electrophysiological investigations of the auditory pathway. Some experiments with AM-sounds will be reviewed in the next sections. This paper focuses on a set of experiments that have been completed with squirrel monkeys. This species seems a good model to study the general mechanisms of envelope encoding in the brain. The monkeys

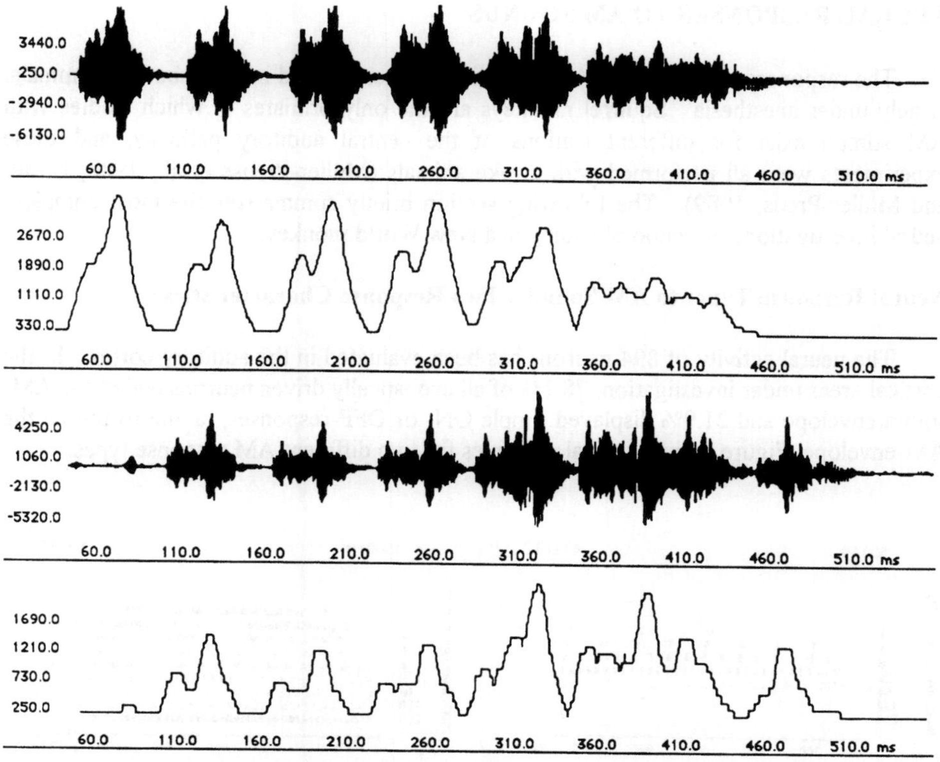

Figure 1. Oscillograms of two squirrel monkey calls and their amplitude envelopes. The envelopes are root mean square (RMS) envelopes calculated with a window width of 10 ms.

use vocalizations with amplitude envelopes between 4 and 64 Hz (Schott, 1975; Fastl et al., 1991) and behavioral studies have demonstrated the information-bearing character of amplitude changes (Maurus et al., 1984). For more details about information processing of time-critical features of complex sounds the reader should consult Langner, 1992, Evans, 1992, and Javel and Mott, 1988.

METHODS

Experiments were performed on awake adult squirrel monkeys with conventional microelectrode recording techniques (see Müller-Preuss et al., 1988 for details). Stimuli were amplitude-modulated sounds and tone/noise bursts. The quantitative analysis of the modulation-frequency transfer was determined by cross-correlating the post-stimulus time histogram (PSTH) with the amplitude envelope of the stimulus. The modulation frequency at which the correlation was highest is referred to as the "**best modulation frequency" (BMF)**. The BMF expresses the best synchronization of the amplitude envelope with the neural firing rate. Additionally, the total sum of spikes of a particular PSTH has been considered by the analysis (=spike rate). The spike-rate modulation transfer function gives the neuron´s tuning properties to AM sounds independent of the temporal pattern of the spike train. The AM sound frequency that gives the highest value for the spike rate is named **SR-max** in the following sections.

NEURAL RESPONSES TO AM SOUNDS

The responses of neurons to AM stimuli have been studied in a number of mammals, largely under anesthesia. Squirrel monkeys are the only primates in which studies with AM stimuli exist for different stations of the central auditory pathway, and these experiments were all performed with awake animals (Müller-Preuss et al., 1988; Bieser and Müller-Preus, 1989). The following section briefly summarizes the most complete neural investigations of temporal coding in a New World monkey.

Neural Response Types to AM Sounds: Two Response Characteristics

The neural activity of 894 neurons has been evaluated in the auditory cortex. In the cortical areas under investigation, 78.1% of all acoustically driven neurons coded the AM-sound envelope and 21.9% displayed simple ON- or OFF-responses, failing to follow the AM-envelope. Figure 2 gives typical examples for two different AM response types.

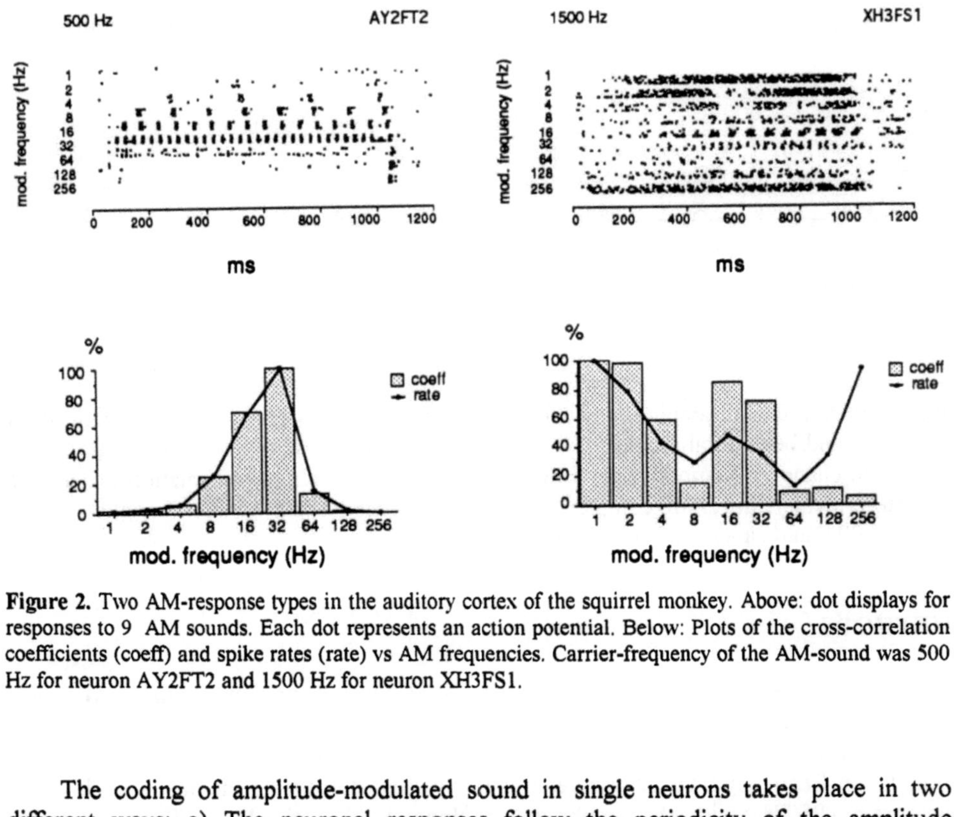

Figure 2. Two AM-response types in the auditory cortex of the squirrel monkey. Above: dot displays for responses to 9 AM sounds. Each dot represents an action potential. Below: Plots of the cross-correlation coefficients (coeff) and spike rates (rate) vs AM frequencies. Carrier-frequency of the AM-sound was 500 Hz for neuron AY2FT2 and 1500 Hz for neuron XH3FS1.

The coding of amplitude-modulated sound in single neurons takes place in two different ways: a) The neuronal responses follow the periodicity of the amplitude modulation envelope, displaying synchronization maxima (**BMF**); the synchronization-transfer function is expressed in normalized coefficients of the cross-correlation. b) The sum of the spikes during the stimulus presentation changes for each modulation frequency without coupling the neural response to the temporal properties of the stimulus (**spike rate**). Both neural responses can be largely independent. For example, the spike rate may be constant for all modulation frequencies, but the phase-locked neural coupling to the AM-envelope expresses a band-pass filter-characteristic with a BMF. The most common neural response to AM sound is a tuning to certain modulation frequencies by both the phase-locked synchronization discharge (BMF) and the spike rate function (SR-max). In 38.7% of all tested neurons, frequency values of BMF and SR-max were identical. 45.8% displayed BMF values below SR-max and only 15.4% of all cortical neurons had a BMF value above SR-max.

TEMPORAL RESOLUTION IN DIFFERENT CORTICAL AREAS

The auditory cortex of mammals shows several more or less complete representations of the cochlear partition. The general functions of the auditory cortex and its subdivisions are still not certain. For many mammalian species, the extent and location of the primary auditory cortex (AI) is defined by its cytoarchitecture and connections to the medial geniculate body (ventral division). For those animals where AI was fine-grain mapped with microelectrodes, an orderly frequency organization was found. Most neurons in this area are tuned to narrow ranges of frequency. Outside AI the location and extent of auditory responsive areas is not well known. Most physiological studies have been conducted in domestic cat, but the characterization of the auditory areas in this species is still incomplete. Up to now four tonotopic fields with systematic frequency organization have been described: AI (primary auditory field), AAF (anterior auditory field), PAF (posterior auditory field) VPAF (ventroposterior auditory field). These areas are surrounded by a peripheral belt where the frequency organization is more complex; most neurons are broadly tuned and the responsiveness to auditory stimuli is often weak. These areas are AII (second auditory field), DP (dorsoposterior auditory field), V (ventral auditory field) and T (temporal auditory field) (Imig and Reale, 1980).

For primates, the available data are limited. For macaque monkeys, six different parcellations of the auditory cortex have been published in the last two decades (Morel et al., 1993). Even the width of the primary field (AI) is uncertain in these Old World monkeys. More progress has been made in understanding the organization of the auditory cortex in New World monkeys. In the squirrel monkey, seven separate auditory areas were distinguished in a cytoarchitectonic and electrophysiological exploration of the auditory cortical areas (Bieser and Müller-Preuss,1992). Besides the primary auditory field AI, a rostral field (R) and an anterolateral field (AL) have also been defined. Medial from these fields are the parainsular field (Pi) and the rostral parainsular field (RPi). Posterior to AI is the postauditory field (Pa). The first temporal field (T1) covers a broad band below the Sylvian fissure on the gyrus temporalis superior (see Figure 3).
The cytoarchitectonic and physiological separation of the different auditory fields in the squirrel monkey is similar to what has been described in another New World monkey, *Aotus trivirgatus* (owl monkey, see Imig et al., 1977; Morel and Kaas, 1992). Despite the fact that for both monkey species the available data are limited, there are striking similarities in the organization of auditory cortex. There exists a core of 2 or perhaps 3 fields, AI, R and RT (the latter found only in Aotus), a surrounding belt consisting of posterior Pa (C in Aotus), Pi and RPi (in Aotus: CM, RM and MRT) located between AI, R and the insula, and T1 and part of AL on the gyrus temporalis superior (in Aotus: PL, AL and LRT). But again, the boundaries of these areas are uncertain. Further studies are necessary, using a combination of fine-grain microelectrode mapping and neuroanatomical tracing methods.

Beyond the tonotopic frequency representation, the functional organization of the mammalian auditory cortex is poorly understood. In cat, differences in sound level coding were found between AI, AAF and PAF (Phillips and Irvine, 1981, 1982) and a specialization for AM coding in different cortical areas was reported by Schreiner and Urbas (1988). To a certain degree, auditory cortical functions in microchiropteran bats can be attributed to information processing of biologically relevant sound features. This is possible because of the relatively stereotyped sounds these animals use. The emitted biosonar pulses and the returning echoes lead to an acoustic information system which can be divided into a few components. In the mustached bat, for example, Suga and colleagues divided the sonogram of the emitted pulses into only eight components: four harmonics with a sequential constant frequency (CF) and frequency-modulated (FM) component for each harmonic. Together with the echo, the animal has to analyze only 16 sound components (Suga,1990). For successful foraging, the bat has to relate particular echo properties to particular target properties. For example, the echo delay gives information about the distance of an object, while the amount of Doppler shift is a

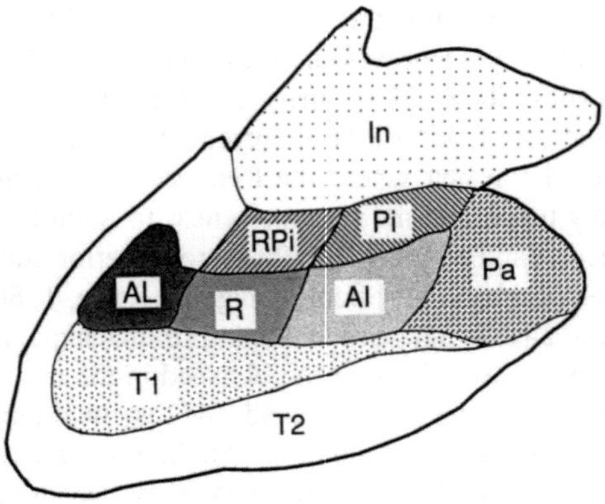

Figure 3. Unfolded view of the cortex lying in and adjacent to the lateral and superior temporal sulci from a squirrel monkey.

measure of relative target velocity. The multiple auditory cortex areas are able to extract such information, i.e., the neurons in these areas are highly specialized to extract the meaningful features. In monkeys, it is not known if different cortical areas are specialized for the analysis of different aspects of complex vocalizations or their features. As mentioned above, encoding of the amplitude envelope is an essential aspect of acoustical information processing. Therefore, this common feature of complex sounds was used in our study as an analytical tool to segregate the cortical areas functionally in the time domain.

Table 1 shows the mean BMF, the SR-max and the related minima and maxima encountered in the auditory cortex and the multi-modal insula (Sudakov et al.,1971) of the squirrel monkey. Best temporal resolution was shown by the primary auditory cortex (AI) and Pi, bordering AI medially, whereas the anterior fields are characterized by a weak temporal resolution. Table 2 indicates which cortical areas have significant differences in their mean BMF. SR-max had higher modulation frequencies (mean: 34-72 Hz) but showed statistically significant differences only between AI and T1 (p=0.0216) and between T1 and R (P=0.0011) in the cortical areas investigated (all statistics were performed using Fisher´s PLSD ANOVA).

Table 1. Mean BMFs and mean SR-max together with their minimal and maximal values in eight different auditory areas.

AREA	mean - BMF (Hz)	minimum (Hz)	maximum (Hz)
AI	17.84	1	350
AL	2.83	1	8
R	7.66	1	64
RPi	2.72	1	16
Pi	22.3	1	128
Pa	11.56	1	128
T1	12.81	1	64
In	9.88	1	64
AREA	**mean SR max (Hz)**	**min SR-max (Hz)**	**max SR-max (Hz)**
AI	50.49	1	512
AL	47.41	1	256
R	34.55	1	256
RPi	67.79	1	256
Pi	53.02	1	256
Pa	52.78	1	256
T1	72.06	1	256
In	59.62	1	256

Table 2. ANOVA-statistics of the mean BMF differences ($p< 0.05$).

AREAS	p	AREAS	p
AI-R	0.0001	Pi-In	0.0148
AI-AL	0.0035	Pi-AL	0.0129
AI-In	0.0450	Pi-R	0.0005
AI-RPi	0.013	T1-RPi	0.050
Pi-RPi	0.006	T1- Pi	0.0252

BMF and SR-max display the modulation frequency where the AM sound envelope is encoded best. In contrast, the **filter shape** of an AM response type describes the AM frequency range which can be encoded by a single neuron. Neuron AY2FT2 in Figure 2, for example, encodes the modulation frequencies from 8 Hz to 32 Hz only. Higher modulation frequencies lead to simple OFF responses. Consequently, these response characteristics show a band-pass filter for the phase-locked response and the spike rate. The response characteristics of neuron XH3FS1 are more complex than AY2FT2. The phase-locked envelope coding of 1, 2, 4, 16 and 32 Hz is very similiar, whereas 8 Hz and the AM frequencies higher than 32 Hz where not encoded by phase-locked firing (in Figure 2 the synchronization is expressed in the cross-correlation coefficients). The spike rate follows the phase-locked shape up to 64 Hz, but at 256 Hz the spike rate increases dramatically, losing the synchronization to the envelope phase of the sound (in Figure 2: rate). Therefore, neuron XH3FS1 shows complex filter characteristics: a band-stop filter between 1 Hz and 64 Hz with a notch at 8 Hz. Within this filter shape the temporal patterning of the sound is encoded. Additionally, a complex filter type is expressed in the spike rate values. The latter filter shape describes the neuron's ability to code AM sounds with a change in the spike frequency without coupling to AM sound phase. Like the BMFs and the SR-max, the filter types for phase synchronization and spike rate often fall apart. That is, the filter cutoff range and the filter shapes are different for both measurements. The synchronization to the AM sound displays band-pass (BP), low-pass (LP) or complex multipeaked (MP), including band-stop, filter characteristics. The spike rate exhibits BP, LP, MP and high-pass (HP) filter characteristics.

The occurrence and distribution of the two most frequent filter type combinations for both the phase-locking and the spike rate in each cortical field is shown in Figure 4.

AI and T1 displayed the same distribution of filter types. In both fields, the BP filters dominate. In the anterior fields, R and AL, LP filters join the BP filters and in Pi the BP filters dominate, but the second most common response type displays no filter characteristics (indicated as X in Figure 4). In Pa, RPi and the insula, neurons unable to follow the amplitude envelope or to express a spike maximum at a certain modulation frequency dominate (designated as X: Pa: 44.7%; RPi: 40.0%; In: 40.6%). Of the small number of neurons tested with AM sounds in T2, none followed the amplitude envelope.

Besides the significant differences in BMFs and SR-max, the filter types also separate the auditory cortex into specialized areas for envelope encoding. The different filter types refer to different computing of the AM sound in different cortical neurons. The distribution of these filter types gives evidence that envelopes of the natural calls are extracted in specialized cortical areas. AI, Pi and T1 represent areas where the envelope fluctuations of the monkey's natural calls may be encoded, whereas the other fields have an insufficient time resolution to manage such information. Similar differences in the temporal resolution between auditory cortical areas have been reported by Schreiner and Urbas (1988) for domestic cat. They reported the highest mean BMF values for AAF (31.1 Hz) and for AI (14.2 Hz). AII, VPAF and PAF displayed mean BMFs between 5.2 and 7.0 Hz. The values for AI in the cat are close to those found in our study for AI. PAF in the cat is thought to be a homologue to field R in New World monkeys and RL in macaques (Luethke et al., 1988). Also the mean values for BMFs in PAF (6.8 Hz) and R in the squirrel monkey (7.7 Hz) are close together. The postauditory field Pa seems to correspond to field C of the owl monkey (Morel and Kaas, 1992; Imig et al., 1977) and is possibly homologous to field AII of the cat (see discussion in Imig et al.,1977). The mean BMFs in cat are lower than our data for Pa (7.0 Hz in cat; 11.6 Hz in monkey). This may be due to the fact that the border between AI and Pa in monkeys is difficult to delineate either electrophysiologically or cytoarchitectonically. Consequently, some units from AI with higher time resolution may have erroneously been attributed to Pa and as a result contributed to the slightly higher mean BMF in the monkey. To directly compare the two investigations, one must consider the relatively high number of neurons in monkey's Pa that are unable to follow the AM envelope. Besides AI, field Pi showed the best temporal resolution. However, we have no data to ensure that Pi is homologous to the "high resolution field" (AAF) described for cats by Schreiner and Urbas.

Values of SR-max have higher modulation frequencies than the synchronized BMFs in all investigated areas. With one exception (T1), there were no statistically significant differences between cortical areas. Therefore, envelope encoding in the auditory cortex leads only for the phase-locked response to functionally differentiated maps.

Taken together, an important result of the squirrel monkey study is the evidence that different auditory areas of the cortex show significantly different patterns of response to changes in intensity of the AM sound. Areas AI, Pi, and T1 are areas with a temporal resolution for AM sound which is in the same range as the envelope fluctuations of most

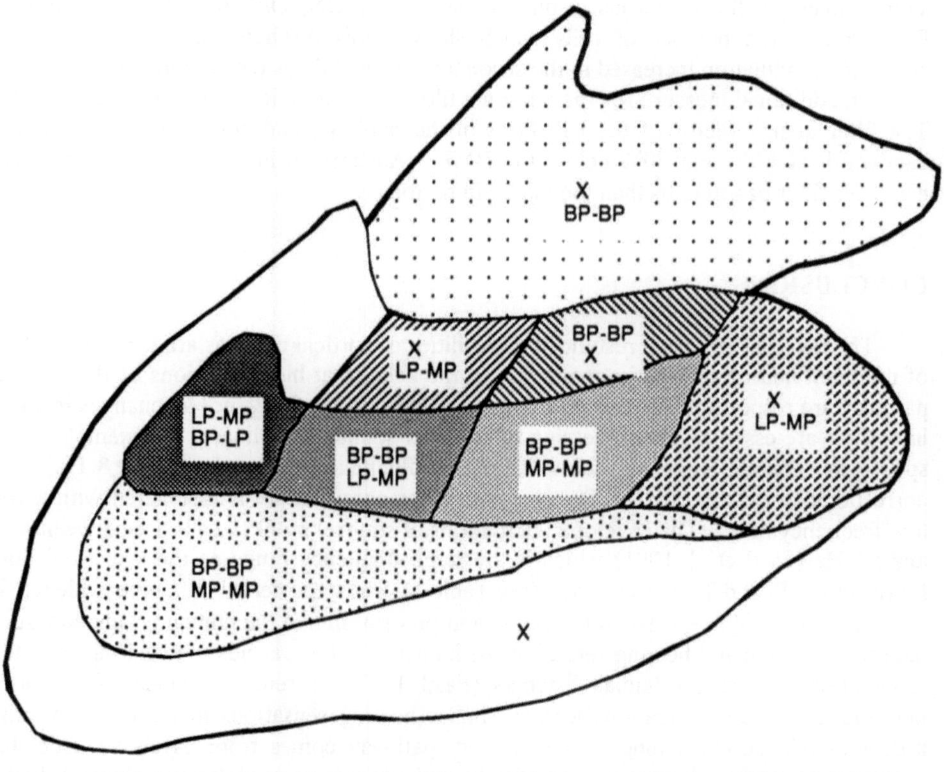

Figure 4. Distribution of the two most frequent filter type combinations in the different auditory fields. BP: band pass; LP: low pass; MP: multipeaked filter; x: simple on, off or on/off responses.

squirrel monkey vocalizations. In contrast, the time resolution of the other fields (AL, RPi, R and Pa) is insufficient to manage such information. Thus, in addition to signal representation in the frequency domain (tonotopicity), the auditory cortex of the squirrel monkey represents information in the time domain.

Additionally, the filter characteristics in the areas with high temporal resolution show the BP filter type to be dominating. It is noteworthy that the psychoacoustic investigations performed with humans showed also a BP-characteristic for the hearing sensation "fluctuation strength" (Fastl,1977), indicating strong similarities between the hearing sensations in humans and in monkeys.

TEMPORAL RESOLUTION OF AUDITORY CORTEX COMPARED TO OTHER STATIONS OF THE AUDITORY PATHWAY

As in other mammals, the primate data show a degradation of temporal resolution in the auditory cortex compared to the more peripheral stations of the auditory pathway (Müller-Preuss et al.,1988). Mean BMFs in the midbrain were between 32 and 64 Hz (Müller-Preuss et al.,1988) and in the medial geniculate body 16-32 Hz (Preuss and Müller-Preuss, 1990). In addition, the filter characteristics of neurons for AM sounds tended to be more complex in the cortex than in the midbrain. The modulation transfer functions for synchronism represented complex BP-filters (MP-filters) and LP filters more often in the cortex than in the midbrain. In the midbrain, all auditory neurons responded to AM sound. In contrast, not all auditory neurons in the cortex were able to follow the AM sound envelope. Such neurons displayed only simple ON, OFF or ON/OFF responses. Furthermore, the number of cells which showed this discharge characteristic to the amplitude modulation increased in the secondary cortical fields (see Figure 4).

An additional measure for the neuron´s filter characteristic is the filter quality factor. The filter quality factor of the BP-filters in the midbrain and cortex was calculated as bandwidth at 40% and 70% below the BMF. All cortical areas had, on the average, narrower filter bandwidths than the midbrain regions.

CONCLUSIONS

The fact that temporal resolution capabilities of cortical neurons are lower than those of midbrain neurons seems in opposition to the belief that higher stations of the auditory pathway are responsible for complex sound coding. Alternatively, not all changes in sound intensities are essential for processing of complex sounds. In humans, for example, fluent speech in different languages shows amplitude fluctuations between 2 and 8 Hz which correspond to syllable rhythm. So the relevant temporal feature is associated with a very low frequency range. The envelope fluctuations of squirrel monkey calls are between 4 Hz and 64 Hz (Fastl et al.,1991) which is the same frequency range as the synchronization BMFs in AI, Pi and T1 of the cortex (see Table 1). Thus, the neural data give evidence for a similar correlation between vocalization and perception like the human psychoacoustical data for language and hearing sensation. At higher AM-frequencies, the hearing sensation "roughness" occurs for human listeners (Fastl 1977). It remains unclear if the faster periodic amplitude fluctuations lead to similar hearing sensations in monkeys. Another aspect of AM sound coding in the auditory pathway comes from psychoacoustic data evaluated in human listeners. Amplitude and frequency modulations show a higher detectability when combined to a mixed modulation (Moore and Sek, 1992). Mixed modulations are sounds with simultaneous amplitude and frequency modulation. It seems that amplitude modulation is used by the auditory system for spectral integration and for grouping those components in a complex sound pattern which share a common source. This procedure is necessary for the separation of information-bearing sounds against the acoustically noisy environment (Bregman et al, 1985). The pronounced amplitude envelopes found in squirrel monkey calls (Schott, 1975; Fastl et al., 1991) may contribute to this hearing mechanism. Therefore, the coincidence of natural call envelope frequencies with the neural discharge properties for this AM-frequency range may be interpreted as an audio-vocal mechanism to compete successfully against the constant insect noise in the natural environment of squirrel monkeys.

FUTURE WORK AND SPECULATIONS

The finding that most of the investigated neurons were able to code the AM sound envelope in a phase-locked manner and/or with a change in the spike rate confirms the importance of this sound feature in acoustic communication. The complex neural response patterns in the cortex in comparison to the simple BP-filtering in the midbrain focusses on the auditory cortex as an area where these features are extracted and where complex species-specific vocalizations are processed. There is also evidence from behavioral lesioning studies, that the auditory cortex of the monkey plays a role in sound detection and location and is essential for perception of species-specific vocalizations (Heffner and Heffner,1990). No data exist about the role of distinct auditory cortex areas in the perception of species-specific calls. The differences in temporal resolution of the cortical areas investigated suggest a specialization in such areas regarding salient aspects of a sound. In order to explain how and where information processing of complex acoustic patterns takes place, one might propose a multiple filter mechanism which extracts meaningful aspects. For the envelope, for example, the midbrain limits the coding of AM sounds up to nearly 300 Hz AM-frequency. The limitation is realized by broad BP-filtering. The cortical areas show complex MP-filter types, LP- filters and BP-filters with a higher quality factor. These filter types are separated in different cortical areas leading to anatomical locations of sound features. The interpretation of data only in terms of filter characteristics, however, is not completely satisfying. Those neurons in the cortex which respond to envelope fluctuations often display complex activity patterns tuned to a BMF or SR-rate. Considering all neurons together, no response type corresponds to a specific modulation frequency or modulation depth. This suggests that, on the cortical level, the envelope fluctuations are treated in a more complex transformation mechanism than at the midbrain level. Another argument for a complex transformation mechanism in the cortex is that the modulation transfer functions for synchronization and rate often fall apart at the cortical level. A single cortical neuron can display different filter types, depending on the AM-frequency, for both the phase-locked response and the spike rate. These results are in keeping with the hypothesis that a stimulus is extracted by the collective properties of the firing of many neurons (Eggermont,1990). Depending on the stimulus, a neuron can respond as a part of different cell assemblies in a phase-locked manner or with non-phase-locked spike frequencies.

For the electrophysiologist, single unit studies do not lose their importance. Our results together with recently published data on the physiology of auditory cortex makes it very likely that acoustic information is coded in a spatiotemporal pattern of neural activity and not in a highly specialized neuron. One can assume, therefore, that information processing in the cortex is guided by topographic maps in the frequency and the time domain leading to a complex computational map. This computational map should show high plasticity depending on the stimulus and the context. The single neuron should participate in different computional maps. Such mechanisms would allow the neural circuits to deal with new acoustic information. To prove this in practice, the design of the experimental approach has to be modified. More emphasis has to be placed on the active participation of the animal. A change in the frequency-time pattern of a sound which is meaningful to the animal should change the activity pattern of neurons in a different way than so-called neutral sounds with the same physical parameters. Up to now, the embedding of single unit studies in behavioral tasks has only been tested by a few groups (Miller et al., 1972; Pfingst et al., 1977; Pfingst and O'Connor, 1981;Weinberger and Diamond, 1987; Weinberger et al., 1990; Edeline and Weinberger, 1993) with simple sound stimuli. Because of the high nonlinearity of the auditory system, satisfactory results probably will be more readily obtained with complex sound patterns. Such data, together with knowledge of the connections of the area investigated, should provide the requisite material for modelling the neural nets presumably involved in pattern discrimination and recognition.

Abbreviations

AI= Primary auditory cortex
AAF= anterior auditory field
DP = dorsoposterior auditory field
Pa = postauditory field
Pi = parainsular auditory field
R = rostral auditory field
RT = rostrotemporal field
T1 = first temporal field
V = ventral auditory field
AII = second auditory field
AL = anterolateral field
In = Insula
PAF = posterior auditory field
PL = posterolateral field
RPi = rostral parainsular auditory field
T = temporal auditory field
T2 = second temporal field
VPAF = ventroposterior auditory field

REFERENCES

Bieser, A. and Müller-Preuss, P., 1989. Temporal processing through auditory neurones:Periodicity decoding and spike rate, *Euro. J. Neurosci. Suppl.* 2: 260.

Bieser, A. and Müller-Preuss. P, 1992, Organization of the auditory cortex in the squirrel monkey, *in:* "Rhythmogenesis in Neurons and Networks", N. Elsner and D.W. Richter, eds., Georg Thieme Verlag, Stuttgart, New York, pp. 228-220.

Bregman, A.S., Abramson, J., Doehring, P., and Darwin, C.J., 1985, Spectral integration based on common amplitude modulation, *Percept. Psychophys.* 37: 483-493.

Cheney, D.L. and Seyfarth, R.M., 1990, "How Monkeys See the World: Inside the Mind of Another Species", University of Chicago Press, Chicago.

Clarey, C.C., Barone, P., and Imig, T., 1992, Physiology of thalamus and cortex, *in:* "The Mammalian Auditory Pathway: Neurophysiology", A.N. Popper and R.R. Fay, eds., Springer-Verlag, Berlin, pp. 233-334.

Edeline, J.M. and Weinberger, N.M., 1993, Receptive field plasticity in the auditory cortex during frequency discrimination training - selective retuning independent of task difficulty, *Behav. Neurosci.* 107: 82-103.

Eggermont, J. J., 1990. "The Correlative Brain - Theory and Experiment in Neural Interaction", Springer Verlag, Berlin, Heidelberg.

Evans, E.F., 1992, Auditory processing of complex sounds - an overview, *Phil. Trans. Royal Soc. Lond. Series B, Biol.* 336: 295-300.

Fastl, H., 1977, Roughness and temporal masking patterns of sinusoidally amplitude modulated broadband noise, *in:* "Psychophysics and Physiology of Hearing", E.F., Evans and J.P.Wilson, eds., Academic Press, London, pp. 403-414.

Fastl, H., Widmann, U., and Müller-Preuss, P., 1991, Correlations between hearing and vocal activity in man and the squirrel monkey, *Acustica* 73: 35-36.

Heffner, H.E. and Heffner, R.S., 1990, Role of primate auditory cortex in hearing, *in*: "Comparative Perception-Vol.II: Complex Signals", W.C. Stebbins and M.A. Beckley, eds., Wiley, New York, pp. 279-310.

Imig, T.J. and Reale, R.A., 1980. Patterns of cortico-cortical connections related to tonotopic maps in cat auditory cortex, *J. Comp. Neurol.* 192: 293-332.

Imig, T.J., Ruggero, M.A., Kitzes, L.M., Javel, E., and Brugge, J.F., 1977, Organization of auditory cortex in the owl monkey (*Aotus trivirgatus*), *J. Comp. Neurol.* 171: 111-128.

Javel, E. and Mott, J.B., 1988, Physiological and psychophysical correlates of temporal processes in hearing, *Hear. Res.* 34: 275-294.

Langner, G., 1992, Periodicity coding in the auditory system, *Hear. Res.* 60: 115-142.

Luethke, L.E., Krubitzer, L.A., and Kaas, J.H., 1988, Cortical connections of electrophysiological and architectonically defined subdivisions of auditory cortex in squirrels, *J. Comp. Neurol.* 268: 181.

Maurus, M., Streit, K.-M., Geissler, B., Barclay, D., Wiesner, E., and Kuehlmorgen, B., 1984, Categorical differentiation in amplitude changes of squirrel monkey calls, *Lang. Commun.* 4: 195-208.
Miller, J.M., Sutton, D., Pfingst, B., and Beaton, R., 1972, Single cell activity in the auditory cortex of rhesus monkeys: Behavioral dependency, *Science* 177: 449-451.
Moore, B.C.J. and Sek, A., 1992, Detection of combined frequency and amplitude modulation, *J. Acoust. Soc. Amer.* 92: 3119-3131.
Morel, A. and Kaas J.H., 1992, Subdivisions and connections of auditory cortex in owl monkeys, *J. Comp. Neurol.* 318: 27-63.
Morel, A., Garraghty, P. E., and Kaas, J.H.,1993, Tonotopic organization, architectonic fields, and connections of auditory cortex in macaque monkeys, *J. Comp. Neurol.* 335: 437-459.
Müller-Preuss, P., Bieser, A., Preuss, A., and Fastl, H., 1988, Neural processing of AM-sounds within central parts of the auditory pathway, *in*: "Auditory Pathway: Structure and Function", J. Syka and B. Masterton, eds., Plenum, New York, pp. 327-33.
Newman, J.D., 1988, Primate hearing mechanisms, *in*: "Comparative Primate Biology, Vol.4: Neurosciences", H.D. Steklis and J. Erwin, eds., Alan R. Liss, New York, pp. 469-499.
Pfingst, B.E. and O'Connor, T.A., 1981, Characteristics of neurons in auditory cortex of monkeys performing a simple auditory task, *J. Neurophysiol.* 45: 16-34.
Pfingst, B.E., O'Connor, T.A., and Miller, J.M., 1977, Response plasticity of neurons in auditory cortex of the rhesus monkey, *Exp. Brain. Res.* 29: 393-404.
Phillips, D.P. and Irvine, D.R.F., 1981, Responses of single neurons in physiologically defined primary auditory cortex (AI) of the cat: frequency tuning and responses to intensity, *J. Neurophysiol.* 45:48-58.
Phillips, D.P. and Irvine, D.R.F., 1982, Properties of single neurons in the anterior auditory field (AAF) of cat cerebral cortex, *Brain Res.* 248:237-244.
Preuss, A. and Müller-Preuss, P., 1990, Processing of amplitude-modulated sounds in the medial geniculate body of squirrel monkeys, *Exp. Brain Res.* 79: 207-211.
Pickles, J.O., 1982, "An Introduction to the Physiology of Hearing", Academic Press, London.
Schott, D., 1975, Quantitative analysis of the vocal repertoire of squirrel monkeys (*Saimiri sciureus*), *Zeitschr. Tierpsychol.* 38: 225-250.
Schreiner, C.E. and Urbas, J.V., 1986, Representation of amplitude modulation in the auditory cortex of the cat. I. The anterior auditory field (AAF), *Hear. Res.* 21: 227-241.
Schreiner, C.E. and Urbas, J.V., 1988, Representation of amplitude modulation in the auditory cortex of the cat. II. Comparison between cortical fields, *Hear. Res.* 32: 49-64.
Sudakov,K., MacLean, P.D., Reeves, A., and Marino,R., 1971, Unit study of exteroceptive inputs to claustrocortex in awake, sitting squirrel monkey, *Brain Res.* 28:19-34.
Suga, N., 1990, Cortical computational maps for auditory imaging, *Neural Networks* 3: 3-21.
Symmes, D., 1981, On the use of natural stimuli in neurophysiological studies of audition, *Hear. Res.* 4: 203-214.
Symmes, D., Newman, J.D., Talmage-Riggs, G., and Katz-Lieblich, A., 1979, Individuality and stability of isolation peeps in squirrel monkeys, *Anim. Behav.* 27: 1142-1152.
Terhardt, E., 1974, On the perception of periodic sound fluctuations (roughness), *Acustica* 30: 201-213.
Pfingst, B.E. and O'Connor, T.A., 1981, Characteristics of neurons in auditory cortex of monkeys performing a simple auditory task, *J. Neurophysiol.* 45: 16-34.
Weinberger, N.M. and Diamond, D.M., 1987, Physiological plasticity in auditory cortex: Rapid induction by learning, *Progr. Neurobiol.* 29: 1-55.
Weinberger, N.M., Ashe, J.H., Metherate, R., McKenna, T.M., Diamond, D.M., and Bakin, J., 1990, Retuning auditory cortex by learning:A preliminary model of receptive field plasticity, *Concepts Neurosci.* 1: 91-132.
Zwicker, E. and Fastl, H., 1990, "Psychoacoustics-Facts and Models", Springer-Verlag, Berlin, Heidelberg.

REPRESENTATIONAL SIGNALING IN NON-HUMAN PRIMATE VOCAL COMMUNICATION

Harold Gouzoules[1,2], Sarah Gouzoules[2] and Jennifer Ashley[1,2]

[1]Department of Psychology
[2]Yerkes Regional Primate Research Center
Emory University
Atlanta, GA 30322, USA

INTRODUCTION

Knowledge about non-human primate vocal communication has grown substantially in recent years and has resulted in a radically different view of the role vocalizations play in animals' lives. Some 25 years ago, anthropologist Jane Lancaster summarized the then prevalent view by noting that, " .... field and laboratory workers have emphasized that vocalizations do not carry the major burden of meaning in most social interactions, but function instead either to call visual attention to the signaller or to emphasize or enhance the effect of visual or tactile signals" (Lancaster, 1968). Downplaying the role of vocal communication even further, Lancaster went on to speculate that "a blind monkey would be greatly handicapped in his social interactions whereas a deaf one would probably be able to function almost normally" (Lancaster, 1968, p.442). In general, the communication systems of monkeys and apes were thought to express the emotional or motivational states of the animals (reviewed in Marler, 1985, 1992; Cheney and Seyfarth, 1990a).

Though Struhsaker (1967) had described acoustically distinct alarm calls in vervet monkeys (*Cercopithecus aethiops*), given in response to different kinds of predators, that elicited adaptively different responses, relatively little significance was attributed to the finding, and the ability to convey information about the external world was thought to be rare and unusual (Lancaster, 1975, p.65). It was not until after Seyfarth, Cheney, and Marler (1980) refocused attention on the relevance of vervet monkey alarm calls that studies of additional call classes in vervets and other primate species explored the possibility of a more widespread occurrence of such signals.

In this chapter, we review representative studies that have reported external reference in the natural vocal communication of non-human primates. We focus on the criteria used for establishing that vocalizations index objects or events in the external world, the available evidence for representational signaling, the conditions under which such signaling is likely to evolve, and the logical framework from which the problem might best be understood.

Current Topics in Primate Vocal Communication
Edited by E. Zimmermann *et al.*, Plenum Press, New York, 1995

REPRESENTATIONAL SIGNALING : DESCRIBING THE PROBLEM

The term "representation" has been used in a variety of ways in psychology and philosophy (e.g., Lloyd, 1989; Newell, 1990). In one sense it may refer to specific knowledge or information an individual possesses or acquires and the manner in which that information is stored, organized or coded in the brain (Mandler, 1983). A second usage relates to symbolism: representation, here, refers to the use of something to stand for or to objects or events in the world (i.e., the "referents"). This capacity has been called "representational" signaling (Gouzoules et al., 1984), "symbolic" communication (Scherer, 1988), "semantic" communication (Seyfarth et al., 1980), or "referential" signaling (Macedonia and Evans, 1993) in the non-human primate literature. As is true of "representation," these other terms have been used in different ways by psychologists, philosophers and linguists, and precise definitions remain elusive or controversial. It is generally agreed that semantics, for example, deals with meaning in language. However, after years of debate, from Plato to the present, there is still disagreement over the meaning of meaning (e.g., Ogden and Richards, 1923; Miller and Johnson-Laird, 1976; Lakoff, 1987; Tiles, 1987). It is the referential sense of the term "representation" that has received most attention in studies of non-human primate communication; however, as has been the case with studies of children (Mandler, 1983), referential representation has been used by some to assess the existence of knowledge as well (e.g., Cheney and Seyfarth, 1990a). Thus, of the various expressions that have been used in studies of communication to discuss relationships between signals, on one hand, and knowledge, behavior, states, objects and events, on the other, we prefer "representation" over others because it covers a wider range of phenomena, from the coding of stimuli in an animal's nervous system to the link between a word and its referents. As Pearce (1987, p. 33) suggests, internal representations can be considered to be the fundamental units of animal cognition.

One relevant level of representation, then, involves the naming of referents; a label, such as a word, stands for the referent and can substitute for it in its absence. Vocalizations that index referents external to the signaller may qualify as representational in this manner. Non-human primate vocalizations were not thought to operate at this level, however. As noted above, the referents of primate vocal signals, until recently, were believed to be concerned with either the internal state or the behavior of the signaller. Would such signals qualify as representational? Conditions can be specified under which they would not, at least in the symbolic, naming sense. It is possible to imagine very simple representational systems: an adaptive response that followed a call could well be a function of evolved properties of the nervous system that link signal detection to motor response without involvement of any representation of information about either the animal that produced the vocalization or any entity or event that caused it. There must, however, even here, be occurrent representation (in the sense of coding) of the signal in the receiver's nervous system. In this case, response is mediated by a representation of the signal and not by one of the individuals that produced the vocalization or the object or event that prompted the signaller to do so. Moreover, a signal that is a manifestation of (in the sense of being derived from) an emotional state, or conveys information about the intensity of that emotional state, or about the likelihood of an impending behavior, does not always provide the receiver with sufficient information to make a choice between incompatible responses (Macedonia and Evans, 1993). Such signals may, however, provide information not only about the signaller but, through inference or association and contextual cues, about entities external to the signaller (e.g., Premack, 1975). For example, a call associated with a euphoric, excited, or pleasurable state might be given by a monkey when it encounters food. If other group members are hungry or foraging when they hear such a call, they may respond to it as if it indexed food directly, as opposed to indirectly through other representations (knowledge or information). Such signals would appear to have external referents. Scherer (1988) goes further to suggest that if listeners are able to identify a particular emotional state of the sender on the basis of acoustical features of the vocalization, and therefore infer the nature of the emotion-producing event, then a symbolic function for emotional vocalizations might be claimed. Clearly, we require evidence that enables us to distinguish among the different ways in which responses can be elicited from vocalizations. However, it should be noted that all biological representational systems, from "simple" detection by the nervous system of environmental features to human language, share certain features and should not be thought of as having sharp boundaries that separate one type from another.

Operationally, the first step in demonstrating representation in the symbolic sense occurs when a vocalization is shown to be capable of eliciting an appropriate response in a receiver deprived of any additional information about the referent. Typically, researchers decide what constitutes an appropriate response on the basis of adaptive or functional analyses or interpretations of the situation. The alarm calls of vervet monkeys are excellent examples of calls that have demonstrably adaptive and functional responses associated with them. The nature of the representation, in the first sense mentioned above, is not easily established through such an approach, however. The exact knowledge or information that either the signaller possesses or the receiver acquires is difficult to assess because this involves a thorough understanding of how the species classifies its world (see Cheney and Seyfarth, 1990a). While it was relatively easy for Seyfarth et al. (1980) to establish through playback experiments with alarm calls that vervet monkeys behaviorally classify leopards and martial eagles differently, it is not known what features distinguish for the monkey a leopard from an eagle, or a leopard from a lion or a cheetah. Thus, Cheney and Seyfarth (1990a, p.171) argue that "the external referents of primate vocalizations are best defined in relation to the external referents of other calls within the animals' repertoire." Nonetheless, the exact nature of the external referents for vervet alarm calls (and, as will be noted, other non-human primate calls) remains elusive, an important issue to which we shall return.

STUDIES OF REPRESENTATIONAL SIGNALING

Vervet Monkey Alarm Calls

Evidence concerning the representational nature of non-human primate vocalizations is most abundant and convincing for the alarm calls of vervet monkeys (summarized in Cheney and Seyfarth, 1990a). Initially, Seyfarth et al. (1980) determined, through field experiments in which tape recordings of the three types of alarm calls were broadcast from hidden speakers, that these vocalizations were sufficient to elicit the adaptive, predator-specific, escape responses. This sort of evidence demonstrates what Marler et al. (1992) and Macedonia and Evans (1993) call "functional" reference and what Cheney and Seyfarth (1992) refer to as semanticity in the "weakest sense" because nothing is revealed about the nature of the representation that the vervets are using. It is possible that upon hearing a leopard alarm a vervet conjures up a mental image of a large spotted quadruped armed with sharp teeth and claws, and runs for the nearest tree. On the other hand, these data alone would also be consistent with a representation of a far simpler kind (Lloyd, 1989; Bickerton, 1990): the representation of the alarm call in the vervet monkeys auditory system might, via connections in the brain, merely trigger motor neurons that control escape behavior.

In additional experiments designed to test whether vervets are able to assess calls according to the things they denote, Cheney and Seyfarth (1988) obtained results that support a notion of semanticity in what they argue is a stronger sense. Two sets of acoustically different vocalizations were used in a habituation/dishabituation procedure.

One pair of calls used included the leopard and eagle alarms, while the other pair was composed of two calls that are given in response to an encounter with another group of vervets. "Wrrs" and "chutters" are acoustically different calls, the former typically given when another group has been spotted, and the latter used during aggressive contact with a different group. Baseline responses of ten subjects to recordings of both the chutters and eagle alarms of one of the group's females were measured. The subjects were then presented with two separate series of eight repetitions of other recorded calls of that same female. In one series the female's wrrs were played and in the other, her leopard alarm. Presentations were separated by 20 minute intervals. Then, 20 minutes following the eighth presentation, the other member of the pair of calls was tested: the repeated wrrs were followed by a chutter, while after the leopard sequence, an eagle alarm was broadcast. Subjects rapidly habituated to both of the repeated calls. Habituation was transferred in the case of the subsequently played chutter, but not to the eagle alarm that followed the leopard alarms. These results are consistent with the view that the monkeys were assessing calls on the basis of their referents rather than their acoustical structure. Cheney and Seyfarth (1990a, pp.154-155) suggest that transfer of habituation occurred in the case of the wrr-chutter pair because these calls had "roughly the same referent," while eagle and leopard alarms denoted different things. Interestingly, when the chutters of one animal were tested in a similar fashion with a sequence of wrrs from a different monkey, habituation was not transferred to the chutter. Thus, call meaning and the identity of the signaller were important determinants of the subjects' responses (see also Owren, 1990). Cheney and Seyfarth (1990a, p.155) argue that the results of these experiments cannot be due to the possibility that alarm calls are simply too costly to ignore because habituation to alarms was no different from that to the intergroup calls. Another alternative explanation, that the transfer of habituation with the intergroup calls occurred because they were more likely than leopard and eagle alarms to occur together (i.e., sensory preconditioning), was rejected because during the period during which the experiments were conducted wrrs and chutters did not, in fact, occur together often. One modification that might have yielded a different set of results in these experiments would be to focus on somewhat stricter response criteria for the alarm call pairings. For wrrs and chutters, looking in the direction of the hidden speaker was the measure used, while for the alarm calls, looking toward the speaker was combined with running to the trees for leopard alarms and looking up (or running into bushes) for eagle alarms. Had only the clearly adaptive escape strategies been used a different picture of the habituation process might have emerged. Looking toward the monkey producing an alarm call and observing the direction of its gaze may provide information about predator whereabouts and, thus, the safest escape route, but it seems reasonable to interpret such behavior when it occurs alone as indicating a lower level of response.

Cheney and Seyfarth's habituation/dishabituation experiments provide us with a glimpse of vervet categorical hierarchies. At the level of basic categories (Rosch et al., 1976), vervet monkeys presumably use wrrs and chutters to designate different external referents. Cheney (1987) reported that intergroup encounters occurred roughly every other day and that high- ranking females were both more active and aggressive participants in intergroup encounters and were more likely to produce wrrs. Vervets were more responsive to wrrs from high- ranking members of neighboring groups than to those of low-ranking members (Cheney and Seyfarth, 1990a, p.66). Playback of wrrs from neighboring groups elicited little reaction when they were broadcast from the appropriate territory, but when the calls were heard coming from an inappropriate direction, they were attended to much more strongly and vigorously.

The wrr is, with very little doubt, associated with the sighting of another group of vervets. Clearly, not only is other information made available through the call, but other representations (knowledge) the monkeys have about their external world, such as the usual location of other groups, is linked to these calls. The vervets give chutters to snakes, members of their own group, to other groups, and to familiar human observers. Though the vervets' responses to the chutters given in these different contexts have not been tested with playback experiments, it appears that they are used to denote aspects of the monkeys' external environments (Cheney and Seyfarth, 1990a, p.120). The chutter used during intergroup encounters is associated with aggressive interactions. Most intergroup chutters (95%) were given during these aggressive episodes while wrrs were used much less frequently (40%) (Cheney and Seyfarth, 1990a, p.152).

Wrrs and chutters seem to be members of a superordinate categorical class that links events concerning other vervet groups, though the precise nature of the similarities is not clear. That habituation does not transfer from an individual monkey's leopard alarm to her eagle alarm suggests that these two basic level categories are not members of a superordinate class even though leopard alarms are occasionally given to stooping eagles about to grasp a victim in its talons (Cheney and Seyfarth, 1990a, p.107). Leopard alarms are sometimes also given to attacking snakes (M. Hauser, cited in Macedonia and Evans, 1993). Macedonia and Evans (1993, p. 183) interpret the relative lack of production specificity for leopard alarms as an indication that the calls are "less specific and hence might be placed further from the referential pole of a motivational-to-referential continuum' (Marler et al., 1992) than eagle and snake alarms." Another possible explanation for the more varied use of leopard calls might be that the call's referent might still be primarily external but what is denoted is something other than the spotted cat that is the most frequent elicitor of the vocalization. In another series of experiments, Seyfarth & Cheney (1990b) conducted habituation/dishabituation tests using vervet eagle and leopard alarms paired with the terrestrial predator and raptor alarms of the sympatric superb starling (*Spero superbus*) to test whether the monkeys assessed starling alarm calls on the basis of their meaning. If this were the case, Seyfarth and Cheney hypothesized that vervet habituation to the raptor (or terrestrial predator) alarm call of one species should transfer to the corresponding call of the other. Thus, habituation of vervets to the starling's raptor alarm should extend to their own eagle alarm; habituation to the eagle alarm should transfer to the starling's raptor call. The same predictions would hold for the vervet leopard alarm and the starling's terrestrial predator call. The starling's alarms, especially its terrestrial predator call, are given to a much wider range of species than are vervet alarm calls, though among the species are those of concern to the vervets. Thus, Seyfarth and Cheney suspected that transfer of habituation might occur more readily with the raptor/eagle pairing.

Results of the starling/vervet alarm call playback experiments suggested that the monkeys respond to starling raptor calls as relatively precise signals that are similar in meaning to their eagle alarms. All eight subjects transferred habituation from one species' alarm call to the other's when the calls broadcast had the same referent (avian predator) but did not extend habituation when the comparison involved vervet leopard alarms paired with starling raptor calls. Transfer of habituation also occurred for tests pairing starling terrestrial predator calls and vervet leopard alarms, though the response disparity between control and test calls was not as great as it was for the avian predator pairing. Seyfarth and Cheney (1990) interpret this difference as an indication that, while vervets treat starling terrestrial predator calls and their own leopard alarms as similar, the calls are judged to be less synonymous than starling raptor calls and their own eagle alarms. Again, it is possible that results with the two sets of calls might have been more similar if response measures were limited to the respective adaptive escape strategies for the calls rather than including looking toward the speaker.

Finally, one other interesting outcome of these experiments deserves attention because it reveals something about how the information attended to in calls may vary in different situations. Recall that for the experiment involving habituation to wrrs, transfer of habituation extended to chutters when the identity of the caller was the same for both vocalizations, but did not occur when one monkey's wrr was tested with another individual's chutter. Cheney and Seyfarth (1990a, p. 201) note that subjects "did not transfer their skepticism to different individual's calls, suggesting that doubts about one individual's reliability did not generalize to include other group members." In the case of the starling/vervet alarm call tests, however, habituation was transferred, even though members of two different species were involved. The monkeys appear not to be attending to information about the identity of the caller when generalizing across species. Why the monkeys should show skepticism to the alarm call of a group member who has not shown herself to be unreliable following their exposure to false starling alarms is not clear.

Alarm Calls in Other Species

Although alarm calls have been studied in other non-human primates, relative to vervet alarms, much less is known about them. While alarm calls have been described for most species, for only a few has it been determined that acoustically distinct calls are used for different kinds of predators (e.g., Kloss's gibbon, *Hylobates klossii* (Tenaza and Tilson, 1977); ringtailed lemurs, *Lemur catta*, and ruffed lemurs, *Varecia variegata* (Macedonia, 1990); Goeldi's monkeys, *Callimico goeldi* (Masataka, 1983). Macedonia and Evans (1993) have reviewed the issue of the meaning of mammalian alarm calls, focusing on a comparison of vervet, lemur, and ground-dwelling sciurid vocalizations.

Alarm call usage in ringtailed lemurs and ruffed lemurs has been well documented in comparative studies (Macedonia, 1990; Macedonia and Yount, 1991; Pereira and Macedonia, 1991; Macedonia, this volume). "Gulps," low amplitude calls produced during a variety of startling circumstances, may be given when ringtailed lemurs initially spot an aerial predator or a terrestrial carnivore. They also may produce a short noisy "rasp" when a raptor has been sighted and loud "shrieks," when the bird is within attack range. The acoustic structure of the shrieks varies with the severity of the incident. Aerial alarms are not elicited by perched raptors: unlike vervet monkey eagle calls, the lemur call is used only for flying raptors. For terrestrial carnivores, ringtail lemurs follow gulps with calls that appear to orient animals to the predator's location ("clicks" and "click series") and others that elicit mobbing of the predator ("yaps"). Macedonia and Evans (1993) consider yaps to be "ground alarm calls." The ground and raptor alarm calls of ringtailed lemurs are given almost exclusively to terrestrial and aerial predators respectively; response urgency (i.e., variation in the imminency of the danger) did not influence the use of these calls. Playback experiments with ringtailed lemur alarms elicited responses similar to those shown when actual predators were encountered. For raptor alarms this meant downward movement when the animals were in trees and looking skyward if they were on the ground. Terrestrial alarms caused the animals to climb into or remain in trees. This response is noteworthy given the authors' description of the use of yaps as "carnivore- mobbing calls" (p.189).

Ruffed lemurs produce "abrupt roars" when a raptor is sighted and "growl-snorts" and, or, "growls" (disturbance vocalizations) followed by "pulsed squawks" (which Macedonia and Evans refer to as "mobbing calls" or "ground alarm calls") when terrestrial predators are encountered, however, occasionally pulsed squawks were given to aerial predators. Terrestrial predator alarms appear to grade structurally with another vocalization, the "wail," that Macedonia and Evans (1993, p. 191) describe as serving an "all clear" function after major disturbances. Macedonia and Evans (1993, p.191) suggest that ruffed lemur terrestrial predator calls may serve primarily as a rallying cry and that the animals' ability to identify the nature of the danger is derived from the probabilistic association of the call with carnivores. Playback experiments indicated that neither abrupt roars nor pulsed squawks were sufficient to elicit appropriate escape responses for the predators with which they are typically associated.

Macedonia and Evans (1993) suggest that these results argue for functional reference in the case of ringtailed lemur alarm calls, but not for those of the ruffed lemur; the two species differ in their alarm call behavior with respect to the two criteria specified for functional reference, production specificity and generation of adaptively different response in playback experiments. They reason that the incompatibility of responses required for different predator classes may account for the evolution of predator-class-specific alarm calls. Relatively terrestrial species like vervet monkeys and ringtailed lemurs (as opposed to the mostly arboreal ruffed lemur) may require escape tactics that are in some ways conflicting in order to deal with different classes of predators. Additionally, it may be that the role mobbing appears to play in the defense tactics of lemurs also distinguishes them from the vervets. The manner in which the alarm calls of the lemurs elicit mobbing may be critical to understanding the information made available by the calls.

Food Call

Vocalizations associated with food have been examined in several species for evidence of referential signaling. Dittus (1984) described toque macaque (*Macaca sinica*) vocalizations that were most often given when a monkey discovered abundant and high-quality food. Individuals hearing the calls ran to the vicinity of the discoverer and began to eat. Dittus suggested that these vocalizations indexed the abundance and quality of the food rather than its specific nature. Such calls might have been selected for because they conveyed pertinent information about feeding conditions to related group members. Dittus rejected an arousal interpretation of these calls, arguing that the monkeys did not call early in the morning, when hunger was likely to be greatest, unless a rich food source was encountered.

The account Dittus provides of toque macaque responses indicates that the monkeys clearly associate the call with the location of food. Importantly, these calls are not used exclusively in the context of finding food. For example, the monkeys sometimes produce them when the sun breaks through the clouds on a cool, overcast day. Furthermore, provisioned and captive macaques predictably produce food calls in chorus as the time they are routinely fed approaches, i.e., they call in anticipation of food and when the entire group is assembled and present. Macaques also call in response to the vocalizations of others (Green, 1975; Dittus, 1988; Masataka, 1988) as well as in response to the arrival of other group members, especially the dominant males, at the food site (Dittus, 1988). Dittus' account of food calls in wild toque macaques might not fail the production specificity criterion for external reference used by Marler et al. (1992) and Macedonia and Evans (1993) because the majority of calls were given in the context of finding food. However, on the basis of observations indicating that these calls are given in a variety of situations that involve known reinforcers for monkeys (warmth, companionship, food), it seems likely that the calls given by toque macaques finding an abundant, high-quality, food source are a manifestation of pleasure, euphoria, or excitement. An alternative possibility, that there is an, as yet unidentified, external referent common to these various situations, seems unlikely. Nonetheless, for wild toque macaques, their occurrence in the context of foraging renders the call sufficient to provide group members information about food.

Hauser and Marler (1993a,b) have conducted extensive research on the food associated calls of rhesus monkeys (*Macaca mulatta*) living on the island of Cayo Santiago, Puerto Rico. Five acoustically distinct calls were noted to occur in food related contexts, the most common of which was a "coo" vocalization much like that given by Japanese macaques (*Macaca fuscata*) in similar situations (Green, 1975). Coos, characterized by a series of harmonic bands with most of the energy located in the fundamental frequency, were also produced during group progressions, mother-infant interactions, as well as other settings, and these vocalizations were not acoustically different from those associated with food (Hauser, 1991).

Another rhesus monkey call associated with food, a grunt, was acoustically very different from the coos. Grunts are pulsatile, atonal vocalizations given in a variety of nonaggressive situations in addition to those involving food, such as when a monkey approaches another (the calls are often given to human observers as they approach or pass by a monkey) or when a female approaches another animal, especially one of high rank, with a young infant. In the latter case, the call is usually delivered in a rapid series accompanied by a vigorous swishing of the tail. Hauser and Marler (1993a) statistically compared 78 food associated grunts with 157 recorded during social interactions and found significant differences. Unlike Hauser's (1991) analysis of coos, however, where the various coo classes were assessed individually for structural differences, the analysis performed on grunts collapsed nonfood grunts into a single class, leaving open the possibility that food grunts are not acoustically distinct from all other kinds of grunts.

The other rhesus food calls described by Hauser and Marler (1993a) were the "warble," "harmonic arch," and the "chirp," all of which were given only by animals actually in possession of food (p. 202). These calls were also more often associated with coconut, a highly preferred food, than with the commercial chow that is provided daily (Hauser and Marler, 1993b). (Though common on the island, coconuts are not available to most of the monkeys because they cannot break open the shell; human observers breaking into a coconut attract gatherings of monkeys.) Coos and grunts were associated more with commercial chow. Group members were more likely to approach individuals producing warbles, harmonic arches, or chirps than to those giving coos or grunts.

Hauser and Marler (1993b) found that the rate of producing food-associated calls was highest from hungry monkeys who had discovered a high quality food source and lowest from satiated ones finding chow. Adult females called more than did adult males and dominance rank had no significant effect on either the rate or likelihood of calling. Hauser and Marler (1993b) interpret their results in a novel manner: they suggest that "call production represents an 'announcement' of food possession or ownership" (p. 212). Food calling monkeys received less aggression than did silent animals with food who were detected; for females, callers and undetected silent monkeys obtained more food than did those who were silent and detected. Of note, some 55% of animals discovering high quality food failed to call, and the authors suggest that the costs of detection are insufficient to deter call suppression, which is interpreted as a form of deception.

Food-associated vocalizations have also been studied in laboratory colonies of golden lion tamarins, *Leontopithecus rosalia*, by Benz (1993) and cotton-top tamarins, *Saguinus oedipus*, by Elowson et al. (1991). In cotton-top tamarins, these calls appear to reflect an individual's interest in a food item: preference hierarchies for various foods differed greatly among animals and a positive correlation was found between the number of calls produced and the preference status of the food. Elowson et al. (1991) suggest that the calls likely reflect motivation for food rather than index some quality about the food itself. Benz (1993) found that six golden lion tamarins also had food preferences that were reflected in their vocal behavior. Some foods were preferred by all subjects (e.g., raisins and mealworms), others were generally less preferred (e.g., nuts, seeds), while some kinds (e.g., crickets) were highly preferred by some but were less desirable to others. Vocalization rate was significantly associated with food preference. Benz also used discriminant function analysis to classify calls according to food class and found that while only 32% of calls were correctly classified when five foods were considered separately, the proportion rose to 52% when the foods were grouped into three larger categories (prey, dried fruit, or grape). Benz concluded that calls conveyed information about the nature of the food encountered.

Chimpanzees (*Pan troglodytes*) sometimes vocalize upon finding food, and the calls have been observed to recruit others to the site (Wrangham, 1977). Hauser and Wrangham

(1987) and Hauser et al. (1993) found that captive chimpanzees were more likely to call when they had found relatively large quantities of food. Hauser et al. (1993) indicated that two distinct calls were given in the context of food, "rough grunts" and "pant-hoots." They tested four captive chimpanzees with different quantities of watermelon (5, 10, and 20 piece conditions, plus one condition involving one large piece equivalent in size to the 20 piece; 10 trials per condition) and reported that the discoverer called on 45.5% of the trials. Calling was more likely when the relative divisibility of the resource was high (e.g., 20 piece condition); rough grunts were given during all trials in which vocalizations were produced, while pant hoots were given only for trials involving the 20 piece and large piece conditions. Rough grunts appear to be food specific (Marler and Tenaza, 1977) while pant-hoots given upon discovery of a food tree do not differ acoustically from those given later during feeding bouts (Clark and Wrangham, 1993). This latter finding led Clark and Wrangham to conclude that pant hoots were not likely to index a specific ecological context ("arrival at food sources," p. 103) but, rather, might convey "an unusual interest in recruiting other individuals," (p. 107). It would seem conceivable, however, that these calls are referencing the qualities of the food rather than the discoverer's arrival at the food site. In Hauser et al.'s (1993) study, rough grunts were produced during all trials where subjects vocalized, but pant hoots were given only to larger quantities of food. The subjects in the study were all unrelated, and the extent to which this may have influenced calling behavior is not known.

While there are indications from these studies that food-associated calls have the potential to convey information about food or aspects of food, failures to conform to the criterion of context specificity or, in some cases, the absence of data concerning this issue, as well as the lack of playback studies make conclusions about the nature of external reference difficult at present.

Agonistic Screams

In many species of non-human primates screams are a prominent feature of aggressive interactions. Early studies of these calls in rhesus monkeys in the then recently established groups at Madingley, Cambridge, revealed pronounced acoustic variation which was interpreted as a graded reflection of the vocalizer's emotional state (Rowell, 1962; Rowell and Hinde, 1962). Later studies of agonistic aiding in free-ranging rhesus (Kaplan, 1977, 1978) and numerous other species (reviewed in: Walters and Seyfarth, 1987; Cheney and Seyfarth, 1990a, pp. 25-44) revealed a complex behavioral system in which alliance formation and recruitment was critical to both the development and maintenance of dominance position in the social group. Vocalizations were frequently noted to play a significant role in eliciting support from allies within the group. Successful recruitment might occur because of vocal recognition of the caller and information provided by the scream about the caller's emotional state or arousal level. Another hypothesis, one that is not mutually exclusive with the previous account, is that screams have external referents and make available more specific information about the agonistic event, information that might be of use to spatially distant allies in making decisions about intervention tactics. Our research has focused primarily on this latter possibility.

Gouzoules et al. (1984) found that in rhesus monkeys on Cayo Santiago, victims of attack usually made use of one of five acoustically distinct screams. Each scream type was associated with a particular class of opponent, defined in terms of relative dominance rank and matrilineal relatedness to the signaller, as well as the nature of the aggression, differentiated as to whether or not physical contact had occurred. Acoustically atonal "noisy" screams were most often given when the opponent was a higher-ranking group member and the incident involved contact aggression such as biting. Noisy screams were least likely to be associated with non-contact aggression from lower-ranking opponents.

"Arched" screams were typically given to lower-ranking opponents during non-contact aggression. They were highly unlikely to occur during serious aggression with higher-ranking opponents or during aggression from matrilineal kin. Both "tonal" and "pulsed" screams were significantly associated with aggression from matrilineal kin; tonal screams were also given to higher-ranking opponents in the absence of physical contact. "Undulated screams were most often produced during non-contact encounters with higher-ranking opponents. These calls were highly unlikely to be given during non-contact aggression with lower-ranking opponents.

Importantly, we found that screams were most often delivered in pure bouts during agonistic episodes, i.e., calls in a series tended to be acoustically similar, a finding that has been reported for a number of non-human primate species (e.g., Todt, 1988, p.92, p.102). Such a pattern would not be expected if screams primarily reflected the arousal level of the caller. We suggest this because agonistic interactions often wax and wane in intensity during a particular incident, and it seems reasonable to suppose that the internal state of the animal does likewise. Additional insight came from sequelae analyses that revealed that the different types of screams did not provide a good prediction of the caller's future behavior. Screams, more than any other type of call in the macaque vocal repertoire, can be heard over considerable distances, and this led us to reason that selection had promoted the evolution of call features that enable messages to be conveyed to distant group members; given the energy costs and potential danger in attracting predators, such features would not be predicted for these calls if the primary receivers were opponents at close range.

Pigtail macaques (*Macaca nemestrina*) also use acoustically different screams when confronting opponents. Interestingly, though pigtail and rhesus macaques can interbreed, producing fertile hybrids (Yerkes colony records), and are very similar in terms of patterns of social organization, the two species employ acoustically very different screams in essentially identical agonistic contexts. For example, during contact aggression from higher ranking opponents rhesus monkeys use atonal noisy screams, pigtails use tonal, frequency-modulated calls. This also suggests that, at least for these screams, there is no direct and simple relationship between motivational state and the acoustical structure of the vocalization (cf. Morton, 1977).

Pigtail and rhesus macaque screams would fail a strict application of the production criterion for external reference suggested by Macedonia and Evans (1993). For both species, screams are often used in agonistic contexts other than those with which they are primarily associated. However, there are several sources of variation in the use of calls that can account for an apparent lack of specificity and yet allow the possibility of external reference. For example, our analyses of pigtail macaque screams (Gouzoules and Gouzoules, 1989) suggest that juvenile monkeys under three years of age are far more likely than older individuals to make mistakes in the contextual use of calls. Production specificity was significantly greater for adult pigtail macaques. Another factor, mentioned above, that may also account for an apparent lack of production specificity is a failure by researchers to accurately describe the objects or events that are indexed by calls. What might initially appear to be different eliciting stimuli could well share common and relevant features to the monkeys.

Playback experiments with rhesus (Gouzoules et al., 1984) and pigtail macaques (Sharp and Gouzoules, unpublished data) revealed that mothers responded most strongly to screams associated with contact aggression from higher-ranking opponents. Responses were measured in terms of latency to look to, and the duration of looking in the direction of, a concealed speaker.[1] Screams associated with non-contact aggression from lower-ranking opponents elicited the next strongest response from mothers. This finding is pivotal because serious aggression with a risk of injury (i.e., fights involving contact with the opponent) were extremely rare for both species and it does not seem likely that caller fear levels should be high. For rhesus, only 2.4% of screams occurred in the context of contact aggression received from lower-ranking opponents, while for pigtails the proportion was 2.9%. Thus the physical consequences for a victim of attack from a lower-ranking opponent were minimal in both species and the high level of response elicited from callers' mothers cannot be explained by reference to protection from injury. Instead, preservation of the matrilineal dominance rank seems a more probable interpretation of the allies' behavior and the pertinent information, namely the relative dominance position of the opponent, appears to be provided by the scream. Screams given primarily to matrilineally related opponents elicit the least strong response from mothers in rhesus monkeys.

Playback experiments on rhesus monkeys were used to examine whether subjects respond differently to calls of dependent offspring as opposed to those from comparably aged collateral kin, e.g., sisters, nieces, cousins, (minimal coefficient of relatedness, through maternal lines, ranged from r = 1/4 to r = 1/16) (Gouzoules et al., 1986). The results supported our earlier conclusions that, for allies, the key information provided by the screams and influencing the likelihood of intervention concerns the nature of the aggression and opponent. While subjects responded more strongly to the noisy screams of their offspring compared to those of more distantly related juveniles, no such difference emerged from a comparison of playbacks involving arched screams from the same callers. When responses to only collateral kin were examined, the same patterns were found: relatedness influenced response to noisy screams but not to arched screams. It may be that the intervention on behalf of a more distant relative whose opponent is higher-ranking (noisy screams) is more costly. We also found that the sex of the caller influenced the intensity of response during playback experiments with the noisy and arched screams of collaterally related juveniles. While there were no differences in response strength to noisy screams of male and female kin, arched screams from young female kin elicited significantly stronger responses than did those from male relatives. This sex difference may be related to male and female life history patterns: pubescent males generally emigrate from the natal group, whereas females remain for life and are dependent on members of their matriline for rank preservation.

Recent work on another cercopithecine species, the sooty mangabey (*Cercocebus atys*), reveals the potential for comparative studies to contribute to an understanding of the evolution of signals that have external referents (Santee, 1992; Santee and Gouzoules, in preparation). Like most other monkeys, sooty mangabeys scream when attacked by group members. Several intriguing differences between macaques and sooty mangabeys have emerged from recent studies, however. Unlike pigtail and rhesus macaques, attacks almost always came from higher-ranking opponents. In sharp contrast to the macaques, rarely did other mangabeys intervene on behalf of the victim of attack. Injury resulting from aggression was infrequent: most bites received appeared to be nominal, almost ritualized in form, and were delivered to the base of the tail of the victim (Gust and Gordon, 1993). In contrast to macaques, grooming and other social behaviors do not correlate well with kinship in this species (Ehardt, 1988). The striking differences in affiliative and aggressive behavior between pigtail and rhesus macaques, on the one hand, and sooty mangabeys, on the other, lead to predictions about the functions of vocal communication during fights. There appears to be less likelihood of serious wounding in sooty mangabey fights and kinship ties (for as yet unknown reasons) seem to be weaker than in macaques. Perhaps related to these conditions, agonistic aiding is not a prevalent part of aggressive interaction. These differences lead us to predict that sooty mangabey screams would be acoustically less heterogeneous than those used by pigtail or rhesus macaques.

Santee (1992) analyzed the acoustic and contextual features of 2438 sooty mangabey screams recorded from 488 agonistic bouts in a large 85 member group maintained at the Yerkes Regional Primate Research Center's Field Station. Dominance relationships and matrilineal kinship were known for all members of the group which had been assembled in early 1969 (Bernstein, 1971). Methods of data collection and analysis for screams and associated behavior was much the same as for previous macaque studies in our lab (e.g., Gouzoules and Gouzoules, 1989). Santee found that the vast majority of screams (96%) were broadband atonal vocalizations, similar to what we termed noisy screams in rhesus monkeys. A small number of screams showed greater acoustical complexity, some with energy concentrated into denser bands, others having frequency-modulated tonal components, while a few showed harmonics above a tonal fundamental frequency. This smaller subset of screams indicates that sooty mangabeys are capable of producing sounds similar to those used by macaques. We hypothesize that the infrequent occurrence of agonistic aiding, the lack of a pronounced influence of matrilineal kinship on behavior, and the apparently less severe nature of aggression in sooty mangabeys result in a social environment that has not been conducive for the evolution of representational screams.

CONCLUSIONS: REFERENTIAL SIGNALS RECONSIDERED

Initially, researchers became interested in representational signals in non-human primate vocal communication simply to demonstrate that, in contrast to the view prevalent until about 1980, the animals could, in fact, communicate about objects and events in their external worlds (e.g., Seyfarth et al, 1980; Cheney and Seyfarth, 1982; Gouzoules et al., 1984). In order to do this, alternative explanations of the caller's behavior and that of receivers had to be eliminated. Arguments and data centered around whether or not, for example, the caller's future behavior or its arousal level were indexed by calls instead of things external to the signaller (e.g., Smith, 1981). It became clear that even when calls indicated something about the internal state or arousal level of a signaller, receivers could nonetheless derive information about the environment based on previous knowledge and context (e.g., some food calls). Eventually the focus of interest shifted to the nature of the representation involved in communication about the external environment and the inferences that could be drawn from different kinds of data. This led Cheney and Seyfarth (1992) to make a distinction between "semanticity in the weakest sense," where data do not reveal anything about the mechanisms that underlie responses to calls, and "semanticity in a stronger sense," where data permit some assessment of the cognitive bases for the behavior. Their clever use of the habituation-dishabituation paradigm provided stronger evidence about whether or not monkeys were responding on the basis of the meaning of calls that apparently had external referents.

Most recently, Marler (1992; Marler et al., 1992) and colleagues have emphasized that no animal signal that has been identified as having an external referent is devoid of motivational content (and the same is true, they argue, for human speech). Interestingly, though earlier studies on vervet alarm calls and rhesus macaque screams stressed that claims for external reference were not incompatible with the possibility that certain acoustical features of the calls were simultaneously mapping aspects of the signaller's internal state, it now seems necessary to recast thinking about animal signals into a framework that is more comprehensive (cf., Gouzoules et al., 1985). Smith (1981, 1990) was clearly concerned about too narrow a focus on the issue of external reference in non-human communication and noted that vocalizations could have multiple referents. Some of the vocalizations of vervets and macaques for which representational claims have been made clearly convey information about the signaller: individual variation in call production permits caller identification. Vervets transferred habituation in some experiments as a function of caller identification. Female macaques identify their relatives' calls when responding to recruitment screams. Pigtail macaque screams have the potential to convey information about which matriline a caller belongs to (Gouzoules and Gouzoules, 1990), though it is not yet certain that pigtail macaques attend to this content. Hauser (1993) has compared a wide range of non-human primate vocalizations with respect to correspondence to Morton's motivational-structural rules (Morton, 1977), which predict a relationship between motivational state of a caller and the acoustical properties of the vocalization it produces, and found, for certain taxa, appreciable fit. Hauser suggests that, for at least some species, acoustic structure is associated more with an external referent than with the caller's internal state and concludes that the acoustic features of non-human primate vocalizations have been designed to convey information about motivational states and aspects of the external environment. As Marler et al. (1992) have concluded, the distinction between the terms "referential" and "motivational," as applied in a categorical fashion to signals and their modes of functioning, cannot be sustained. Instead, they postulate varying degrees of referentiality, as in points on a continuum.

Are Monkeys Fuzzy Logicians?

In concluding this review of representational signals in non-human primates, we focus attention on the form of logic typically used to study the question of external reference. Most of the studies we have reviewed have followed what Marler et al. (1992) have indicated is the essential first step for determining the meaning of a signal: investigate correspondence between acoustic characteristics and external events. These data are then used to assess whether or not the vocalization meets the production specificity criterion for external reference that Marler et al. (1992) and Macedonia and Evans (1993) have recommended. However, in order for such an assessment to be made properly, a correct decision has to have been arrived at regarding the nature of the external referent(s). If, for example, two objects or events presumed to be members of different categories by the investigator, actually share features or properties that link them for the animals, then production specificity may be deemed low when, in fact, it is not. Macedonia and Evans (1993) suggest that vervet monkey leopard alarms might be further from "the referential pole of a 'motivational-to-referential continuum' than eagle or snake alarms" because the former are elicited occasionally by other types of predators (e.g., attacking snakes or eagles about to seize a monkey). However, if, as Marler et al. (1992) argue, alarm calls convey both motivational and referential information, it is not clear why, under supposedly more emotionally charged conditions, the animals should have switched to an acoustically different call, rather than have modified the acoustic features that reflect emotional state. It may be, instead, that the incidents involving leopard alarms that appear inappropriate (described by Cheney & Seyfarth, 1990a, p.108, and by Macedonia & Evans, 1993, p.182) actually share critical features with the more typical conditions under which the calls are produced: we do not know what these might be at the present, however the "inappropriate" use of leopard alarm calls appears to be restricted to conditions of imminent danger.

Perhaps the way we have traditionally categorized the objects and events that are indexed by particular non-human primate vocalizations predisposes us to find cases of "inappropriate" usage of calls. All attempts to categorize the putative things that animals communicate about have assumed that the animals employ a logic system in which there are only two truth values, 1 or 0 ("true" and "false"). For example, in our studies, we have classified opponents in agonistic encounters as either higher-ranking or lower-ranking than the victim. We classified the intensity of encounter by distinguishing between contact aggression and non-contact aggression. Studies of food calls have categorized food as high quality or low quality, divisible or not, preferred or not. It is clear that some species react to objects in their worlds in a manner consistent with this simple view. Prey recognition in toads is one example (Lloyd, 1989). Yet it is possible that such bivalent logic is not the basis for the categorization of many things relevant to and communicated about by monkeys. Perhaps a multi-valued or "fuzzy" logic system, one with more than two truth values, would be a more appropriate way to conceptualize this problem.

Fuzzy logic has been proposed as a means to handle the concept of partial truth and one of its first applications was to model the uncertainty of natural language (for reviews of fuzzy logic see McCawley, 1981; Lakoff, 1987; Massaro, 1987; Zadeh, 1992; Kosko, 1993). It employs truth values between 0 and 1 as a way to deal with inexact concepts. McCawley, (1981, p.380) illustrates one benefit of such an approach by noting that "with intermediate truth values one is not forced to draw an arbitrary distinction between tall individuals and not-tall individuals but can recognize individuals who are somewhat tall, slightly tall, and so on without being unqualifiedly tall or unqualifiedly not tall." A fuzzy set approach to categorizing the dominance of opponents, the severity of aggression, or aspects of discovered food would hold that members of the set of objects or events are members to some degree, in contrast to classical set theory where membership is all or none. Some members of a fuzzy set belong to it to a greater degree than others. Thus, for example, an attack from the alpha male or alpha female of the group, for a rhesus monkey female, would (under most circumstances) have a truth value of close to 1 for the simian proposition "attack from a higher-ranking opponent." It is entirely possible that the truth value for such proposition would be considerably less for an incident involving a different monkey, one which ranks closer to the victim on the ordinal form of dominance hierarchy typically constructed by observers to assess dominance relationships. In the same way, aggression that resulted in wounding and injury would likely have a truth value close to one for the proposition "severe aggression," but an interaction involving a slap or a push might not. Figure 1 shows a hypothetical set of truth values for these two classification problems involving social interactions in monkeys. It is important to recognize that the truth values shown are not probability values for group membership. Thus a truth value of .5 does not mean that the particular object or event has a 50% probability of belonging to a class, but instead reflects the idea that fuzzy set members belong to a class to different degrees.

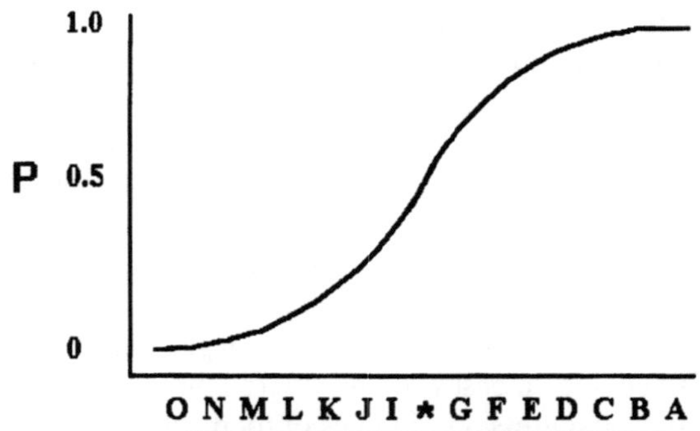

Figure 1. Hypothetical function showing the truth or certainty values for elements of two fuzzy subsets, dominance of opponent ("opponent is higher-ranking") and severity of aggression ("aggression is severe"). Horizontal axis: (top) lists 15 group members with ordinal dominance rank of individual A equal to 1 and that of individual O equal to 15; (bottom) lists a series of aggressive behaviors experienced (increasing severity). Vertical axis shows P (0< P <1), the degree to which entries on horizontal axis are true or certain for group member H (*).

Macaque agonistic interactions might thus be thought of as fuzzy subsets in that particular encounters are members to different degrees based on functions that have as input elements relating to the opponent's dominance status, the severity of the aggression and, for rhesus macaques, relatedness. Relatedness, too, appears to fit well with the notion of a fuzzy subset: kinship correlated behavior in non-human primates does not manifest itself in a way consistent with a bivalent "kin" or "nonkin" (1,0) logic. Instead, individual relatives are members of the class to different degrees; in some instances, the coefficient of relatedness may sometimes be seen as a shorthand to the truth value of membership.

While there has always been an awareness of the utility of concepts such as kinship and dominance as tools for researchers trying to understand non-human primate behavior, there is growing appreciation and acceptance of the likelihood that the animals themselves make use of such concepts in their natural behavior (see reviews in Cheney and Seyfarth, 1990a,b). How we study such concepts in animals will depend very much on the system of logic we employ to study them. Variation is inherently part of as well as the basis for the biological world, something often acknowledged, but rarely incorporated into thinking about the problems animals face in dealing with their physical and social environments. To different extents, many species have evolved brains that enable individuals to cope with ubiquitous variation through representations of biologically relevant items. Representations permit animals to partition the world into manageable components. They vary from the coding of stimuli in the nervous system that permit the detection of important features of the environment, to complex forms that abstract to what we call knowledge, to systems of communication that represent representations. At all levels, representational systems involve reduction. Representations are not equivalent to their precursors; they are necessary simplifications that reduce variability. Representational systems permit animals to effectively deal with a fuzzy world, with natural selection determining the relevant level of extraction and synthesis, or categorization.

Representational signaling in non-human primates accomplishes the same thing, and while one goal of research in this area should continue to be to try to understand the meaning of these vocalizations, perhaps greater progress can be achieved by incorporating a "fuzzy" perspective into thinking about the problem. If the fuzzy view we are suggesting here has validity, and we feel the behavior of well-studied species is consistent with this possibility, one fact for researchers attempting to understand the meaning of calls that correlate with events and objects in the social and physical environments of non-human primates seems clear. Conclusions that particular calls do not have external referents, or proposals specifying the nature of the external referents, that are drawn from analyses employing bivalent classification schemes, should be viewed with caution.

ACKNOWLEDGEMENTS

We thank U. Neisser, M. Tomasello, E. St. Andre, D. Cheney, M. Hauser and E. Zimmermann for comments on a previous draft of this paper. Our research was supported by grants from the National Science Foundation (BNS 84-06435, BNS 87-19230 and IBN 92- 09844), the H.F. Guggenheim Foundation, and the Emory University Research Committee to H.G. and S.G. and by NIH grant RR 00165 to the Yerkes Regional Primate Research Center. The Yerkes Center is fully accredited by the American Association of Accreditation of Laboratory Animal Care (AAALAC).

REFERENCES

Benz, J.J., 1993, Food-elicited vocalizations in golden lion tamarins: design features for representational communication, *Anim. Behav.* 45:443-455.

Bernstein, I.S., 1971, The influence of introductory techniques on the formation of captive mangabey groups, *Primates* 12:33-44.

Bickerton, D., 1990, "Language and Species," University of Chicago Press, Chicago.

Cheney, D.L., 1987, Interactions and relationships between groups, *in*: "Primate Societies," B.B. Smuts, D.L. Cheney, R.M. Seyfarth, R.W. Wrangham, and T.T. Struhsaker, eds., University of Chicago Press, Chicago.

Cheney, D.L. and Seyfarth, R.M., 1982, How vervet monkeys perceive their grunts: field playback experiments, *Anim. Behav.* 30: 739-751.

Cheney, D.L. and Seyfarth, R.M., 1988, Assessment of meaning and the detection of unreliable signals by vervet monkeys, *Anim. Behav.* 36:477-486.

Cheney, D.L. and Seyfarth, R.M., 1990a, "How Monkeys See the World," University of Chicago Press.

Cheney, D.L. and Seyfarth, R.M., 1990b, The representation of social relations by monkeys, *Cognition* 37:167-196.

Cheney, D.L. and Seyfarth, R.M., 1992, Meaning, reference, and intentionality in the natural vocalizations of monkeys, *in:* "Topics in Primatology, Vol. 1, Human Origins," T. Nishida, W.C. McGrew, P. Marler, M. Pickford, and F.B.M. de Waal, eds., University of Tokyo Press, Tokyo.

Clark, A.P. and Wrangham, R.W., 1993, Acoustic analysis of wild chimpanzee pant hoots: do Kibale forest chimpanzees have an acoustically distinct food arrival pant hoot? *Amer. J. Primatol.* 31: 99-109.

Dittus, W.P.J., 1984, Toque macaque food calls: semantic communication concerning food distribution in the environment, *Anim. Behav.* 32:470-477.

Dittus, W., 1988, An analysis of toque macaque cohesion calls from an ecological perspective, *in:* "Primate Vocal Communication," D. Todt, P. Goedeking, and D. Symmes, eds., Springer-Verlag, New York.

Ehardt, C.L., 1988, Absence of strongly kin-preferential behavior by adult female sooty mangabeys (*Cercocebus atys*), *Amer. J. phys. Anthropol.* 76:233-243.

Elowson, A.M., Tannenbaum, P.L., and Snowdon, C.T., 1991, Food associated calls correlate with food preferences in cotton-top tamarins, *Anim. Behav.* 42:931-937.

Gouzoules, H. and Gouzoules, S., 1989, Design features and developmental modification of pigtail macaque, *Macaca nemestrina*, agonistic screams, *Anim. Behav.* 37:383-401.

Gouzoules, H. and Gouzoules, S., 1990, Matrilineal signatures in the recruitment screams of pigtail macaques, *Macaca nemestrina*, *Behaviour* 115:327-347.

Gouzoules, S., Gouzoules, H., and Marler, P., 1984, Rhesus monkey (*Macaca mulatta*) screams: representational signalling in the recruitment of agonistic aid, *Anim. Behav.* 32:182-193.

Gouzoules, H., Gouzoules, S., and Marler, P., 1985, External reference and affective signaling in mammalian vocal communication, *in:* "The Development of Expressive Behavior (Biology-Environment Interactions)," G. Zivin, ed., Academic Press, New York.

Gouzoules, H., Gouzoules, S., and Marler, P., 1986, Vocal communication: a vehicle for the study of social relationships, *in:* "The Cayo Santiago Macaques," R.G. Rawlins and M.J. Kessler, eds., State University of New York Press, Albany, N.Y.

Green, S., 1975, Variation of vocal pattern with social situation in the Japanese monkey (*Macaca fuscata*): a field study, *in:* "Primate Behavior, Vol. 4," L.A. Rosenblum, ed., Academic Press, New York.

Gust, D.A and Gordon, T.P., 1993, Conflict resolution in sooty mangabeys, *Anim. Behav.* 46:685-694.

Hauser, M.D., 1991, Sources of acoustic variation in rhesus macaque vocalizations, *Ethology* 89 29-46.

Hauser, M.D., 1993, The evolution of non-human primate vocalizations: effects of phylogeny, body-weight, and social context, *Amer. Nat.* 142:528-542.

Hauser, M.D. and Marler, P., 1993a, Food-associated calls in rhesus macaques (*Macaca mulatta*): I. Socioecological factors, *Behav. Ecol.* 4:194-205.

Hauser, M.D. and Marler, P., 1993b, Food-associated calls in rhesus macaques (*Macaca mulatta*): II. Costs and benefits of call production and suppression, *Behav. Ecol.* 4:206-212.

Hauser, M.D. and Wrangham, R.W., 1987, Manipulation of food calls in captive chimpanzees: a preliminary report, *Folia Primatol.* 48: 207-210.

Hauser, M.D., Teixidor, P., Field, L., and Flaherty, R., 1993, Food-elicited calls in chimpanzees: effects of food quantity and divisibility, *Anim. Behav.* 45:817-819.

Kaplan, J., 1977, Patterns of fight interference in free-ranging rhesus monkeys, *Amer. J. phys. Anthropol.* 47:279-288.

Kaplan, J.R,. 1978, Fight interference and altruism in rhesus monkeys, *Amer. J. phys. Anthropol.* 49:241-250.

Kosko, B., 1993, "Fuzzy Thinking: the New Science of Fuzzy Logic," Hyperion, New York.

Lakoff, G., 1987, "Women, Fire, and Dangerous Things: What Categories Reveal about the Mind," University of Chicago Press, Chicago.

Lancaster, J., 1968, Primate communications and the emergence of human language, *in:* "Primates," P.C. Jay, ed., Holt, Rinehart and Winston, New York.

Lancaster, J.B., 1975, "Primate Behavior and the Emergence of Human Culture," Holt, Rinehart and Winston, New York.

Lloyd, L.B., 1989, "Simple Minds," M.I.T. Press, Cambridge, MA.

Macedonia, J.M., 1990, What is communicated in the antipredator calls of lemurs: evidence from playback experiments with ringtailed and ruffed lemurs, *Ethology* 86:177-190.

Macedonia, J.M. and Evans, C.S., 1993, Variation among mammalian alarm call systems and the problem of meaning in animal signals, *Ethology* 93:177-197.

Macedonia, J.M. and Yount, P.L., 1991, Auditory assessment of avian predator threat in semi-captive ringtailed lemurs (*Lemur catta*), *Primates* 32:169-182.

Mandler, J.M., 1983, Representation, *in:* "Handbook of Child Psychology: Vol. 3. Cognitive Development," J.H. Flavell and E.M. Markman, eds., Wiley and Sons, New York.

Marler, P., 1985, Representational vocal signals of primates, *Fortschr. Zool.* 31:211-221.

Marler, P., 1992, Functions of arousal and emotion in primate communication: a semiotic approach, *in:* "Topics in Primatology, Vol. 1, Human Origins," T. Nishida, W.C. McGrew, P. Marler, M. Pickford, and F.B.M. de Waal, eds., University of Tokyo Press, Tokyo.

Marler, P. and Tenaza, R., 1977, Signaling behavior of apes with special reference to vocalization, *in:* "How Animals Communicate," T. Sebeok, ed., Indiana University Press, Bloomington.

Marler, P., Evans, C.S., and Hauser, M.D., 1992, Animal signals: motivational, referential, or both? *in:* "Nonverbal Communication: Comparative and Developmental Approaches," H. Papousek, U. Jürgens, and M. Papousek, eds., Cambridge University Press, Cambridge.

Masataka, N., 1983, Categorical responses to natural and synthesized alarm calls in Goeldi's monkeys (*Callimico goeldii*), *Primates* 24:40-51.

Masataka, N., 1988, Temporal and sequelae analysis of food calling behavior in Japanese macaques, *Research Reports of the Arashiyama West and East Groups of Japanese Monkeys:* 51-55.

Massaro, D.W., 1987, Categorical partition: a fuzzy-logical model of categorization behavior, *in:* "Categorical Perception," S. Harnad, ed., Cambridge University Press, Cambridge.

McCawley, J.D., 1981, "Everything That Linguists Have Always Wanted to Know about Logic," University of Chicago Press, Chicago.

Miller, G.A. and Johnson-Laird, P.N., 1976, "Language and Perception," Harvard University Press, Cambridge.

Morton, E.S., 1977, On the occurrence and significance of motivation-structural rules in some bird and mammal sounds, *Amer. Nat.* 111:855-869.

Newell, A., 1990, "Unified Theories of Cognition," Harvard University Press, Cambridge.

Ogden, C.K. and Richards, I.A., 1923, "The Meaning of Meaning," Routledge and Kegan Paul, London.

Owren, M.J., 1990, Acoustic classification of alarm calls by vervet monkeys (*Cercopithecus aethiops*) and humans (*Homo sapiens*): I. Natural calls, *J. Comp. Psychol.* 104:20-28.

Pearce, J.M., 1987, "Introduction to Animal Cognition," Erlbaum Associates, Hove and London.

Pereira, M.E. and Macedonia, J.M., 1991, Response urgency does not determine antipredator call selection by ringtailed lemurs, *Anim. Behav.* 41:543-544.

Premack, D., 1975, On the origins of language, *in:* "Handbook of Psychobiology," M.S. Gazzaniga and C.B. Blackemore, eds., Academic Press, New York.

Rosch, E., Mervis, C.B., Gray, W.D., Johnson, D.M., and Boyes-Braem, P., 1976, Basic objects in natural categories, *Cogn. Psychol.* 8:382-439.

Rowell, T.E., 1962, Agonistic noises of the rhesus monkey (*Macaca mulatta*), *Symp. Zool. Soc. Lond.* 8:91-96.

Rowell, T.E. and Hinde, R.A., 1962, Vocal communication by the rhesus monkey (*Macaca mulatta*), *Proc. Zool. Soc. Lond.* 138: 279-294.

Santee, D.P., 1992, "Aspects of the Vocal Communication of Sooty Mangabeys (*Cercocebus atys*): Agonistic Screams and `Whoopgobbles'," Ph.D. Dissertation, Emory University, Atlanta.

Scherer, K.R., 1988, On the symbolic functions of vocal affect expression, *J. Lang. Soc. Psychol.* 7:79-100.

Seyfarth, R.M. and Cheney, D.L., 1990, The assessment by vervet monkeys of their own and another species' alarm calls, *Anim. Behav.* 40:754-764.

Seyfarth, R.M., Cheney, D.L., and Marler, P., 1980, Vervet monkey alarm calls: semantic communication in a free-ranging primate, *Anim. Behav.* 28:1070-1094.

Smith, W.J., 1981, Referents of animal communication, *Anim. Behav.* 29:1273-1275.

Smith, W.J., 1990, Animal communication and the study of cognition. *in:* "Cognitive Ethology: the Minds of Other Animals (essays in honor of Donald R. Griffin)," C.A. Ristau, ed., Lawrence Erlbaum Associates, Hillsdale, N.J.

Struhsaker, T.T., 1967, Auditory communication among vervet monkeys (*Cercopithecus aethiops*). *in:* "Social Communication among Primates," S.A. Altmann, ed., University of Chicago Press, Chicago.

Tenaza, R. and Tilson, R., 1977, Evolution of long-distance alarm-calls in Kloss'gibbon, *Nature* 268:233-235.

Tiles, J.E., 1987, Meaning. *in:* "The Oxford Companion to the Mind," R.L. Gregory, ed., Oxford University Press, Oxford.

Todt, D., 1988, Serial calling as a mediator of interaction processes: crying in primates. *in:* "Primate Vocal Communication," D. Todt, P. Goedeking, and D. Symmes, eds., Springer-Verlag, New York.

Walters, J.R. and Seyfarth, R.M., 1987, Conflict and cooperation, *in:* "Primate Societies," B.B. Smuts, D.L. Cheney, R.M. Seyfarth, R.W. Wrangham and T.T. Struhsaker, eds., University of Chicago Press, Chicago.

Wrangham, R.W., 1977, Feeding behaviour of chimpanzees in Gombe National Park, Tanzania, *in:* "Primate Ecology," T.H. Clutton-Brock, ed., Academic Press, New York.

Zadeh, L.A., 1992, Knowledge representation in fuzzy logic, *in:* "An Introduction to Fuzzy Logic Applications in Intelligent Systems, R.R. Yager and L.A. Zadeh, eds., Kluwer Academic Publishers, Norwell, Ma.

Note

[1] Such measures reflect, most directly, the ability of the calls to attract the attention of the subject. Naturally occurring screams frequently elicit approach and intervention by an ally and, for playback experiments, this might lead to habituation. Thus short exemplars of calls were used to minimize the likelihood that a subject would approach the concealed speaker. Calls were broadcast at similar amplitudes and because rhesus and pigtail macaques employ acoustically distinct screams in the same contexts, there is nothing in the call structure itself that could account for the observed differences in response. Competing hypotheses that predict subjects' interest in the broadcast calls are then used to assess the information made available by the vocalization.

LINGUISTIC CAPACITY: AN ONTOGENETIC THEORY WITH EVOLUTIONARY IMPLICATIONS

John L. Locke

Neurolinguistics Laboratory
Massachusetts General Hospital
and Harvard Medical School
Boston, MA 02114, USA

What forces propel the young child into language use? After an unprecedented surge in research over the last two decades or so we are still no nearer to providing anything even approaching a definitive answer to this question (Schaffer, 1989).

INTRODUCTION

In this chapter, I will present a theory on the infant's development of language that was inspired, in part, by research on non-human primates and other animals. According to the theory, long before infants speak or even know about the existence of linguistic rules and representations they store utterances and supporting contextual information. Utterance storage is carried out primarily by the right cerebral hemisphere, which is heavily involved in the processing of vocal intonation and affect at the time storage operations commence. Many of these prosodically organized utterances are reproduced, giving listeners the impression that the infant is using language. Lexical forms accumulate at a gradual rate until a point at which - for reasons not well understood - rate of word learning accelerates. The consequent proliferation of words threatens to overload the infant's small capacity store for intonationally or prosodically organized material. Overloading is avoided by activation of a mechanism, typically sited in the left hemisphere, that analyzes utterances into their component parts. When infants discover the patterning of these units, they learn something about the organization of the ambient sound system. Organizational learning permits infants to compute novel word forms, the first evidence of grammatical behavior. While non-human primates seem unable to analyze and compute utterances in this fashion, they may be capable of preliminary operations that in the human appear to facilitate lexical acquisition and use.

Most linguists that avail themselves of non-human primate data do so for the purpose of constructing a theory on the evolution of linguistic capacity in the species. My own use of findings from studies of non-human primates has been different. It has begun with the assumption that the infant's progress toward language does not begin with its first word, but arises from earlier capabilities that are non-linguistic by definition. Seeking inspiration as to the nature of these capabilities, it has made sense to me to examine the early social and emotional development of non-human primates that, like our own infants, are also non-linguistic. As a result of this approach, the ontogenetic theory that I will present here has readily discernible implications for the evolution of language.

This approach may surprise primatologists who have thought of language as a capability that exceeds the reach of their research subjects, and therefore falls outside their own purview as investigators and scholars. And that would be a very predictable reaction because the mainstream of modern linguistics has been more concerned with theories of grammar than social communication and ethologically valid models of language use. Language also has been defined in very abstract terms, and treated by many linguists as though it were synonymous with generative morphology and syntax. But primate studies have much to contribute to theories of language *development*, as well as to speculation on the nature and evolution of linguistic capacity.

Recent shifts in the *Zeitgeist* in developmental psycholinguistics make non-human primates more interesting than ever. Ever since Chomsky's attack on Skinner over 30 years ago, psycholinguists with any hope for a successful career had to think twice before using the word "learning" in connection with language development, at least if the learning involved was not specifically restricted to linguistic operations. But it is now recognized that spoken language rests on several different types of vocal and phonetic learning and a range of potentially contributory precursive behaviors (cf. Bullowa, 1979; Golinkoff, 1983; Lindblom and Zetterstrom, 1986; Speidel and Nelson, 1989). As a consequence, it is now "legal" to ask about infants' cognitive and neural development, and social and perceptual experiences, in the quest to understand how and why they begin to speak.

This approach is also a theoretical necessity. For if infants are engaged in behaviors that facilitate language before they possess the cognitive capability to fully appreciate its existence, then their behaviors must be motivated by one or more "non-linguistic" factors (Locke, in submission). Merely owning the genes of a species known to have linguistic capacity would be insufficient.

It has been argued that language requires specialized mechanisms that are encapsulated or dissociated from other, more general processing systems, but most discussions to date have centered on perception (cf. Fodor, 1983; Mattingly and Studdert-Kennedy, 1991). If our species has a "language module," however, it would have to perform the structural analyses and computations that are required by grammar. I assume that brain mechanisms that are structurally capable of carrying out such analyses cannot activate or become fully operable without stored utterances - the candidates for analysis - anymore, say, than visual analytical mechanisms can develop without light. A grammatical module thus *have to be fed.* That is, it could only become functional if the child first paid attention to, and absorbed, the superficial behaviors of talking people (Locke, 1994b, in press, b). There is no reason to attribute the attentional biases responsible for storing utterances to the mechanism that will analyze them once stored (Locke, 1992). It is thus ironic that the human language module, for all its important capabilities, has no way to take in knowledge about language.

PRECURSORS TO SPOKEN LANGUAGE

In the last 20 years there has been a steady accumulation of evidence that spoken language in children is continuous with earlier vocal and gestural behaviors. Findings now link frequency and quality of spontaneous vocalization, joint attention, vocal imitation, and maternal response to the emergence of words (Hardy-Brown et al., 1981; Locke, 1989; Nelson, 1985; Snow, 1989; Tomasello et al., 1986; Veneziano, 1988). Although some of the research merely identifies a simple chronological progression, other studies seem to have identified specific causal factors.

Comparative studies (cf. Parker and Gibson, 1990) suggest the existence of cognitive parallels in the development of human and non-human primates. The capabilities we share with primates are, generally speaking, the first to develop in the human infant. Much of non-human primate socialization revolves around affective signaling by voice and face, and so do the interactions of the human young. Both groups of primates retain their capacity to communicate at that level, but humans additionally take on more arbitrary and codified means of communicating.

Language develops in a social context, through interactions between the infant and people who are talking. The infant's social environment is defined by physical cues, especially the facial movement and vocal variations of caregivers and other interested parties (Locke, 1993a, 1993b, 1994b). I would suggest that when infants orient to these cues, and respond in kind, they start down a developmental growth path that leads to spoken language. Let us examine the initial capabilities that deposit infants on this path.

SPECIALIZATION FOR SOCIAL COGNITION

The ability to infer dispositions and intentions is a highly evolved social capability. In primates, such socially cognitive acts depend on neural specializations and modular properties that stand somewhat apart from other forms of cognition (Brothers, 1990; Brothers and Ring, 1992). If we wish to explore the potential for vocal learning in primates, and the biological foundations of spoken language in human primates, there are several things we must know about these specializations.

One of the first things we must learn is how the brain responds to faces and facial movements, and we must appreciate the neurophysiology of demonstrated functional links between face perception and social vocalization. In humans, behavioral and physiological studies reveal a greater involvement of the right cerebral hemisphere during the display and interpretation of facial emotion (cf. Etcoff, 1989). This effect also has been obtained in monkeys (Hamilton and Vermeire, 1988; Hauser, 1993). Moreover, in non-human primates, single unit recordings from temporal cortex have revealed cell ensembles with differential reactions to faces and other visually complex objects and cells that respond more vigorously to upright and intact faces than to inverted or rearranged faces (cf. Desimone, 1991; Perrett and Mistlin, 1990). There also are cells that appear to be sensitive to the directionality of facial view, line of gaze of the stimulus face, facial expression, facial orientation, specific facial features, expressions, and a variety of movements. Interestingly, cells that code for facial identity are distinct from cells coding for facial expression.

We also need to learn what goes on in the brain when listeners attempt to extract from the human *voice* information that is relevant to the speaker's identity, intentions in speaking, and attitude toward his addressees. There is evidence that when human listeners analyze such cues, they activate processing machinery in the non-dominant, right cerebral hemisphere (cf. Locke, 1993b). Evidence suggests that with similar cues, non-human primates also may favor the right hemisphere, although they, too, may prefer the left hemisphere when processing conspecific calls (Hauser and Andersson, in press; Heffner and Heffner, 1984, this volume; Petersen et al., 1978, 1984).

As one might suppose, there is support for a functional linkage between face and voice processing. At present, much of it comes from clinical studies showing that in adult patients a lesion in the right hemisphere tends to impair interpretation and execution of both facial expressivity *and* vocal prosody and emotion (Ross, 1981). Among the developmentally disabled, there is now a small literature indicating that autistic children characteristically show aversion to or disinterest in the human face, atypical face scanning patterns, and deficits in the interpretation of facial affect (Hobson et al., 1988; Langdell, 1978). Malformations associated with autism have been found in the cerebellum and limbic system, including the amygdala (Bauman and Kemper, 1985), a structure that has come to be identified with affective processing (Kling and Brothers, 1992).

Many of the linguistic precursors documented in humans may also be observed in non-human primates who, like humans, have large brains, precocial sensory systems, neotenous development and nurturant social structure. Comparative studies can therefore be helpful in several broad areas, as I will attempt to demonstrate. One pertains to the human infant's specialization for social cognition, including the infant's ability to respond to supporting social experiences that occur between the infant and its primary caregivers. A second area relates to the capacity to analyze utterances and apprehend phonetic structure. A third area relates to the motor control mechanisms required for self-expression by way of modulated voice. As we will see, capabilities in these areas emerge during fairly distinct phases in the development of language (Locke, 1994a, in press, a).

THE VOCAL LEARNING PHASE

It might seem that one achieves linguistic capability and then sets out to use it, but in order to develop the capacity for language the human infant must learn a particular language.

An infant learns a language by observing people who are speaking that language (Locke, 1994b). There are many sights and sounds in the infant's environment that are unrelated to talking. How does the infant come to separate speech from non-speech activities and thus pay preferential attention to the linguistically relevant actions of people who are speaking?

In the first phase in the development of linguistic capacity, infants learn and respond to properties of the human voice. Evidence is converging to suggest that vocal learning begins prenatally, when the fetus learns enough about its mother's voice to recognize it after birth. The 3-day-old neonate's preference for maternal voice (DeCasper and Fifer, 1980) is presumed to occur because the mother's airborne voice resembles her voice as transmitted prenatally (DeCasper and Spence, 1986) through the soft tissues and fluids of her body. At 2 to 4 days of age, there also is a documented preference for the voice of an adult female that resembles the mother's because it is speaking the same language that the mother spoke prenatally (Mehler et al., 1988; Moon et al., 1993).

Although mature vervet and squirrel monkeys apparently can identify juvenile and infantile offspring from their vocalizations (Cheney and Seyfarth, 1980, 1990; Symmes and Biben, 1985; Waser, 1977), there is very little evidence on the parallel question - whether infant monkeys and apes respond preferentially to the sound of their mother's voice. I also am unaware of any efforts to explore the effects of maternal voice on the non-human primate fetus, although investigators have found cardiac accelerations to conspecific threat calls in neonatal chimpanzees that were not found for other conspecific calls and noises and not obtained in orangutans (Berntson and Boysen, 1989). At 3 and 4 months, isolation-reared chimpanzees showed different cardiac responses to conspecific screams and laughter, suggesting specialized perceptual processing mechanisms in these species (Berntson et al., 1990).

That human neonates prefer the language heard in utero at the same age that they prefer the voice heard prenatally suggests that, early on, maternal and linguistic preference may be one and the same. It also suggests that a single set of cues permits distinctions among individuals and among spoken languages. Given the filtering of soft tissues, these cues would be of the relatively low frequency type that reveal little about specific consonants and vowels but preserve linguistic *prosody*, i.e., the frequency and intensity of the voice as well as temporal information relating to the rhythm and rate of speech. This is shown by recordings of intrauterine maternal voice transmission (Querleu et al., 1985) and by electronic (low-pass) filtering of speech which preserves differences between vocal emotions (Ross et al., 1973; Scherer et al., 1972) and languages (Cohen and Starkweather, 1961). Prosody thus conveys to the fetus a great deal of information about the identity and language of its host.

In the present context, it is important to consider the indexical significance of language. The way a person speaks a language may, and frequently does, contain clues to the individual's age, social class, education, region of birth or residence, politics, and national affiliation. That is to say, upon hearing individuals speak we learn something about their *group*.

The mother speaks a particular language and also has her own unique vocal signature. The infant's listening preference thus rests, potentially, on a single set of vocal characteristics that distinguish its mother from (a) all other speakers of her language and (b) speakers of all other languages. The infant's initial preference for the mother's "voice" is, thus, a preference for *the sound pattern of her language* as well as any personally identifying information.

According to this view, the infant's preference for a language - a general characteristic of his or her social group - follows from familiarization with a specific individual, the infant's mother or primary caregiver. Moreover, this individual-to-group sequence conforms to a general pattern that applies to avians as well as primates (Bolhuis, 1991). For example, ducklings appear to respond preferentially to conspecific calls because of their similarity to the duckling's own vocalizations (Gottlieb, (1991a,b).

The 3-day-old human also looks preferentially at its mother, relative to other mothers, but facial learning can only occur postnatally, and does so in the first few minutes and hours following birth (Bushnell et al., 1989; Field et al., 1984). Occurring so quickly, one assumes that the face has been of such importance phylogenetically that newborns are preadapted to look at it, and thus to extract information from the face. It is an ability that we share with non-human primates (Sackett et al., 1970), who prefer the faces of conspecifics (whether they prefer the face of their mother seems to be unknown at present). They also share the human infant's interest in the eyes, a preference that is undiminished by the existence of mouth movements (Haith et al., 1977; Keating and Keating, 1982, 1993; Mendelson et al., 1982).

In the second 6 months of life, when brain developments enable human infants to inhibit attention to irrelevant stimulation (Chugani et al., 1987) - and thus, presumably, to listen selectively - they acquire a preference for the sound of the ambient language over a foreign one (Jusczyk et al., 1993a,b), and demonstrate awareness of the relationship between prosodic cues and co-varying linguistic units (Goodsit et al., 1993; Hirsh-Pasek et al., 1987). During this time, listening experience strengthens some adult-like perceptual categories that are evident in the first few weeks of life and weakens others. In effect, native language listening experience has the effect of shifting the perceptual boundaries of vowel-like sounds at about 6 months (Kuhl et al., 1992), and modifies the boundaries of consonant-like sounds at about 10 to 12 months (Werker and Polka, 1993).

This might suggest that the year-old human infant is already deeply involved in *linguistic learning*, but this is not the case. In a model of phonetic perceptual development, Werker and Pegg (1992) proposed that infants progress through four different types of perceptual learning on their way to full linguistic capacity. In the first two types the infant is influenced by ambient stimulation, but it is primarily *auditory* and relatively nonspecific. Since in the case of consonant-like sounds, these phases occur by or before the age of 10 months - an age in which lexical comprehension is usually demonstrable - the learning of words antedates the kinds of perceptual learning that are most straightforwardly linguistic.

As for production, infants may occasionally reproduce maternal intonation contours in the first 6 months of life (Masataka, 1992), and reveal in spontaneous utterances their assimilation of ambient sound patterns in the second 6 months (Boysson-Bardies et al., 1992). During the vocal learning phase, infants thus become acquainted with vocal cues that identify people, convey emotion, regulate social behavior, and superficially characterize their native language.

At present, it is not clear whether non-human primates' development of species-typical vocal patterns involves ambient learning A cross-fostering study by Masataka and Fujita (1989) suggests that it does, at least in Japanese and rhesus monkeys, but a similarly designed and well controlled study of the same species by Owren et al. (1992) produced little evidence for vocal learning.

Vocal Turn Taking

Interaction is obviously critical to social learning, and an important format for interaction involves mother-infant alternations of behavior. Infants and their mothers tend to take turns vocalizing. This practice probably facilitates acquisition of native language, because to perceive and reproduce sound patterns infants must have good perceptual access to the material to be reproduced. And to respond contingently, a demonstrably helpful process, mothers must hear the infant's reproductions of their speech with some clarity.

Several investigators have found a marked increase in vocal turn taking between 12 and 18 weeks of age (Ginsburg and Kilbourne, 1988). Studies suggest that at this age, infants begin to forestall vocalization if their mother is speaking and to fall silent if their mother starts to speak (Moran et al., 1987), a practice that intensifies in subsequent months (Papousek and Papousek, 1989).

Vocal turn taking may be related to maternal attachment. For one thing, infants are more likely to turn off their own voice if that increases perceptual access to the voice of a person they want to hear (Barrett-Goldfarb and Whitehurst, 1973). Moreover, Bowlby (1969) suggests that attachment begins in earnest when infants' preference for their mother's company is satisfied either facially or vocally, i.e., not just facially. Since the phase in which this happens usually begins at about 12 weeks, attachment may commence at about the same time as vocal turn taking. A second reason for a relationship between vocal turn taking and attachment is that both behaviors involve emotional interpretation and inhibition, functions made possible by development of prefrontal cortex. Studies suggest that prefrontal cortex becomes unusually active as early as 6 to 7 months (Bell and Fox, 1992; Chugani et al., 1987), perhaps beginning sometime after 3 months (Chugani and Phelps, 1986).

Turn taking also occurs in non-human primates, at least in adult animals. In the vocal exchanges between squirrel monkeys, there is evidence for a conversation-like temporal structure (Biben et al., 1986), and what may be a turn-taking cue was reported recently by Hauser (1992b). In vervets and rhesus macaques, he found that a decline in the fundamental frequency of certain call types was associated with vocal interruption by another animal.

Of ontogenetic significance, there is evidence for turn-taking in the nonvocal play behaviors of chimpanzee mother-infant dyads (Plooij, 1979), and adult squirrel monkeys appear to use several distinctive call types only or primarily in certain caregiving contexts (e.g., while inspecting or retrieving their infant; Biben et al., 1989).

Do neonatal monkeys and apes react selectively to maternal voice to the same degree as human newborns? No answer seems to be available at the moment. But it is important to find out, for if young primates are less reactive to maternal voice, one might not expect that with increasing fearfulness they would be selectively reassured by her voice. And if the sound of her voice failed to function in this way there would be less reason for infant apes and monkeys to monitor maternal vocal activity.

Infants prefer the company of their mother over that of a stranger, and thus we may assume that they prefer to hear, and therefore actively listen to, her voice. For this reason alone, one might expect infants to store more maternal speech than non-maternal speech. But they also may have more opportunity to learn from their mother, since her presence would encourage quiet alertness - which other research indicates may be necessary for efficient vocal learning (Kugiumutzakis, 1993) - rather than an agitated or fretful state.

THE STORAGE PHASE

We saw earlier that vocal storage begins as early as the prenatal period, when experience begins to tune vocal learning mechanisms to the prosody of the mother and, incidentally, her language. This is followed by a phase in the development of linguistic capacity that is massively dedicated to storage of utterances (Locke, in press, a). Typically, this phase is revealed in the form of lexical comprehension in infants as young as 8 months. At this age, the normally developing infant responds appropriately to an average of 36 words and produces fewer than 2. At 10 months, infants comprehend about 67 words, but they will not produce that many words until at least 6 months later (Bates et al., 1994).

The true number of words that are stored may be even greater than this, for to be credited with lexical storage infants must respond appropriately to words. Appropriate responding cannot occur unless supporting contextual and referential information also was stored, along with the words, and the infant is sufficiently responsive (e.g., by looking or pointing at named objects) that its comprehensions can be inferred by adult observers. There is, thus, a long lexical storage period that precedes speech. A somewhat analogous period has been proposed in the vocal development of songbirds (Marler and Peters, 1982).

It is appropriate to note that one non-human primate - the bonobo, Kanzi - was capable of storing and comprehending a fairly large vocabulary of spoken words that were not specifically taught or trained, but used in the context of his everyday social interactions with language researchers (Savage-Rumbaugh et al., 1993). This behavior may have been facilitated by a human-like enculturation that encouraged joint attention to objects (Tomasello et al., 1993).

Are other maternal behaviors absorbed at the age that utterances are first comprehended? Uzgiris et al. (1989) studied imitation of a variety of motor activities by eighty infants in 9- to 12-minute face-to-face interactions. Infants rarely matched maternal behavior at 2.5 and 5.5 months, but a significant increase between 5.5 and 8.5 months. As it turned out, every imitative act involved the voice, and in about a third of the matchings the voice was the only behavior reproduced. However, in the remaining cases infants imitated activities of the voice and the face or the voice and the hands. This suggests that during the storage phase infants may take on a number of vocal and other behaviors, not just those having linguistic status.

Inhibition and Perceptual Learning

As we have seen, vocal and lexical learning advances significantly during the second six months of life. Maternal facial monitoring and, we may presume, attention to affective and indexical properties of the mother's voice, reaches a peak between 7 and 12 months. Words are first evidently stored during this period. Perceptual learning first conspicuously reorganizes to listening experience during this interval. Vocal production is first influenced by specific ambient stimulation. Why do all these important preparations for, and stages in, spoken language occur during this particular period?

The development of prefrontal cortex seems to play a major role in these ontogenetic processes and events. It is now understood that this area figures importantly into the development of social and emotional responding, as is revealed in attachment behaviors, both in non-human primates and humans (Kalin and Shelton, 1989; Kalin et al., 1991). It also makes possible inhibition of attention to irrelevant stimulation and is therefore responsible for selective attention (Diamond, 1990; Goldman-Rakic, 1987). Since, as seen earlier, activity of the prefrontal cortex increases greatly during the 7 to 12 months interval, this area may play a major role in selective attention to speech, thus enabling utterance storage and the attentional tuning of perceptual mechanisms (Werker and Pegg, 1992).

VOCAL ACCOMMODATION

Stored material begins to become expressible through a process I have termed vocal accommodation, an infantile variant of what social psychologists call "speech accommodation". Vocal accommodation refers to minor adjustments, usually unintentional, in one's vocal output that cause it to be more like that of an influential individual with whom one is interacting. This process requires very little modulatory control since it mainly affects a range of phonatory qualities, including the pitch, loudness, and length of utterances.

There is evidence that vocal accommodation initially occurs in the human infant before talking begins (Boysson-Bardies et al., 1989; Kuhl and Meltzoff, 1989). Exposure to stereotyped utterances in contextually limited circumstances typically produces an initial stock of word-like vocal behaviors that antedate development of phonology and morphology by many months. Vocal accommodation is therefore linguistically significant in the human, even though it does not qualify as an example of linguistic behavior according to strict definitions of language (e.g., Terrace et al., 1980; see below).

Vocal accommodation also seems to occur in non-human primates. For example, when a squirrel monkey engages in "dialogue" with a conspecific cagemate, the frequency and morphology of its vocalizations may be influenced by immediately preceding vocal behaviors of the other animal (Maurus et al., 1987). It may be relevant to theories of linguistic evolution that some of the vocal parameters on which there is evidence of accommodation in non-human primates, e.g., fundamental frequency modulations (Maurus et al., 1988), also are used contrastively in spoken languages.

In a study of coo calls in free-living rhesus monkeys, Hauser (1992a) documented some behaviors that in my judgment may qualify as vocal accommodations. By noting which animals vocalized, and their genetic relationship to other animals, Hauser was able to discover something quite remarkable. His spectrographic analyses revealed that coo calls from the members of one matriline differed systematically from the coos of other matrilines. The acoustic basis for differences lay in the spectral prominences above the second harmonic, proportion of missing harmonics, intensity of the fundamental frequency and second harmonic, and energy between harmonics. The result of these variations was that, to the human listener, one matriline sounded "nasal" and the others did not. Although Hauser could not absolutely rule out an organic explanation such as a heritable palatal deficiency, these effects may well have been the result of infants' accommodations to their mother's nasal quality. Familial studies of this sort are of value both to primatology and developmental psycholinguistics.

Prosodic Speech

The first detectable infant speech may be a word or what sounds like a phrase (Peters, 1983). Phrases are usually a rote or nearly rote copy of word sequences that have been heard frequently in the speech of others, e.g., "shoes and socks" and "time to go to bed." Although articulation may be imprecise, the prosody of this material is usually well preserved (Nelson, 1981). Often, phrases contain articles and pronouns, as well as words that are otherwise late to develop and may not occur elsewhere in the speech of the child (Hickey, 1993). These phrases therefore give pregrammatical children an air of linguistic sophistication.

The word sequences described above are called *formulas*. Formulas are vocal material that has not been subjected to, and whose comprehension and production do not require, analysis into constituent syllables and sounds. But formulas are not rare. Although their frequency seems to vary with children's learning style, in the second year of life about a fifth of infant's lexical entries are formulaic phrases (Lieven et al., 1992).

From a functional standpoint, these units enable young children to speak grammatically even before they know the words - or know that there *are* words - embedded in their vocal material. There is also no indication that young children are unconsciously aware that the words they do know have constituent elements such as morphemes or phonemes, and they appear not to understand the concept of word or lexical meaning, or the taxonomic principle underlying word use (Markman and Hutchinson, 1984). For this reason, it is parsimonious to suppose that most or even *all* of infants' speech is formulaic. The shorter utterances are heard by listeners as words, the longer utterances as multisyllabic words or phrases.

The neural mechanisms that support storage operations may be regarded as components of this utterance acquisition system that captures and reproduces prosodic patterns as shaped by the infant's perceptual biases and interests. The ostensibly "non-linguistic" right hemisphere predominates in the processing of underanalyzed material when adults listen to idioms (Van Lancker, 1987) and pregrammatical children listen to words that they do not comprehend (Mills et al., 1993).

Speech Without Language

The significance of the above is that the human infant need not possess grammatical capability either to comprehend or to produce speech. Moreover, the first 10 or 12 months of talking occur before development of grammatical functions, and before the cognitive growth needed to use stored utterances propositionally, i.e., to convey thoughts they have and others do not. If infants are to identify perceptual cues and functional units, and analyze the structure of language, they will need to see some value in the activities that linguistic competence permits. Even if their vocal tract and vocal-motor control system are perfectly normal, infants are unlikely to work out all the articulatory routines that are needed for expression of linguistic knowledge unless they anticipate some functional benefits. In the normally developing child, the efforts to analyze and execute spoken language seem to be driven by a desire to communicate, a desire that arises from our species specialization in social cognition. Whether non-human primates have an equivalent urge to communicate by way of learned gestural or vocal behaviors is unclear (Terrace et al., 1980).

The phrase "urge to communicate" may seem disturbingly inexact. How, after all, could one test for this attribute, or precisely describe it? Fortunately, this concept rests upon other mental abilities that are susceptible to evaluation. Chief among these abilities is the child's awareness that other individuals have mental experiences that differ from his own. If children do not recognize discrepancies between their own and other people's knowledge and feelings, they will be able to say words, i.e., to talk, but they will be unable to use words strategically and flexibly to express thoughts. When a child excitedly attempts to speak, one surmises that he has just experienced what he believes to be *his own* thoughts. This "theory of other minds", as it is called, can be tested for and quantified in human children (Baron-Cohen et al., 1985).

The child's first appreciation of other minds seems somehow to be related to, and may be preceded by, intentional deception and pretend play. To some developmentalists, effective deception implies the ability to perceive and to act upon the dispositions and intentions of others. Human children usually reveal this capacity in their second year of life (Kuczynski et al., 1987; Watson and Fischer, 1977).

Non-human primates seem to lack an awareness of other individuals' mental activity (Premack and Woodruff, 1978), and there is reason to believe that some animals even have difficulty recognizing their own existence. Consider the results of research using the mirror test. In that paradigm, a spot of rouge is put on the nose or forehead of the subject. When given the chance to look in a mirror, it is assumed that animals who know that the image is their own will touch the spot or try to rub it off. Normally developing infants do this by about 22 months of age (Mans et al., 1978); infants with Down syndrome do not react in this way until somewhat later (Hill and Tomlin, 1981).

A review of the literature (Gallup, 1982) suggests that chimpanzees and orangutans recognize themselves in the mirror but that the vast majority of primate species, including gorillas and several species of gibbons and monkeys, seem incapable of figuring out that the movements they see in the mirror were initiated by them, and therefore that the animal they see *is* them. There is just a handful of published accounts (Cheney and Seyfarth, 1990; Steklis and Raleigh, 1979; Whiten and Byrne, 1988) in which one animal seemingly deceived another intentionally. But in many of these cases it is not clear that the deceiver intended the deception, and attributed intentions to its victim (Byrne and Whiten, 1991; Heyes, 1993).

Normally developing 18-month-olds are thus able to talk but they give no evidence of grammatical capability. By this I mean that their speech is relatively unconstrained by application of linguistic rules, and is relatively uncreative. They say little that they have not heard before. Terrace and his colleagues have commented that "if a child's initial utterances and his responses to his parents' utterances constituted the only evidence of his linguistic ability, there would be little reason to conclude that a child's production and comprehension of words are governed by a grammar." (Terrace et al., 1980, p. 434).

While their utterance-reproduction capability allows pregrammatical children to talk, and thereby to be regarded as talkers, it will not enable them to achieve a large lexicon. For there are presumed to be absolute storage limitations on material that is prosodically well specified but underspecified at the level of component sounds (Lindblom et al., 1983). Lacking the ability to break utterances into smaller pieces of speech, there exists no possibility of a grammar forming in the storage phase. Nonetheless, this phase does pave the way for other developmental linguistic steps by providing systems still to be activated with an enormous sample of linguistically relevant data.

The 12- to 18-month-old's capacity to store and recognize prosodic material, and to relate it to external referents, exists to some degree in other primates. Kanzi, the bonobo mentioned earlier, participated in a language project from an early age and now comprehends hundreds of words, and responds appropriately to numerous sentences that have never been heard previously (Savage-Rumbaugh et al., 1993). Is Kanzi linguistic? If research on non-human primates is truly to be *comparative*, it is essential that we adopt a common definition of language. Terrace et al. (1980) observed that their own research subject - Nim, a young chimpanzee who was taught sign language - used novel, multi-word combinations, but failed to "express a semantic proposition in a rule-governed sequence of signs." (p. 401). Normally developing 18-month-old human infants produce many communicative behaviors that are not linguistic and many language-like behaviors that fail to express a thought. Are they linguistic? By the definition of Terrace et al., they too lack the capacity for human language, and will continue to lack it until - barring motoric limitations - they first demonstrate the capability of expressing semantic propositions using rule-governed sequences of words.[1]

THE ANALYTIC PHASE

Phase 3 in the development of linguistic capacity involves operations - initially the analysis, and then the computation, of structure. These operations are performed on material that has been acquired in the preceding storage phase. I assume the mechanisms carrying out these operations are experience-dependent; if the infant has no stored utterances, there will be nothing for its analytical mechanisms to work on, and no stored forms to reconfigure by application of computational rules. The development of grammatical capability therefore depends on prior success in the storage phase. Where the second phase is affected by external factors, including the availability of appropriate stimulation, functions instated in the third phase are affected primarily by internal factors. That is, the ability to analyze utterances is affected by environmental stimulation only insofar as that stimulation is correctly relayed to a structure analyzer by subordinated processing systems. To understand language development, it is therefore not enough to know what has been said in the presence of the child; one must know about the success of internal operations, that is, which utterances were perceived, stored, and submitted to a constituency analysis by the child.

The structure analysis system locates recurring elements within and across utterances, and thereby learns the rules by which utterances are to be synthesized and parsed. In effect, it presents the child with the units needed for all linguistic domains - morphology, syntax, phonology, and the lexicon. The initial, and perhaps most conspicuous, form of internal evidence that an analyzer is at work comes from children's regularization of irregular verbs. When this happens, words like "went" may be temporarily and inconsistently expressed as "goed" (Marcus et al., 1992). These forms are not merely heard and reproduced by children, they are *generated*.

I believe that a rapidly expanding lexicon presses analytical mechanisms into operation because of storage limitations on underanalyzed vocal material. At about 16 to 18 months, there is a rapid acceleration in the rate at which new words and phrases are comprehended (Reznick and Goldfield, 1992). These words will begin to compete for available space if the infant does not discover some internal organization - which presupposes discovery of the units that comprise phonology. Using a looking time measure of language comprehension, it was found recently that infants begin to discriminate phonetic segments that distinguish words in their own receptive vocabulary, i.e., develop classically defined phonemic categories, at about 19 months of age (Werker et al., in preparation). This advance is attributable to an analytical mechanism in the left hemisphere, for using the same paradigm, other researchers have discovered a rightward looking bias that develops between 16 months and 19 months (Mount et al., 1989).

A parallel argument has been advanced by Marchman and Bates (in press), who have suggested that a certain number of words must be known if the infant is to discover morphological regularities. In either case, one would expect, and does find, a positive relationship between vocabulary and grammatical measures (Bates et al., 1988).

Are non-human primates able to carry out analytical operations resembling or analogous to those performed on speech by 2- to 3-year-old children? Although a spirited

---

1 Another argument that the initial vocabulary of children is not, in and of itself, a strictly linguistic matter is suggested by a study of the vocabulary development of gorillas and children who were exposed to sign language (Bonvillian and Patterson, 1993). Though differences were found in the rate of learning and the role of iconicity, frequency analyses produced identical word category profiles in the two species, extending from general nominals, which predominated in gorillas and children, to action words, modifiers, specific nominals, personal-social terms, and function words.

argument has been made for a language-like organizational ability in non-human primates (Greenfield, 1991), there is no evidence that they possess the ability to first decompose highly structured vocalizations into component parts and then resequence such phoneme-size pieces of speech into novel patterns.

MORE VOCAL CONTROL

Much attention has been devoted to the role of vocal tract anatomy and peripheral effector systems in relation to the hominid's emerging capacity for spoken language. However, the importance of the pharynx may have been overstated (cf. Lieberman, 1984). There are documented cases of children who have learned to speak without a tongue (MacKain, 1984), and it is possible to learn language without a left cerebral hemisphere. But it is essential that the vocal tract be hooked up to a reasonably competent control center somewhere in the brain.

Children increasingly develop vocal control over the first half-dozen years of life. The process seems to take a step forward when babbling begins, usually at about 6 to 7 months. Reduplicated babbling refers to the alternate depression and elevation of the mandible, with points of oral contact, while phonating. To the ear, this activity produces syllabic elements, e.g., dadada, that resemble speech. Although a behavior referred to as "lipsmacking" has been described in the non-human primate literature (Anthoney, 1968; Kenney et al., 1979; Marler and Tenaza, 1977; Redican, 1975), it is not clear that this behavior, which seeming lacks phonatory support, is quite the same thing as babbling.

Conditioning studies suggest that non-human primates can be able to learn when and with what duration to vocalize, as participants in carefully controlled experiments (Sutton et al., 1973, 1978), but there is little evidence that they are able to make fine adjustments in their vocal quality in response to feedback or reinforcement (although see Masataka, this volume). When efforts have been made to condition vocal behavior, and then to bilaterally lesion areas of the neocortex that are homologous to Broca's, no change has occurred either in the number or duration of learned (or spontaneous) vocalizations. Such changes as these occurred only when lesions were made in the area of the anterior cingulate gyrus (Sutton et al., 1981; Trachy et al., 1981), a structure associated with volitional speech in humans. If neocortex plays little or no role in conditional vocalization, as the work of Sutton and his colleagues suggests, it seems unlikely that non-human primates would be capable of the rampant vocal learning that occurs in language development.

Because the left Sylvian fissure may be longer than the right in great apes (LeMay and Geschwind, 1975) and Old World monkeys (Falk, 1978), it is unclear how to interpret similar asymmetries in the human brain. What, then, about the neurophysiology of phonation in non-human primates? Fortunately, we are beginning to obtain some preliminary information on which parts of the primate brain control laryngeal and respiratory activity associated with phonation. Research by Larson (Larson, 1988; Larson and Kistler, 1986) and by Jürgens and his associates (this volume) suggests that much of phonatory activity is controlled by the midbrain periaqueductal gray, which receives projections from the limbic system. This is not surprising, considering the close relationship between emotionality and vocal behavior, and Sutton's work suggesting that spontaneous vocalization survives bilateral lesions of the cortex. It would be interesting to see what parts of monkey brain are activated by conspecific and heterospecific vocalizations of various types.

PHYLOGENY: LEXICAL PRESSURES TO ORGANIZE

To summarize the argument offered here, the child's use of words presupposes lexical comprehension, which depends on storage of utterances and supporting contextual information. This is typically carried out mainly by the right cerebral hemisphere, which is predominantly involved in prosody and vocal affect and is better developed than the left hemisphere at the time these events occur. My claim is that lexical storage is performed by infants long before they begin to speak or even know about the existence of linguistic rules and representations, on a social motivation that arises, in part, from a functional need to monitor caregiver vocalization. Monitoring of maternal voice is particularly functional in securely attached infants who value her vocal stimulation both for its indexical value and its ability to convey estimates of infant risk.

Stereotyped utterances that infants store and reproduce are recognized by familiar listeners as speech. Lexical forms accumulate at a gradual rate until a point at which - for reasons not well understood currently - rate of word learning accelerates. After some months, this proliferation of stored forms threatens to overload their small capacity prosodic store. Overloading is avoided by activation of a mechanism, typically sited in the left hemisphere, that analyzes utterances into their component parts (Locke, 1992). When the infant detects the patterning of these units, it is then in a position to learn something about the distributional rules of the ambient phonology. From these operations, computational capacity emerges in the form of words whose surface morphology could only have been generated by rule.

The evidence we have seen to this point, whether it relates to the social and cognitive capabilities of non-human primate adults or human infants, suggests that hominid young would have been oriented to the face and voice, and would have been able to send and interpret affective signals through these modalities. If they possessed sufficient phonatory and oral control to vary pitch, loudness, duration, and vocal quality, they could have talked, and perhaps achieved an expressive vocabulary of several hundred utterances or more.

It can be assumed that the right cerebral hemisphere of hominids controlled facial activity and processed vocal activity associated with the expression of emotion because such seems to be true both of non-human primates and humans. It also can be assumed that the left cerebral hemisphere of hominids processed species-significant vocal contrasts, as has been found in non-human primates and humans.

Pressures to elaborate on this lexical capability, e.g., a felt need to work cooperatively (Pinker and Bloom, 1990), could have exerted an effect only because the young already had, and were regularly exercising, the capabilities to store, process, and produce utterances. Speculations on evolutionary pressures for language are, thus, arguments in favor of enhanced *lexical* communication. One might speculate that with burgeoning vocabulary there was increasing need to subject prosodically stored material to analysis. Grammatical capability would thus have emerged as a neurocognitive, and ultimately neurolinguistic, response to the rampant proliferation of phonetic representations.

A SHARED RESEARCH AGENDA

If non-human primates lack the vocal-analytic capacity required by linguistic grammar, and demonstrate relatively less vocal-motor control, non-human primates still may possess one or more of the capacities that support lexical comprehension and thus feed grammatical systems in the human infant. Primatologists can help determine which social, cognitive, and neural developments support and lead to grammatical capability ontogenetically and in the species.

It would be extremely productive if primatologists and developmental psycholinguists were to pursue a shared agenda of biolinguistic research. From my viewpoint, it would be helpful if this agenda included the following areas of inquiry:

Early Learning

We have seen a great many cognitive and social similarities between human and non-human primate infants, especially in the first few years of life. This suggests that very young monkeys and apes closely resemble the human fetus and neonate in ways that may be relevant to vocal learning. One question that might be asked, with appropriate laboratory controls, is how the non-human primate fetus reacts to externally presented alarm calls and other vocalizations that it can hear but the mother cannot? Other questions include: how does the fetus normally react to, and what does it learn from, the sound of its mother's voice? And, do physiologic measures reveal differential responses to conspecific and heterospecific vocalizations? In a more social context, it should be asked when the non-human primate infant first recognizes or preferentially attends to maternal face and voice. When the attached infant is separated from its mother, does it react differently to the sound of her voice as opposed to the voice of some other adult female? Exploration of these issues would tell us more about development of the potential for vocal learning in human and non-human primates, and shed light on inter-specific differences that are present as early as the prenatal period.

Social Neurophysiology

Research indicates that in several sub-species of squirrel monkeys there is selective response by adult animals to the peeps of their own sub-species (Snowdon et al., 1985). With neuroimaging techniques, either PET or fast-scanning MRI, it is possible to ask which regions and structures of non-human primate brain are activated by conspecific and heterospecific vocalizations of various types. It would be predicted, of course, that some conspecific calls will be lateralized to the left hemisphere, but more specific localization should be possible. Recognizing that humans have cells that respond preferentially to autogenous vocalizations (Creutzfeldt et al., 1989a,b), what is the interconnectivity and functional communication between any clusters of conspecific voice-sensitive cells that are found and cells that control or selectively activate to the animal's own vocalizations?

There seem to be phylogenetic and functional links between vocal and facial signaling in primates generally (Andrew, 1963a,b), and one presumes that neuroanatomic connections must also exist between these communicative systems. In macaques, some studies have located multimodal or polysensory cells that appear to respond to both visual and auditory stimulation (Bruce et al., 1981), and in humans at least one study has reported links between facial and lexical processing (Heit et al., 1988). It would be useful to explore interactions between these affective communication systems that are shared by all primates using a variety of neurophysiological techniques.

In an unusual study of *human* alarm calls, if they may be called that, Comuzzie and Wilcox (in preparation) recently reported that adults localized contrived words more accurately when spoken in an alarmed than in a conversational manner. This result raises a host of interspecific questions including, but certainly not limited to, the human neonate's response to the contact and alarm calls of non-human primates, and the response of monkeys and apes to the range of affective qualities that naturally occurs in human speech. It also would be of some interest to know the responses of human and non-human primates to vocal acts in which the affective information on the face supports or conflicts with the information that is heard.

Joint Efforts

The developmental linguistic theory presented here was inspired by particular types of non-human primate research. Those more concerned with the evolution of language have proceeded differently than I, some placing a higher value on hemispheric motor asymmetries (MacNeilage et al., 1987) or vocal tract anatomy (Lieberman, 1984) - issues relating to the human sound producing capability - others exploring the ability of non-human primates to learn spoken, signed, or specially contrived languages (cf. Savage-Rumbaugh et al., 1993; Terrace et al., 1980). Regardless of the approach, with comparisons between species and collaborations between researchers our understanding of the human faculty for spoken language is certain to expand in interesting ways.

ACKNOWLEDGMENTS

The author wishes to thank Marc Hauser for many inspiring discussions, and to Anne Fernald and Robert Hinde for helpful criticisms of the manuscript.

REFERENCES

Andrew, R. J., 1963a, Evolution of facial expression, *Science* 141:1034-1041.

Andrew, R. J, 1963b, The origin and evolution of the calls and facial expressions of the primates, *Behaviour* 20:1-109.

Anthoney, T. R., 1968, The ontogeny of greeting, grooming and sexual motor patterns in captive baboons (superspecies *Papio cynocephalus*). *Behaviour* 31:358-372.

Baron-Cohen, S., Leslie, A. M., and Frith, U., 1985, Does the autistic child have a "theory of mind"?, *Cognition* 21:37-46.

Barrett-Goldfarb, M. S. and Whitehurst, G. J., 1973, Infant vocalizations as a function of parental voice selection, *Dev. Psychol.* 8:273-276.

Bates, E., Bretherton, I., and Snyder, L., 1988, "From First Words to Grammar: Individual Differences and Dissociable Mechanisms", Cambridge University Press, New York.

Bates, E., Marchman, V., Thal, D., Fenson, L., Dale, P., Reznick, J. S., Reilly, J., and Hartung, J., 1994, Developmental and stylistic variation in the composition of early vocabulary, *J. Child Lang.* 21:85-123.

Bauman, M. and Kemper, T. L., 1985, Histoanatomic observations of the brain in early infantile autism, *Neurology* 35:866-874.

Bell, M. A. and Fox, N. A., 1992, The relations between frontal brain electrical activity and cognitive development during infancy, *Child Develop.* 63:1142-1163.

Berntson, G. G. and Boysen, S. T., 1989, Specificity of the cardiac response to conspecific vocalizations in chimpanzees, *Behav. Neurosci* .103:235-245.

Berntson, G. G., Boysen, S. T., Bauer, H. R., and Torello, M. S., 1990, Conspecific screams and laughter: cardiac and behavioral reactions of infant chimpanzees, *Develop. Psychobiol.* 22:771-787.

Biben, M., Symmes, D., and Bernhards, D., 1989, Contour variables in vocal communication between squirrel monkey mothers and infants, *Develop. Psychobiol.* 22:617-631.

Biben, M., Symmes, D., and Masataka, N., 1986, Temporal and structural analysis of affiliate vocal exchanges in squirrel monkeys (*Saimiri sciureus*), *Behaviour* 98: 259-273.

Bolhuis, J. J., 1991, Mechanisms of avian imprinting: a review, *Biol. Rev.* 66:303-345.

Bonvillian, J. D. and Patterson, F. G., 1993, Early sign language acquisition in children and gorillas: vocabulary content and sign iconicity, *First Lang.* 13:315-338.

Bowlby, J., 1969, "Attachment. Volume 1. Attachment and loss", Basic Books, New York.

Boysson-Bardies, B. de, Halle, P., Sagart, L., and Durand, C., 1989, A crosslinguistic investigation of vowel formants in babbling, *J. Child Lang.* 16:1-18.

Boysson-Bardies, B. de, Vihman, M.M., Roug-Hellichius, L., Durand, C., Landberg, I., and Arao, F., 1992, Material evidence of infant selection from the target language: a cross-linguistic phonetic study, *in*: "Phonological Development: Models, Research Implication", C.A. Ferguson, L. Menn and C. Stoel-Gammon, eds., York Press, Timonium, Maryland.

Brothers, L., 1990, The social brain: A project for integrating primate behavior and neurophysiology in a new domain, *Concepts Neurosci.* 1:27-51.
Brothers, L. and Ring, B. A., 1992, Neuroethological framework for the representation of minds, *J. Cogn. Neurosci.* 4:107-118.
Bruce, C., Desimone, R., and Gross, C. G., 1981, Visual properties of neurons in a polysensory area in superior temporal sulcus of the macaque, *J. Neurophysiol.* 46:369-384.
Bullowa, M., 1979, "Before Speech: The Beginning of Interpersonal Communication", Cambridge University Press, New York.
Bushnell, I. W. R., Sai, F., and Mullin, J. T., 1989, Neonatal recognition of the mother's face, *Brit. J. Develop. Psychol.* 7: 3-15.
Byrne, R. W. and Whiten, A., 1991, Computation and mindreading in primate tactical deception, *in*: "Natural Theories of Mind", A. Whiten, ed., Basil Blackwell, Oxford.
Cheney, D. L. and Seyfarth, R. M., 1980, Vocal recognition in free-ranging vervet monkeys, *Anim. Behav.* 28:362-367.
Cheney, D. L. and Seyfarth, R. M., 1990, "How Monkeys See the World: Inside the Mind of Another Species", University of Chicago Press, Chicago.
Chugani, H. T. and Phelps, M. E., 1986, Maturational changes in cerebral function in infants determined by 18FDG positron emission tomography, *Science* 231:840-843.
Chugani, H. T., Phelps, M. E., and Mazziota, J. C., 1987, Positron emission tomography study of human brain functional development, *Ann. Neurol.* 22:487-497.
Cohen, A. and Starkweather, J., 1961, Vocal cues to language identification, *Am. J. Psychol.* 74:90-93.
Comuzzie, D. K. C. and Wilcox, K. A., in preparation, The localizability of human alarm calls.
Creutzfeldt, O., Ojemann, G., and Lettich, E., 1989a, Neuronal activity in the human lateral temporal lobe. I. Responses to speech, *Exp. Brain Res.* 77:451-475.
Creutzfeldt, O., Ojemann, G., and Lettich, E., 1989b, Neuronal activity in the human lateral temporal lobe. II. Responses to the subjects own voice, *Exp. Brain Res.* 77:476-489.
DeCasper, A. and Fifer, W. P., 1980, On human bonding: Newborns prefer their mothers' voices, *Science* 208:1174-1176.
DeCasper, A. and Spence, M., 1986, Prenatal maternal speech influences newborns' perception of speech sounds, *Infant Behav. Develop.* 9:133-150.
Desimone, R., 1991, Face-selective cells in the temporal cortex of monkeys, *J. Cogn. Neurosci.* 3:1-8.
Diamond, A., 1990, The development and neural bases of memory functions as indexed by the AB and delayed response tasks in human infants and infant monkeys, in: "The development and neural bases of higher cognitive functions", A. Diamond, ed., *Ann. New York Acad. Sci.* 608: 267-317.
Etcoff, N. L., 1989, Asymmetries in recognition of emotion, *in*: "Handbook of Neuropsychology", F. Boller and J. Grafman, eds., Elsevier, New York.
Falk, D., 1978, External neuroanatomy of Old World monkeys (*Cercopithecoidea*), *Contributions to Primatology* 15, Karger, Basel.
Fernald, A. and Kuhl, P., 1987, Acoustic determinants of infant preference for motherese speech, *Infant Behav. Develop.* 10:279-293.
Field, T. M., Cohen, D., Garcia, R., and Greenberg, R., 1984, Mother-stranger face discrimination by the newborn, *Infant Behav. Develop.* 7:19-25.
Fodor, J., 1983, "Modularity of Mind", MIT Press, Cambridge, Mass.
Gallup, G. G., 1982, Self-awareness and the emergence of mind in primates, *Am. J. Primatol.* 2:237-248.
Ginsburg, G. P. and Kilbourne, B. K., 1988, Emergence of vocal alternation in mother-infant interchanges, *J. Child Lang.* 15:221-235.
Goldman-Rakic, P. S., 1987, Development of cortical circuitry and cognitive function, *Child Develop.* 58:601-622.
Golinkoff, R. M., 1983, "The Transition from Prelinguistic to Linguistic Communication", Erlbaum, Hillsdale, New Jersey.
Goodsit, J. V., Morgan, J. L., and Kuhl, P. K., 1993, Perceptual strategies in prelingual speech segmentation, *J. Child Lang.* 20:229-252.
Gottlieb, G., 1991a, Experiential canalization of behavioral development: theory, *Dev. Psychol.* 27:4-13.
Gottlieb, G., 1991b, Experiential canalization of behavioral development: results, *Dev. Psychol.* 27:35-39.
Greenfield, P. M., 1991, Language, tools and brain: the ontogeny and phylogeny of hierarchically organized sequential behavior, *Behav. Brain Sci.* 14:531-595.
Haith, M. M., Bergman, T., and Moore, M. J., 1977, Eye contact and face scanning in early infancy, *Science* 198:853-855.
Hamilton, C. R. and Vermeire, B. A., 1988, Complementary hemisphere specialization in monkeys, *Science* 242:1691-1694.

Hardy-Brown, K., Plomin, R., and DeFries, J. C., 1981, Genetic and environmental influences on the rate of communicative development in the first year of life, *Develop. Psychol.* 17:704-717.
Hauser, M., 1989, Ontogenetic changes in the comprehension and production of Vervet monkey (*Cercopithecus aethiops)* vocalizations, *J. Comp. Psychol.* 103:149-158.
Hauser, M., 1992a, Articulatory and social factors influence the acoustic structure of rhesus monkey vocalizations: A learned mode of production?, *J. Acoust. Soc. Amer.* 91:2175-2179.
Hauser, M. D., 1992b, A mechanism guiding conversational turn-taking in vervet monkeys and rhesus macaques, *in*: "Topics in Primatology. Vol. 1. Human Origins", Tokyo University Press, Tokyo.
Hauser, M. D., 1993, Right hemisphere dominance for the production of facial expression in monkeys, *Science* 261:475-477.
Hauser, M. D. and Andersson, K., in press, Left hemisphere dominance for processing vocalizations in adult, but not infant rhesus monkeys: field experiments, *Proceed. Nat. Acad. Sci.*.
Heffner, H. E. and Heffner, R. S., 1984, Temporal lobe lesions and perception of species-specific vocalizations by macaques, *Science* 226:75-76.
Heit, G., Smith, M. E., and Halgren, E., 1988, Neural encoding of individual words and faces by the human hippocampus and amygdala, *Nature* 333:773-775.
Heyes, C. M., 1993, Anecdotes, training, trapping and triangulating: do animals attribute mental states? *Anim. Behav.* 46:177-188.
Hickey, T., 1993, Identifying formulas in first language acquisition, *J. Child Lang.* 20:27-41.
Hill, S. D. and Tomlin, C., 1981, Self-recognition in retarded children, *Child Develop.* 52:145-150.
Hirsh-Pasek, K., Kemler Nelson, D. G., Jusczyk, P. W., Cassidy, K. W., Druss, B., and Kennedy, L., 1987, Clauses are perceptual units for young infants, *Cognition* 26:269-286.
Hobson, R. P., Ouston, J., and Lee, A., 1988, What's in a face? The case of autism, *Brit. J. Psychol.* 79:441-453.
Jusczyk, P. W., Cutler, A., and Redanz, N. J., 1993a, Infants' preference for the predominant stress patterns of English words, *Child Develop.* 64:675-687.
Jusczyk, P. W., Friederici, A. D., Wessels, J. M. I., Svenkerud, V. Y., and Jusczyk, A. M., 1993b, Infants' sensitivity to the sound patterns of native language words, *J. Mem. Lang.* 32:402-420.
Kalin, N. H. and Shelton, S. E., 1989, Defensive behaviors in infant rhesus monkeys: environmental cues and neurochemical regulation, *Science* 243:1718-1721.
Kalin, N. H., Shelton, S. E., and Takahashi, L. K., 1991, Defensive behaviors in infant rhesus monkeys: ontogeny and context-dependent selective expression, *Child Develop.* 62:1175-1183.
Keating, C. F. and Keating, E. G., 1982, Visual scan patterns of rhesus monkeys viewing faces, *Perception* 11:211-219.
Keating, C. F. and Keating, E. G., 1993, Monkeys and mug shots: cues used by rhesus monkeys (*Macaca mulatta*) to recognize a human face, *J. Comp. Psychol.* 107:131-139.
Kellogg, W. N., 1968, Communication and language in the home-raised chimpanzee, *Science* 182:423-427.
Kenney, M. D., Mason, W. A., and Hill, S. D., 1979, Effects of age, objects, and visual experience on affective responses of rhesus monkeys to strangers, *Develop. Psychol.* 15:176-184.
Kessen, W., Levine, J., and Wendrich, K., 1979, The imitation of pitch in infants, *Infant Behav. Develop.* 2:93-100.
Kling, A. S. and Brothers, L. A., 1992, The amygdala and social behavior, *in*: "The Amygdala", J. Aggleton, ed., John Wiley and Sons, New York.
Kuczynski, L., Zahn-Waxler, C., and Radke-Yarrow, M., 1987, Development and content of imitation in the second and third years of life: a socialization perspective, *Dev. Psychol.* 23:276-282.
Kugiumutzakis, G., 1993, Intersubjective vocal imitation in early mother-infant interaction, *in*: "New Perspectives in Early Communicative Development", J. Nadel and L. Camaioni, eds., Routledge, New York.
Kuhl, P. K. and Meltzoff, A. N., 1988, Speech as an intermodal object of perception, *in*: "Perceptual Development in Infancy", A. Yonas, ed., Minnesota Symposia on Child Psychology, Erlbaum, Hillsdale, NJ.
Kuhl, P.K., Williams, K.A., Lacerda, F., Stevens, K.N., and Lindblom, B., 1992, Linguistic experience alters phonetic perception in infants by 6 months of age, *Science* 255:606-608.
Langdell, T., 1978, Recognition of faces: An approach to the study of autism, *J. Child Psychol. Psychiat.* 19:255-268.
Larson, C.R., 1988, Brain mechanisms involved in the control of vocalization, *J. Voice* 2:301-311.
Larson, C.R. and Kistler, M.K., 1986, The relationship of periaqueductal gray neurons to vocalization and laryngeal EMG in the behaving monkey, *Exp. Brain Res.* 63:596-606.
LeMay, M. and Geschwind, N., 1975, Hemispheric differences in the brains of great apes, *Brain Behav. Evol.* 11:48-52.

Lieberman, P., 1984, "The Biology and Evolution of Language", Harvard University Press, Cambridge, Mass.

Lieven, E.V.M., Pine, J.M., and Dresner Barnes, H., 1992, Individual differences in early vocabulary development: redefining the referential-expressive distinction, *J. Child Lang.* 19:287-310.

Lindblom, B., MacNeilage, P., and Studdert-Kennedy, M., 1983, Self-organizing processes and the explanation of phonological universals, *in*: "Explanations of Linguistic Universals", B. Butterworth, B. Comrie and D. Dahl, eds., Mouton, The Hague.

Lindblom, B. and Zetterstrom, R., 1986, "Precursors of Early Speech", Stockton Press, New York.

Locke, J.L., 1989, Babbling and early speech: Continuity and individual differences, *First Lang.* 9:191-206.

Locke, J.L., 1992, Neural specializations for language: a developmental perspective, *Sem. Neurosci.* 4:425-431.

Locke, J. L., 1993a, The role of the face in vocal learning and the development of spoken language, *in*: "Developmental Neurocognition: Speech and Face Processing in the First Year of Life", B. de Boysson-Bardies, S. de Schonen, P. Jusczyk, P. MacNeilage, and J. Morton, eds., Kluwer Academic Publishers, The Netherlands.

Locke, J. L., 1993b, "The Child's Path to Spoken Language", Harvard University Press, Cambridge, Mass.

Locke, J. L., 1994a, Gradual emergence of developmental language disorders, *J. Speech Hear. Res.* 37:608-616.

Locke, J.L., 1994b, Development of the capacity for spoken language, *in*: "Handbook of Child Language", P. Fletcher and B. MacWhinney, eds., Blackwell Publishers, Cambridge, Mass.

Locke, J.L., in press, a, Phases in the development of linguistic capacity, *in*: "Discussions in Neuroscience", C.L. Bolis, D. C. Gajdusek and G. McKhann, eds.

Locke, J.L., in press, b, Talking themselves into a language: phases in children's development of linguistic competence, *Amer. Sci.*

Locke, J.L., in submission, Why do infants begin to talk? Language as an unintended consequence.

MacKain, K.S., 1984, Speaking without a tongue, *J. Nat. Stud. Speech Lang. Hear. Assoc.* 12:46-71.

MacNeilage, P.F., Studdert-Kennedy, M., and Lindblom, B., 1987, Primate handedness reconsidered, *Behav. Brain Sci.* 10:247-263.

Mans, L., Cicchetti, D., and Sroufe, L.A., 1978, Mirror reactions of Down's syndrome infants and toddlers: cognitive underpinnings of self-recognition, *Child Develop.* 49:1247-1250.

Marchman, V. A. and Bates, E., in press, Continuity in lexical and morphological development: a test of the critical mass hypothesis, *J. Child Lang.*

Marcus, G. F., Pinker, S., Ullman, M., Hollander, M., Rosen, T. J., and Xu, F., 1992, Overregularization in language acquisition, *Monogr. Soc. Res. Child Develop.* 57 (Serial No. 228).

Markman, E. M. and Hutchinson, J. E., 1984, Children's sensitivity to constraints on word meaning: taxonomic vs. thematic relations, *Cogn. Psychol.* 16:1-27.

Marler, P. and Peters, S., 1982, Long-term storage of learned birdsongs prior to production, *Anim. Behav.* 30:479-482.

Marler, P. and Tenaza, R., 1977, Signaling behavior of apes with special reference to vocalization, *in*: "How Animals Communicate", T.A. Seboek, ed., Indiana Press, Bloomington.

Masataka, N., 1992, Pitch characteristics of Japanese maternal speech to infants, *J. Child Lang.* 19:213-223.

Masataka, N. and Fujita, K., 1989, Vocal learning of Japanese and rhesus monkeys, *Behaviour* 109:191-199.

Mattingly, I. G. and Studdert-Kennedy, M., eds., 1991, "Modularity and the Motor Theory of Speech Perception", Erlbaum, Hillsdale, NJ.

Maurus, M., Barclay, D., and Streit, K.-M., 1988, Acoustic patterns common to human communication and communication between monkeys, *Lang. Commun.* 8:87-94.

Maurus, M., Kühlmorgen, B., Wiesner, E., Barclay, D., and Streit, K.-M., 1987, Interrelations between structure and function in the vocal repertoire of Saimiri: asking the monkeys themselves where to split and where to lump, *Europ. Arch. Psychiat. Neurol. Sci.* 236:35-39.

Mehler, J., Jusczyk, P., Lambertz, G., Halsted, N., Bertoncini, J., and Amiel-Tison, C., 1988, A precursor of language acquisition in young infants, *Cognition* 29:143-178.

Mendelson, M. J., Haith, M. M., and Goldman-Rakic, P. S., 1982, Face scanning and responsiveness to social cues in infant rhesus monkeys, *Dev. Psychol.* 18:222-228.

Mills, D. L., Coffey-Corina, S. A., and Neville, H. J., 1993, Language acquisition and cerebral specialization in 20-month-old infants, *J. Cogn. Neurosci.* 5:317-334.

Moon, C., Cooper, R. P., and Fifer, W. P., 1993, Two-day olds prefer their native language, *Infant Behav. Develop.* 16:495-500.

Moran, G., Krupka, A., Tutton, A., and Symons, D., 1987, Patterns of maternal and infant imitation during play, *Infant Behav. Develop.* 10:477-491.

Mount, R., Reznick, J. S., Kagan, J., Hiatt, S., and Szpak, M., 1989, Direction of gaze and emergence of speech in the second year, *Brain Lang.* 36:406-410.

Nelson, K., 1981, Individual differences in language development: implications for development and language, *Dev. Psychol.* 17:170-187.

Nelson, K., 1985, "Making Sense: The Acquisition of Shared Meaning", Academic Press, New York.

Owren, M.J., Dieter, J.A., Seyfarth, R.M., and Cheney, D.L., 1992, Evidence of limited modification in the vocalizations of cross-fostered rhesus (*Macaca mulatta*) and Japanese (*M. fuscata*) macaques, *in*: "Topics in primatology. Volume 1. Human Origins", T. Nashida, W. McGrew, P. Marler, M. Pickford and F. DeWaal, eds., University of Tokyo Press, Tokyo.

Papousek, M. and Papousek, H., 1989, Forms and functions of vocal matching in interactions between mothers and their precanonical infants, *First Lang.* 9:137-158.

Parker, S.T. and Gibson, K.R., 1990, "'Language and Intelligence in Monkeys and Apes: Comparative Developmental Perspectives", Cambridge University Press, New York.

Perrett, D.I. and Mistlin, A.J., 1990, Perception of facial characteristics by monkeys, *in*: "Comparative Perception: Complex Signals. Volume 2", W.C. Stebbins and M.A. Berkley, eds., John Wiley, New York.

Peters, A.M., 1983, "The Units of Language Acquisition", Cambridge University Press, New York.

Petersen, M.R., Beecher, M.D., Zoloth, S.R., Moody, D.B., and Stebbins, W.C., 1978, Neural lateralization of species-specific vocalizations by Japanese macaques (*Macaca fuscata*), *Science* 202:324-327.

Petersen, M.R., Zoloth, S.R., Beecher, M.D., Green, S., Marler, P.R., Moody, D.B., and Stebbins, W.C., 1984, Neural lateralization of vocalizations by Japanese macaques: communicative significance is more important than acoustic structure, *Behav. Neurosci.* 98:779-790.

Pinker, S. and Bloom, P., 1990, Natural language and natural selection, *Behav. Brain Sci.* 13:707-784.

Ploog, D., 1979, Phonation, emotion, cognition, with reference to the brain mechanisms involved (Ciba Foundation Symposium), *Brain and Mind* 69: 79-98.

Plooij, F., 1979, How wild chimpanzee babies trigger the onset of mother-infant play - and what the mother makes of it, *in*: "Before Speech: the Beginnings of Interpersonal Communication", M. Bullowa, ed., Cambridge University Press, Cambridge.

Premack, D. and Woodruff, G., 1978, Does the chimpanzee have a theory of mind?, *Behav. Brain Sci.* 1:515-526.

Querleu, D., Renard, X., and Versyp, F., 1985, Vie sensorielle du foetus, *in*: "L'Environnement de la Naissance", G. Levy and M. Tournaire, eds., Vigot, Paris.

Redican, W.F., 1975, Facial expressions in non-human primates, *in*: "Primate Behavior: Developments in Field and Laboratory Research", L.A. Rosenblum, ed., Academic Press, New York.

Reznick, J.S. and Goldfield, B.A., 1992, Rapid change in lexical development in comprehension and production, *Develop. Psychol.* 28:406-413.

Ross, E.D., 1981, The aprosodias: Functional-anatomic organization of the affective components of language in the right hemisphere, *Arch. Neurol.*38:561-569.

Ross, M., Duffy, R. J., Cooker, H.S., and Sargeant, R. L., 1973, Contribution of the lower audible frequencies to the recognition of emotions, *Amer. Ann. Deaf* 118:37-42.

Sackett, G.P., Suomi, S.J., and Grady, S., 1970, Species preferences by macaque monkeys. Cited by G. P. Sackett and G. C. Ruppenthal, 1974, Some factors influencing the attraction of adult female macaque monkeys to neonates, *in*: "The Effect of the Infants on its Caregiver", M. Lewis and L. A. Rosenblum, eds., John Wiley, New York.

Savage-Rumbaugh, E.S., Murphy, J., Sevcik, R.A., Brakke, K.E., Williams, S.L., and Rumbaugh, D.M., 1993, Language comprehension in ape and child, *Monogr. Soc. Res. Child Develop. 58* (Serial No. 233).

Schaffer, H. R., 1989, Language development in context, *in*: "The Social and Cognitive Aspects of Normal and Atypical Language Development", S. von Tetzchner, L. Siegel and L. Smith, eds., Springer-Verlag, New York.

Scherer, K.R., Koivumaki, J., and Rosenthal, R., 1972, Minimal cues in the vocal communication of affect: judging emotions from content-masked speech, *J. Psycholing. Res.* 1: 269-285.

Snow, C. E., 1989, Imitativeness: a trait or a skill?, *in*: "The Many Faces of Imitation in Language Learning", G. E. Speidel and K. E. Nelson, eds., Springer-Verlag, New York.

Snowdon, C. T., Coe, C. L., and Hodun, A., 1985, Population recognition of infant isolation peeps in the squirrel monkey, *Anim. Behav.* 33:1145-1151.

Speidel, G.E. and Nelson, K.E., 1989, "The Many Faces of Imitation in Language Learning", Springer-Verlag, New York.

Steklis, H. D. and Raleigh, M. J., 1979, Requisites for language: Interspecific and evolutionary aspects, *in*: "Neurobiology of Social Communication in Primates: An Evolutionary Perspective", H. D. Steklis and M. J. Raleigh, eds., Academic Press, New York.

Sutton, D., Larson, C., Taylor, E. M., and Lindeman, R. C., 1973, Vocalization in rhesus monkeys: conditionability, *Brain Res.* 52:225-231.

Sutton, D., Samson, H. H., and Larson, C. R., 1978, Brain mechanisms in learned phonation of *Macaca mulatta*, *in*: "Recent Advances in Primatology", D. J. Chivers and J. Herbert, eds., Academic Press, London.

Sutton, D., Trachy, R. E., and Lindeman, R. C., 1981, Primate phonation: unilateral and bilateral cingulate lesion effects, *Behav. Brain Res.* 3:99-114.

Symmes, D. and Biben, M., 1985, Maternal recognition of individual infant squirrel monkeys from isolation call playbacks, *Amer. J. Primatol.* 9:39-46.

Terrace, H. S., Petitto, L. A., Sanders, R. J., and Bever, T. G., 1980, On the grammatical capacity of apes, in: "Children's Language: Volume 2", Gardner Press, New York.

Tomasello, M., Mannle, S., and Kruger, A. C., 1986, Linguistic environment of 1- to 2-year-old twins, *Dev. Psychol* .22:169-176.

Tomasello, M., Savage-Rumbaugh, S., and Kruger, A. C., 1993, Imitative learning of actions on objects by children, chimpanzees, and enculturated chimpanzees, *Child Develop.* 64:1688-1705.

Trachy, R. E., Sutton, D., and Lindeman, R. C., 1981, Primate phonation: anterior cingulate lesion effects on response rate and acoustical structure, *Amer. J. Primatol.* 1:43-55.

Uzgiris, I. C., Benson, J. B., Kruper, J. C., and Vasek, M. E., 1989, Contextual influences on imitative interactions between mothers and infants, *in*: "Action in a Social Context: Perspectives on Early Development", J. Lockman and N. Hazen, eds., Plenum, New York.

Van Lancker, D., 1987, Nonpropositional speech: neurolinguistic studies, *in*: "Progress in the Psychology of Language, Volume 3", A. Ellis, ed., Erlbaum, Hillsdale, NJ.

Veneziano, E., 1988, Vocal-verbal interaction and the construction of early lexical knowledge, *in*: "The Emergent Lexicon: The Child's Development of a Linguistic Vocabulary", M. D. Smith and J. L. Locke, eds., Academic Press, New York.

Waser, P. M., 1977, Individual recognition, intragroup cohesion and intergroup spacing: Evidence from sound playback to forest monkeys, *Behaviour* 60:28-74.

Watson, M. W. and Fischer, K. W., 1977, A developmental sequence of agent use in late infancy, *Child Develop.* 48:828-832.

Werker, J. F., Cohen, L. B., and Lloyd, V., in preparation, Infants' ability to associate sounds with objects: a developmental study.

Werker, J. F. and Pegg, J. E., 1992, Infant speech perception and phonological acquisition, *in*: "Phonological Development: Models, Research, Implications", C. A. Ferguson, L. Menn and C. Stoel-Gammon, eds., York Press, Timonium, Maryland.

Werker, J. F. and Polka, L., 1993, Developmental changes in speech perception: new challenges and new directions, *J. Phonetics* 21:83-101.

Whiten, A. and Byrne, R. W., 1988, Tactical deception in primates, *Behav. Brain Sci.* 11:233-244.

WHAT PRIMATE CALLS CAN TELL US ABOUT HUMAN EVOLUTION

Philip Lieberman

Cognitive and Linguistic Sciences
Brown University
Providence RI 02912-1978, USA

INTRODUCTION

An influential, traditional school of linguistics and philosophy that is often identified with Noam Chomsky and his followers, but which is by no means limited to "generative" or "transformational" linguistics, categorically differentiates the neural bases of human language from the vocal communications of other primates (e.g., Bickerton, 1990; Wilkins and Wakefeld, in press). There supposedly is no direct evolutionary link between the brain mechanisms that regulate vocal communication in monkeys and apes and the brain mechanisms that regulate human speech and syntax. Adherents of this position note that whereas the neural substrate regulating human speech allows arbitrary sequences of speech sounds to be produced on a "voluntary" basis, non-human primate's vocalizations are largely "bound" to particular emotional states. Hewes (1973), for example, proposes that the first form of human language involved manual signing rather than vocal signals. Pinker and Bloom (1990) claim that the neural mechanisms that make human syntax possible evolved from some part of the non-human primate brain that had no function. An extreme position is found in Chomsky's (1972, 1976, 1980) repeated claims that human language could not have evolved by means of Darwinian processes.

However, given the many physiological similarities that obtain between the sound-producing capabilities of non-human primates and humans and attendant similar acoustic features, a more probable evolutionary scenario would involve voluntary human speech evolving in stages. It is possible that the roots of human language are to be found in adaptations yielded neural mechanisms that extended or modulated the relationships that hold between particular primate vocalizations, particular affects, and particular emotional states. In other words, the process by which hominids acquired the neural mechanisms that underlie those aspects of present linguistic ability that appear to be "uniquely human" speech and syntax, may have operated in stages. The tight bond between an emotion and a particular vocalization may have been loosened by processes in which different affective states became associated with the same vocalization. Meaning would have been initially "disambiguated" conveyed by context or gestural signals, later by modulations of the vocalization.

Although studies of vocal communication in non-human primates have determined some of the upper bounds of their utterances with respect to human speech as well as many of the similarities, our knowledge is still imperfect. We know, for example, what sounds the non-human primate vocal tract cannot produce. We need to obtain the "body of knowledge" concerning the nature of non-human primate vocalizations that would allow us to answer these questions. We simply don't know enough about the acoustic nature of non-human primate vocal communication, the articulatory maneuvers underlying these vocalizations and the emotional and possible semantic referents of these vocalizations. Although excellent subjective accounts of non-human primate vocal communication are available (e. g., Goodall, 1986) they inherently cannot give the detailed information that is necessary to answer the question of the degree to which non-human primates possess the neural substrate that underlies the human ability to produce "voluntary" speech. Human listeners are inherently tied to the phonetic framework of their native language when they attempt to transcribe speech signals by auditory means. The perception of vowels, for example, is tied to one's native language by age 6 months (Kuhl et al., 1992). Human listeners who have different linguistic backgrounds will "hear" identical acoustic signals as belonging to different vowel categories. We have found that this is the case for phoneticians with years of training. In a series of transcription studies of the utterances of 18-month-old children more than 50 percent of the vowels and consonants transcribed by three phoneticians (native speakers of Danish, English and Hindi) differed. Similar deficiencies limit auditory transcriptions of intonation and stress (Lieberman, 1960). The recent history of research on human speech demonstrates that objective psychoacoustic experiments, acoustic analyses, and detailed studies of speech production using such techniques as cineradiography and electromyography were necessary to obtain even preliminary insights into the manner in which human speech is produced and perceived.

The characteristic that defines voluntary speech should be noted at the outset of this paper. The articulatory gestures that specify particular sounds can be combined in novel sequences to form words that are not simply stereotyped response patterns triggered by our emotional state, or particular events or objects. Non-human primates lack this ability. Behavioral studies consistently show that non-human primates lack the general voluntary control of vocal output that characterizes human speech.

For example, acoustic analyses of chimpanzee calls show that they make use of phonetic "features" that play a linguistic role in human speech (Lieberman, 1968, 1975). Chimpanzees produce falling formant frequency transitions by rounding their lips that are similar to those that specify "labial" sounds like [b] and [w] in English. Chimpanzees also have larynges that are capable of generating a glottal source that though it has somewhat more high frequency content (Lieberman, 1968) than is typical of human speech (Fant, 1960) is adequate for speech production; most chimpanzee calls appear to be phonated (Goodall, 1986). Therefore, chimpanzees could, in principle, produce the labial stop consonants [b] and [p] which are differentiated by means of the timing between lips and laryngeal output (Lisker and Abramson, 1964). Moreover, chimpanzees have speech-producing anatomy that, in principle, could produce nasalized vowels similar to the human vowels [I], [U], [ae], etc. (Lieberman et al., 1972), and some of their calls involve slightly different vowel-like contrasts (Lieberman, 1968). Many chimpanzees have been raised from infancy in close proximity to humans speaking to them and to other humans and have been actively tutored. However, no chimpanzee has ever produced <u>voluntary</u> speech-like vocalizations approximating human words like <u>bad</u>, <u>pad</u>, <u>bit</u>, <u>pit</u>, <u>bat</u>, etc. though intensive attempts have been made to get them to "talk." The Hayes and Hayes (1951) study in which a young chimpanzee was exposed to a similar environment of a human infant acquiring speech was perhaps the final attempt to teach a chimpanzee to talk.

It is apparent that the acoustic features that constitute the chimpanzee calls that, in principle, could be isolated and then melded together to form these human words are <u>bound</u>

together. However, chimpanzees can't voluntarily dissociate the bonds that link the acoustic features that make up their calls. Though their stereotyped vocalizations incorporate elements that could approximate many human words, chimpanzees are not able to "free" these elements and produce voluntary speech. Humans can say whatever they please whether they're happy, hungry, feeding, frightened, etc. In contrast, as many observers of chimpanzees and other non-human primates note, primate calls are part of a stereotyped response pattern that signals a particular emotional state (Goodall, 1986). Although behavioral and field studies have noted instances where non-human primates either change the rate at which they utter a call or suppress a call (Hauser et al., 1992) these modifications do not necessarily constitute "voluntary" speech. Operant conditioning of non-human primates that changes their emotional state would also modify their calls. As the primates' emotional state changes, so would its stereotyped calls. Likewise, changes in the primates' fearful emotional state contingent on its decision to avoid sharing food might result in the suppression of food calls. The suppression of the food call therefore might be intermediate between involuntary, stereotyped calls and truly voluntary calls. Independent measures of emotion or affect, e. g. heart rate or assessment of vocal measures directly related to affect such as pitch perturbations (Lieberman, 1963) would be useful in assessing this possibility. We would expect to find correlations between these measures of emotion and the change in call rate, spectral characteristics of the calls, etc. In any event, we need to have more data to establish the possibility of degrees of control of vocalization intermediate to fully voluntary human speech.

Likewise, the stereotyped vocal responses of adult non-human primates may differ as the result of either maturation or learning, but the end result is a stereotyped call that signals particular threats, rewards, etc. In this regard, the data presented by Cheney and Seyfarth (1990) appear to demonstrate the effects of maturation and associative learning - specific calls have become associated with particular predators by means of either maturational processes or general associative learning or some combination thereof. Hauser (1989) demonstrates that this is the case for rhesus monkey calls. In this sense the calls are not strictly stereotyped, but neither are they fully voluntary. If non-human primates truly had voluntary control of their vocalizations, we would expect to find them imitating human speech when they were raised in human environments. They never produce any speech sounds despite the most intensive training. The deficit cannot be ascribed to lack of intentionality on the part of the chimpanzees or deficient effort on the part of the experimenters since the chimpanzees communicate by means of manual means (Rumbaugh and Savage-Rumbaugh, 1992; Savage-Rumbaugh et al., 1986). In contrast, any normal human child raised in any reasonably normal environment effortlessly acquires speech. We can conclude that chimpanzees and other non-human primates don't acquire any aspect of speech because they lack the requisite brain mechanisms.

The case is also clear against chimpanzee's having syntactic ability similar to that of normal human beings. Many language-trained chimpanzees clearly have limited lexical abilities and can learn about 200 words using various manual-visual signals. However, they are <u>not</u> able to make use of even simple distinctions of word order. They cannot comprehend the semantic distinctions conveyed by word order in sequences like <u>see me</u> versus <u>me see.</u> Many claims to the contrary have been made; in recent years some investigators (Savage-Rumbaugh et al., 1986) have proposed that the linguistic abilities of the bonobo (*Pan paniscus*) surpass those of the "common" chimpanzee (*Pan troglodytes*) with respect to syntax. However, Rumbaugh and Savage-Rumbaugh (1992) in their recent study of the grammatical ability of the bonobo Kanzi conclude, "there is no example in which a difference in [word] order signals a difference in meaning." Chimpanzees do combine words to form new words. They, for example, spontaneously form new words by compounding old words forming <u>waterbird</u> - for a waterfowl, from the semantically distinct words <u>water</u> and <u>bird</u> (Gardner et al., 1989). Chimpanzees, in other words, command some aspects of morphophonemics. However, they cannot handle the syntax of human language with the proficiency of a normal three-year-old human child. Clearly some qualitative neural distinction sets the syntactic abilities (rather the lack thereof) of chimpanzees apart from humans.

SOME ASPECTS OF THE NEUROPHYSIOLOGY OF SPEECH PRODUCTION

The traditional account of the neurophysiological basis of human speech production is largely based on the study of aphasia - damage to the brain that results in deficits in speech production. Paul Broca in 1861 first claimed that damage to part of the neocortex in the dominant hemisphere of the brain can result in aphasia, i. e., a complex of speech and language deficits. The traditional view of Broca's aphasia is that damage localized to this particular area of the brain will result in these deficits while damage to any other part of the brain won't. This belief is reflected in popularized accounts of how the human brain "works", and in the supposition of many linguists that human beings have a specific, localized "language organ" (Chomsky, 1986). However, that supposition is not true. In the past decade new brain imaging and electrophysiological techniques have revealed that permanent aphasia is the consequence of damage to subcortical "pathways" or "circuits" connecting Broca's area to the parts of the brain that directly control muscles, and to the prefrontal cortex. In fact, subcortical damage that disrupts the connections from Broca's area, leaving it intact, can result in aphasia (Alexander et al., 1987).

Damage to the internal capsule (the bundle of nerve fibers that connect the neocortex to the midbrain), the putamen, and the caudate nucleus (some of the structures of the basal ganglia) can yield impaired speech production and agrammatism similar to that of the classic aphasias as well as other cognitive deficits (Alexander et al., 1987; Naeser et al., 1982). Alexander and his colleagues reviewed 19 cases of aphasia with language impairments that ranged from fairly mild disorders in the patient's ability to recall words, to "global aphasia" in which the patient produced very limited "dysarthric" non-propositional speech and was not able to comprehend syntax. Dysarthric speech derives from a loss of the coordinated motor control of the respiratory system and larynx - phonation becomes irregular and "breathy." The patients who had the most extreme language and speech deficits also lost control of their right arms and legs. Lesions in the basal ganglia and the nerve connections that run down from the neocortex were noted for the 19 cases. In general, severe language deficits (oral apraxia, dysarthria, anomia, agrammatism) occurred in patients who had suffered the most extensive subcortical brain damage. These patients also suffered paralysis of their dominant right hands.

Studies of the deficits resulting from neurodegenerative diseases, like Parkinson's disease (PD), also show the effects of damage to subcortical circuits on cognition and language. These diseases cause major damage to the basal ganglia, sparing the cortex. The primary deficits of subcortical disease are motoric; tremor, rigidity and akinesia occur. However, subcortical diseases can also cause linguistic and cognitive deficits. In extreme form, the deficits associated with these subcortical diseases constitute a dementia (Albert et al., 1974; Cummings and Benson, 1984). Deficits in the comprehension of syntax have been noted in several independent studies of Parkinson's disease (Grossman et al., 1991, 1992; Lieberman et al., 1990, 1992; Natsopolous et al., in press). A pattern of speech production, syntax and cognitive deficits similar in nature to those typical of aphasia can occur in even moderately impaired Parkinson's disease (Lange et al., 1992; Lieberman et al., 1992; Morris et al., 1988).

Moreover, studies that make use of CT and PET scans of victims of aphasia clearly show that damage to pathways to prefrontal cortex is implicated in these deficits. Metter et al. (1989), for example, using CT scans found that Broca's patients had subcortical damage to the internal capsule and parts of the basal ganglia. PET scans showed that these patients had vastly reduced metabolic activity in the left prefrontal cortex and Broca's region. Previous studies (Metter et al., 1987) showed a strong correlation between prefrontal and Broca's region metabolic activity and "functional motor loss of the arms and legs, as well as spontaneous speech and writing [and] in normal subjects, a strong correlation between prefrontal cortex and decision making" (Metter et al. 1989, p. 31). They conclude that the behavioral deficits of Broca's aphasia - general "difficulty in motor sequencing and executing motor speech tasks," and "the presence of language comprehension abnormalities" derive from damage to circuits to the prefrontal cortex.

PET scan studies of subcortical disease reach similar conclusions (Metter et al., 1984). Progressive supranuclear palsy (PSP) patients have less metabolic activity in the prefrontal cortex than normal controls. The destruction by PSP of the circuits through the basal ganglia that stimulate the prefrontal cortex is responsible for the reduced activity and concomitant cognitive deficits (D'Antonia et al., 1985). Recent behavioral data show that reduced basal ganglia pathway activity to prefrontal cortex in even moderately impaired PD patients produces cognitive deficits similar in kind to those typical of frontal lesions (Lange et al., 1992; Lieberman et al., 1992; Morris et al., 1988; Sahakian et al., 1988).

ON THE NEURAL REGULATION OF NON-HUMAN PRIMATE CALLS

Although a homologue of Broca's area (the part of the human brain traditionally associated with language) can be identified in monkeys on anatomical grounds, massive lesions in or near this area of the neocortex have no effect whatsoever on their vocalizations. Electrophysiological and ablation studies show that it is nonfunctional with respect to regulation of their vocalizations. It is the case that neocortical lesions in non-human primates have no effect on vocalizations. As MacLean and Newman (1988) note, the effects on vocalizations ascribed to the neocortical lesions noted in some primate studies actually were due to damage to the anterior cingulate gyrus (the "paleomammalian" limbic system, c. f. MacLean, 1985). The vocalizations of non-human primates are not regulated by neocortical structures. Though parts of their neocortex resemble Broca's area anatomically, it is not involved in regulating their vocalizations. Their vocalizations are instead controlled by the cingulate cortex, the "old" motor cortex that evolved with the earliest mammals, the basal ganglia, and midbrain structures. No non-human primate vocalizations studied appear to be regulated by neocortical structures (Sutton and Jürgens, 1988).

WHAT MIGHT DIFFERENTIATE THE HUMAN BRAIN FROM OTHER BRAINS?

Tracer studies that reveal the detailed wiring diagram of the brain cannot be performed on human subjects. However, they have been used to explore the basal ganglia pathways to and from prefrontal cortex in non-human primates and other animals (Parent, 1986). Basal ganglia pathways in mammals carry signals between various parts of cortex. The basal ganglia have become proportionately larger as the cortex has expanded in the course of mammalian evolution; as cortex expands so do basal ganglia (Parent, 1986). Prefrontal cortex also becomes proportionately larger as the phylogenetic scale is ascended (Brodmann, 1909). Prefrontal cortex is about 200 percent larger in humans than in chimpanzee (Deacon, 1988). It enters into all manner of controlled, planned motoric functions as well as in abstract thought (Stuss and Benson, 1986). It is involved in cognitive tasks in which one shifts parameters when classifying patterns, pictures or concepts (Lange et al., 1992).

The basal ganglia have also become more differentiated in the course of mammalian evolution. The basal ganglia of rodents, for example, lack the independent caudate nucleus and putamen found in primates (Parent, 1986). There is no reason to suppose that the basal ganglia of humans are structurally identical to those of apes given the fact that the human basal ganglia are larger than ape's. It is reasonable to propose that the voluntary speech behavior of humans may rest on the enlargement and elaboration of prefrontal cortex and the basal ganglia circuits connecting it to other parts of the brain. The basal ganglia in rodents code the motoric acts that make up sequential grooming patterns (Aldridge et al., 1993). It is possible that the ability of humans to "free up" the stereotyped motor acts that make up non-human primate calls derives from prefrontal cortex "overriding" basal ganglia coded patterns (Lieberman, in press).

A PROGRAM TESTING THESE HYPOTHESES

Studies that determine whether the acoustic cues that chimpanzees actually use when they communicate in natural surroundings are necessary to determining whether some degree of voluntary control may be present in non-human primates. For example, it may be the case that different vocal "dialects" exist analogous to the different gestures that have been observed in different chimpanzee populations, accompanying similar behaviors (Goodall, 1986). This might indicate malleability in the chimpanzee vocal communication system. Different call patterns might, for example, signal similar affects and be transmitted to juvenile chimpanzees. This would be consistent with a gradual evolutionary process for the evolution of the neural substrate for human voluntary speech that had its roots in early hominoids. But we first have to establish the nature of the acoustic cues and articulatory gestures that underlie chimpanzee vocalizations and their referents. Comparative neurophysiological studies using non-invasive electrophysiological techniques presently under development would then be able to address the question of the locus of the functional differences between the human and chimpanzee brain. Given the syntactic deficits of chimpanzees and the proposed link between speech and syntax (Lieberman, 1984, 1985, 1991) we would also be in a position to address the issue of the neural basis and evolution of human syntactic ability and the supposed "modularity" (Fodor, 1983) of the human brain and conflicting theories concerning the evolution of human language. For example, the theories proposed by Bickerton (1990) and Pinker and Bloom (1990) claim that human speech is disjoint from primate vocalizations in contrast to theories such as those proposed by Kimura (1979) and Lieberman (1975, 1984, 1991, in press).

Different types of studies employing different methodologies may be appropriate:

A - Acoustic analyses of the communications of primates in their natural surround that establish correspondences between message and signal characteristics. Analyses that relate message content to acoustic features are a necessary first step. Studies in progress analyzing the vocalizations captured in video records that have been made at Gombe by Jane Goodall, Chris Boehm and their colleagues present one example (Fitch, 1993). Fitch's analysis shows that formant frequency patterns are probably part of the primate signalling system - lower formant frequencies generated by lip-rounding and protrusion form part of the agressive displays. Experiments with humans show that lower formant frequency patterns are associated with larger body size; similar relations may hold for non-human primates. If we can determine the range of acoustic cues that are used in primate communication, we can then go on to determine the degree of possible voluntary control thereof. One obvious question is whether the acoustic cues used by primates form a subset of those used in human speech. In other words, do all non-human primate vocalizations constitute a "primitive" subset of human speech with human speech simply adding additional acoustic cues? That seems unlikely given vocal specializations in certain non-human primate species, e. g., howler monkeys.

B - Perceptual studies establishing the saliency of possible acoustic primate cues. Although acoustic analyses can isolate particular cues that would appear to be potentially used to convey information (e. g., the falling formant transitions of chimpanzee calls similar to those that occur in labial stops), it is necessary to establish whether they actually are functional. Playback experiments modelled on the psychoacoustic techniques commonly used to establish the perceptual saliency of potential acoustic cues for human speech are necessary. Controlled synthesis or computer-implemented manipulation of these cues is almost an absolute necessity. A model experiment would, for example, contrast chimpanzee calls that feature a terminal formant fall with ones that are similar except for the fall. The problem is in achieving a response from the non-human primate subjects. Perhaps~ "bilingual" chimpanzees fluent in sign language and their natural calls would be useful - the "Dr. Doolittle" paradigm.

C - Analyses that determine the articulatory gestures underlying these acoustic cues. The data bases used for acoustic analyses (A) may also prove useful in this regard. Hauser's (Hauser et al., 1993) study relating mandible position to aggressive displays in rhesus macaques is a model. Comparative studies using human subjects are also useful in this regard. Fitch (in preparation) shows that human subjects relate body size to formant frequency variations typical of those that commonly occur in chimpanzee calls.

D - Behavioral tests in which captive apes were taught to vocalize "words" - utterances conveying distinct referential information - that involved permutations and new combinations of a subset of the acoustic features of primate vocalizations might establish partial voluntary control. Although non-human primates clearly lack full voluntary control, it is possible that call distinctions based on modification of the isolation call - intonation and stress - might be malleable given the apparent regulation and malleability of the primate isolation call by the cingulate cortex (MacLean and Newman, 1988). If chimpanzees, for example, could be taught to communicate "words" that (a) conveyed different referential information, and (b) phonetically differed in their fundamental frequency contours, we would have an example of voluntary control of vocalization.

E - Coordinate neurophysiological experiments could establish the neural bases of intermediate levels of voluntary vocalization in non-human primates if such instances could be identified using the techniques noted above.

F - Finally, further cognitive studies establishing the degree to which aspects of cognitive behavior associated with prefrontal cortical activity in humans were present in non-human primates would help to test theories positing a biological and evolutionary relationship between language and cognition (Lieberman, 1984, 1991).

SUMMARY

A traditional view of the evolution of human linguistic ability claims that it is not related to the vocal communications of non-human primates. Whereas the neural substrate regulating human speech allows arbitrary sequences of speech sounds to be produced on a "voluntary" basis, the vocalizations of present-day non-human primates are "bound" to particular emotional states. Therefore, these theories claim that the neural mechanisms that regulate human speech and syntax evolved from neural mechanisms that regulated other aspects of behavior or even nonfunctional neural structures. However, given the many similarities that exist between human speech and primate vocalizations a more probable evolutionary scenario would involve voluntary human speech evolving in stages; adaptations that at first yielded neural mechanisms that extended or modulated the relationships that hold between particular primate vocalizations and particular emotional states. Detailed, precise studies that establish the acoustic cues and articulatory gestures by which living non-human primates communicate and comparative neurophysiologic studies are necessary to determine the nature and course of the evolutionary process that yielded the human capacity for voluntary speech. Since recent studies also implicate the neural mechanisms governing human voluntary speech in other more "cognitive" aspects of language such as syntax and abstract cognition itself, studies of primate vocal communication may address the nature and evolution of these aspects of human behavior.

REFERENCES

Albert, M. A., Feldman, R.G., and Willis, A. L., 1974, The "subcortical dementia" of progressive supranuclear palsy, *J. Neurol. Neurosurg. Psychiat.* 37:121-130.

Aldridge, J. W., Berridge, K.C., Herman, M., and Zimmer, L., 1993, Neuronal coding of serial order: Syntax of grooming in the neostriatum, *Psychol. Sci.* 4:391-395.

Alexander, M. P., Naeser, M. A., and Palumbo, C. L., 1987, Correlations of subcortical CT lesion sites and aphasia profiles, *Brain* 110:961-991.

Bickerton, D., 1990, "Language and Species", University of Chicago Press, Chicago.

Brodmann, K., 1909, "Vergleichende Histologische Lokalisation der Großhirnrinde in ihren Prinzipien dargestellt auf Grund des Zellenbaues", Barth, Leipzig.

Cheney, D. L. and Seyfarth, R. M., 1990, "How Monkeys See the World", University of Chicago Press, Chicago.

Chomsky, N., 1972, "Language and Mind", Extended Edition, Harcourt, Brace and World, New York.

Chomsky, N., 1976, On the nature of language, *in*: "Origins and Evolution of Language and Speech", H. B. Steklis, S. R. Harnad, and J. Lancaster, eds., New York Academy of Sciences, New York.

Chomsky, N., 1980, Rules and representations, *Behav. Brain Sci.* 3:1-61.

Chomsky, N., 1986, "Knowledge of Language: Its Nature, Origin and Use", Prager, New York.

Cummings, J. L. and Benson, D. F., 1984, Subcortical dementia: Review of an emerging concept, *Arch. Neurol.* 41:874-879.

D'Antonia, R., Baron, J. C., Samson, Y., Serdaru, M., Viader, F., Agid, Y., and Cambier, J., 1985, Subcortical dementia: Frontal cortex hypometabolism detected by positron tomography in patients with progressive supranuclear palsy, *Brain* 108:785-799.

Deacon, T. W., 1988, Human brain evolution II. Embryology and brain allometry, *in*: "Intelligence and Evolutionary Biology", NATO ASI Series, H. J. Jerison and I. Jerison, eds., Springer, Berlin.

Fant, G., 1960, "Acoustic Theory of Speech Production", Mouton, The Hague.

Fitch, W. T. III, 1993, Vocal tract length and the evolution of language, PhD dissertation, Brown University, Providence.

Fodor, J., 1983, "Modularity of Mind", MIT Press, Cambridge, Mass.

Gardner, R. A., Gardner, B. T., and Van Cantfort, T. E., 1989, "Teaching Sign Language to Chimpanzees", State University of New York Press, Albany.

Goodall, J., 1986, "The Chimpanzees of Gombe: Patterns of Behavior", Harvard, Cambridge, Mass..

Grossman, M., Carvell, S., Gollomp, S., Stern, M. B., Vernon, G., and Hurtig, H. I., 1991, Sentence comprehension and praxis deficits in Parkinson's disease, *Neurology* 41:160-1628.

Grossman, M., Carvell, S., Stern, M. B., Gollomp, S., and Hurtig, H. I., 1992, Sentence comprehension in Parkinson's disease: The role of attention and memory, *Brain Lang.* 42:347-384.

Hauser, M. D., 1989, Ontogenetic changes in the comprehension and production of vervet monkey (*Cercopithecus aethiops*) vocalizations, *J. Comp. Psychol.* 103:149-158.

Hauser, M. D., 1992, Costs of deception: Cheaters are punished in rhesus monkeys (*Macaca mulatta*), *Proceed. Nat. Acad. Sci.* 89:12137-12139.

Hauser, M.D., Evans, C.S., and Marler, P., 1993, The role of articulation in the production of rhesus monkey (*Macaca mulatta*) vocalizations, *Anim. Behav.* 45: 423-434.
Hayes, K.J. and Hayes, C., 1951, The intellectual development of a home-raised chimpanzee, *Proceed. Amer. Phil. Soc.* 95:105-109.
Hewes, G.W., 1973, Primate communication and the gestural origin of language, *Curr. Anthropol.* 14:5-24.
Kimura, D., 1979, Neuromotor mechanisms in the evolution of human communication, *in*: "Neurobiology of Social Communication in Primates", H.D. Steklis and M.J. Raleigh, eds., Academic Press, New York.
Kuhl, P.K., Williams, K.A., Lacerda, F., Stevens, K.N., and Lindblom, B., 1992, Linguistic experience alters phonetic perception in infants by 6 months of age, *Science* 255:606-608.
Lange, K.W., Robbins, T.W., Marsden, C.D., James, M., Owen, A.M., and Paul, G.M., 1992, L-Dopa withdrawal in Parkinson's disease selectively impairs cognitive performance in tests sensitive to frontal lobe dysfunction, *Psychopharmacol.* 107:394-404.
Lieberman, P., 1961, Perturbations in vocal pitch, *J. Acoust. Soc. Amer.* 33: 597-603.
Lieberman, P., 1965, On the acoustic basis of the perception of intonation and stress by linguists, *Word* 21:40-54.
Lieberman, P., 1968, Primate vocalizations and human linguistic ability, *J. Acoust. Soc. Amer.* 44:1157-1164.
Lieberman, P., 1975, "On the Origins of Language: An Introduction to the Evolution of Speech", Macmillan, New York.
Lieberman, P., 1984, "The Biology and Evolution of Language", Harvard University Press, Cambridge, Mass.
Lieberman, P., 1985, On the evolution of human syntactic ability: Its pre-adaptive bases - motor control and speech, *J. Human Evol.* 14:657-668
Lieberman, P., 1991, "Uniquely Human: The Evolution of Speech, Thought, and Selfless Behavior", Harvard University Press, Cambridge, Mass.
Lieberman, P., 1992, Could an autonomous syntax module have evolved? *Brain Lang.* 43: 768-774.
Lieberman, P., in press, Biologically bound behavior, free-will, and human evolution, *in*: "Conflict and Cooperation in Nature", J. I. Casti, ed., John Wiley, New York.
Lieberman, P., Crelin, E. S., and Klatt, D. H., 1972, Phonetic ability and related anatomy of the newborn, adult human, Neanderthal man, and the chimpanzee, *Amer. Anthropol.* 74:287-307.
Lieberman, P., Friedman, J., and Feldman, L. S., 1990, Syntactic deficits in Parkinson's disease, *J. Nerv. Ment. Dis.* 178: 360-365.
Lieberman, P., Kako, E. T., Friedman, J., Tajchman, G., Feldman, L. S., and Jiminez, E. B., 1992, Speech production, syntax comprehension, and cognitive deficits in Parkinson's disease, *Brain Lang.* 43:169-189.
Lisker, L. and Abramson, A. S., 1964, A cross language study of voicing in initial stops: acoustical measurements, *Word* 20:384-442.
MacLean, P. D., 1985, Evolutionary psychiatry and the triune brain, *Psychol. Med.* 15:219-221.
MacLean, P. D. and Newman, J. D., 1988, Role of midline frontolimbic cortex in the production of the isolation call of squirrel monkeys, *Brain Research* 450:111-123.
Metter, E. J., Riege, W. H., Hanson, W. R., Phelps, M. E., and Kuhl, D. E., 1984, Local cerebral metabolic rates of glucose in movement and language disorders from positron tomography, *Amer. J. Physiol.* 246: R897-R900.
Metter, E. J., Kempler, D., Jackson, C. A., Hanson, W. R., Reige, W. H., Camras, L. M., Mazziotta, J. C., and Phelps, M. E., 1987, Cerebral glucose metabolism in chronic aphasia, *Neurology* 37:1599-1606.
Metter, E. J., Kempler, D., Jackson, C., Hanson, W. R., Mazziotta, J. C., and Phelps M. E., 1989, Cerebral glucose metabolism in Wernicke's, Broca's, and conduction aphasia, *Arch. Neurol.* 46:27-34.
Morris, R. G., Downes, J. J., Sahakian, B. J., Evenden, J. L., Heald, A., and Robbins, T. W., 1988, Planning and spatial working memory in Parkinson's disease, *J. Neurol. Neurosurg. Psychiat.* 51:757-766.
Naeser, M. A., Alexander, M. P., Helms-Estabrooks, N., Levine, H. L., Laughlin, S. A., and Geschwind, N., 1982, Aphasia with predominantly subcortical lesion sites; description of three capsular/putaminal aphasia syndromes, *Arch. Neurol.* 39:2-14.
Natsopoulos, D., Grouios, G., Bostantzopoulou, S., Mentenopoulos, G., Katsarou, Z., and Logothetis, J., in press, Algorithmic and heuristic strategies in comprehension of complement clauses by patients with Parkinson's disease, *Neuropsychol.*
Parent, A., 1986, "Comparative Neurobiology of the Basal Ganglia", John Wiley, New York.
Pinker, S. and Bloom, P., 1990, Natural selection and natural language, *Behav. Brain Sci.* 13:707-784.
Rumbaugh, D. M. and Savage-Rumbaugh, E. S., 1992, Biobehavioral roots of language: Words, apes and a child. Conference paper presented at University of Bielefeld, Germany.

Sahakian, B. J., Morris, R. G., Evenden, J. L., Heald, A., Levy, R., Philpot, M., and Robbins, T. W., 1988, A comparative study of visuospatial memory and learning in Alzheimer-type dementia and Parkinson's disease. *Brain* 111:695-718.
Savage-Rumbaugh, S., McDonald, K., Sevcik, R. A., Hopkins, W. D., and Rubert, E., 1986, Spontaneous symbol acquisition and communicative use by pygmy chimpanzees (*Pan paniscus*), *J. Exp. Psychol. General* 115:211-235.
Stuss, D. T. and Benson, D. F., 1986, "The Frontal Lobes", Raven, New York.
Sutton, D. and Jürgens, U., 1988, Neural control of vocalization, *in*: "Comparative Primate Biology, Volume 4", H. D. Steklis and J. Erwin, eds., Arthur D. Liss, New York.
Wilkins, W. K. and Wakefeld, J., in press, Brain evolution and neurolinguistic preconditions, *Behav. Brain Sci.*

INDEX

MIX
Papier aus verantwortungsvollen Quellen
Paper from responsible sources
FSC® C105338

If you have any concerns about our products,
you can contact us on
ProductSafety@springernature.com

In case Publisher is established outside the EU,
the EU authorized representative is:
Springer Nature Customer Service Center GmbH
Europaplatz 3, 69115 Heidelberg, Germany

Printed by Libri Plureos GmbH
in Hamburg, Germany